MATHEMATICAL SURVEYS
AND MONOGRAPHS

NUMBER 23

Introduction to
VARIOUS ASPECTS
OF DEGREE THEORY
IN BANACH SPACES

E. H. ROTHE

American Mathematical Society
Providence, Rhode Island

1980 *Mathematics Subject Classification* (1985 *Revision*). Primary 55M25;
Secondary 46-02.

Library of Congress Cataloging-in-Publication Data

Rothe, Erich H.
    Introduction to various aspects of degree theory in Banach spaces.
    (Mathematical surveys and monographs, ISSN 0076-5376; no. 23)
    Bibliography: p.
    Includes index.
    1. Degree, Topological. 2. Banach spaces. I. Title. II. Series.
QA612.R68   1986   514'.2   86-8038
ISBN 0-8218-1522-9 (alk. paper)

MS-Stk
SCIMON

Contents

Preface

As its title indicates this book is not meant to be an encyclopedic presentation of the present state of degree theory. Likewise, the list of references is not meant to be a complete bibliography. The choice of subjects treated is determined partly by the fact that the book is written from the point of view of an analyst, partly by the desire to throw light on the theory from various angles, and partly the choice is subjective.

A detailed description of the subjects treated is contained in §10 of the following introduction. At this point we mention, together with relevant references, some subject matter belonging to degree theory which is not treated in this book.

(a) The Leray-Schauder degree theory in Banach spaces may be extended to linear convex topological spaces as already noticed by Leray in [35]. A complete self-contained treatment may be found in Nagumo's 1951 paper [43].

(b) Already, in his 1912 paper [9], Brouwer established degree theory for certain finite-dimensional spaces which are not linear. For later developments in this direction and for bibliographies, see, e.g., the books by Alexandroff-Hopf [2], by Hurewicz-Wallmann [29], and by Milnor [39]. In their 1970 paper [20] Elworthy and Tromba established a degree theory in certain infinite-dimensional manifolds. See also the systematic exposition by Borisovich-Zvyagin-Sapronov [6].

(c) There are various degree theories for mappings which are not of the type treated by Leray and Schauder: a theory by Browder and Nussbaum on "intertwined" maps [11] and a theory of "A-proper" maps by Browder and Petryshyn [12] in which the degree is not integer-valued but a subset of the integers. See also [10]. Moreover, the method used in the Fenske paper [21] mentioned in §8 of the following introduction applies also to mappings which are "α-contractions." Various mathematicians considered degree theory for multiple-valued maps. A survey of these generalizations and further bibliographical data may be found in the book [37] by Lloyd.

(d) Some important theorems based on degree theory (e.g., those contained in §§2–3 of Chapter 5 and parts of Chapters 8 and 9 of this book) and some generalizations thereof can be derived by the use of cohomology theory in infinite-

v

dimensional spaces. See the paper by Geba and Granas [**22**], in particular Chapter IX, and the book by Granas [**25**], in particular Chapter 11.

(e) The axiomatic treatment of degree theory by Amann and Weiss [**3**].

(f) The "coincidence degree theory" of Mawhin [**38**].

(g) For supplementary reading and further references we mention books by the following authors: Krasnoselskiĭ [**31**], 1956, J. Cronin [**14**], 1964, J. T. Schwartz [**51**], 1965, K. Deimling [**15**], 1974, E. Zeidler [**54**], 1976, F. Browder [**10**], 1976, G. Eisenack–C. Fenske [**19**], 1978, N. G. Lloyd [**37**], 1978. We also mention the paper [**52**], 1980 by H. W. Siegbert which contains an interesting exposition of the history of finite-dimensional degree theory from Gauss via Kronecker to Brouwer.

(h) Ever since it was established by Leray and Schauder [**36**], degree theory in Banach spaces has been an important tool for the treatment of boundary value problems (including periodicity problems) for ordinary and partial differential equations, for integral equations, and for eigenvalue and bifurcation problems. These applications to analysis are not treated in the present book, but the reader will find a number of them discussed in many of the books quoted in part (g) of this preface. See, e.g., Chapters III–VI of [**31**], Chapters II and IV of [**14**], and Chapter 9 of [**37**]. The Cronin book [**14**] discusses, in particular, papers up to 1962 using the "Cesari method." For the role of degree theory in the further development of this method, now usually referred to as the "Alternative Method," the reader may consult the bibliography on pp. 232–234 in *Nonlinear Phenomena in Mathematical Science*, ed. V. Lakshmikantham, Academic Press, 1982.

We finish this preface with a word on the organization of this book. In addition to the introduction the book consists of nine chapters. Each chapter (except Chapter 1) is divided into sections and each section into subsections. (The notation "1.2" in Chapter 3 means subsection 1.2 in that chapter; "(1.2)" refers to a formula in that chapter. However §1.2 refers to §2 in Chapter 1.) Each chapter is followed by notes containing some of the proofs and historical remarks. Finally, there are two Appendixes A and B.

My thanks go to my colleagues and friends Professors Lamberto Cesari, Charles Dolph, and R. Kannan for many encouraging conversations.

My thanks also go to Mrs. Wanita Rasey who, with great patience, converted an often not easily decipherable handwritten script into a readable typescript.

I am obliged to the editorial board of the Mathematical Surveys and Monographs for reviewing the typescript and to the editorial staff of the American Mathematical Society, in particular to Ms. Mary C. Lane, Lenore C. Stanoch, and Holly Pappas for their cooperation and labor in transforming the manuscript into a book.

Introduction

1. It is one of the most important tasks of analysis to compute the number $N(f, \Omega, y_0)$ of solutions $x \in \Omega$ of the equation

$$f(x) = y_0, \tag{1}$$

where Ω is a subset of some domain space E and where y_0 is a point of a range space E_1.

If Ω is the open interval (a, b) on an x-axis, if f is a polynomial in x, and if y_0 is a point on a y-axis, then the above problem is completely solved by a classical theorem of Sturm (1835) which gives a procedure to calculate the number $N(f, \Omega, y_0)$.

However, this number may not be continuous in dependence on y_0 or f. If, e.g., Ω is the interval $(-1, +1)$ on the real axis R and f is the map $\Omega \to R$ given by $f(x) = y = x^2$, then for any positive $\varepsilon < 1$

$$N(f, \Omega, y_0) = \begin{cases} 2 & \text{if } y_0 = \varepsilon, \\ 1 & \text{if } y_0 = 0, \\ 0 & \text{if } y_0 = -\varepsilon. \end{cases}$$

Thus our number N is not continuous in y_0 at $y_0 = 0$. It is just as easy to see that $N(f, \Omega, y_0)$ may not change continuously if f is changed continuously. For instance if Ω is as above and $f(x) = x^2 + \alpha$ with α real, then $N(f, \Omega, 0)$ is not continuous at $\alpha = 0$.

2. The basic difference between the root count $N(f, \Omega, y_0)$ and the "degree" $d(f, \Omega, y_0)$ is that the latter has the above-mentioned continuity properties. There are two good reasons for insisting on these: (i) without them a small error in the computation of y_0 or f may lead to a wrong root count; (ii) if $d(f, \Omega, y_0)$ is continuous in a certain class C of mappings f, then this number, being integer-valued, is constant for all $f \in C$. It may happen that C contains an f_0 for which the degree is particularly easy to compute. A case often occurring in the applications is that C contains the identity map $f_0 = I$.

3. We return to the example $y = f(x) = x^2$ of §1 to indicate how a root count $d(f, \Omega, y_0)$ may be obtained which is continuous in y_0 at $y_0 = 0$. For $y_0 > 0$ we note that, with increasing x, the value of $f(x)$ decreases as x crosses the root $x^- = -\sqrt{y_0}$ of (1), while $f(x)$ increases as x crosses the root $x^+ = +\sqrt{y_0}$ of (1). This can be stated as follows: the differentials $Df(x^-; h) = f'(x^-)h = -2\sqrt{y_0}h$ and $Df(x^+; h) = 2\sqrt{y_0}h$ (considered as linear maps in h) both map R onto R in a one-to-one fashion. However, the first one reverses the "orientation" of R while the second one preserves it. For this reason we count x^- as a negative contribution to our root count, and x^+ as a positive one, i.e., we attach to x^- an index $j(x^-) = -1$, set $j(x^+) = +1$, and define $d(f, \Omega, y_0) = j(x^-) + j(x^+)$. Thus, for $0 < y_0 < 1$, $d(f, \Omega, y_0) = 0$. But it is natural to assign to our "degree" the value zero also if $y_0 < 0$ since (1) has no solution for such y_0. We then only have to define $d(f, \Omega, 0) = 0$ to obtain continuity at y_0.

As will be seen later, this method of "root counting" can be vastly generalized and leads to a definition of the degree which has the desired continuity properties of the degree with one important exception: in general, $d(f, \Omega, y_0)$ is not continuous as y_0 crosses a point of the image $f(\partial\Omega)$ of the boundary $\partial\Omega$ of Ω. A simple example is the following one: let Ω again be the interval $(-1, +1)$ and let $f(x) = 2x$. It is then easily seen that

$$d(f, \Omega, y_0) = \begin{cases} 1 & \text{if } |y_0| < 2, \\ 0 & \text{if } |y_0| > 2, \end{cases}$$

if one uses the same method of root counting as in the previous example. Thus ± 2 are points of discontinuity. But these two points are the images of the points $1, -1$ whose union is $\partial\Omega$.

It is for these considerations that in the precise definition of the degree to be given later we always suppose that

$$y_0 \neq f(\partial\Omega). \tag{2}$$

4. The approach to degree theory used for the simple examples of §3 can be extended to mappings in an arbitrary (real) n-dimensional Euclidean space E^n. In the present section we describe the basic steps of this extension.

Let Ω be a bounded open set in E^n, and let f be a map of the closure $\overline{\Omega}$ of Ω into E^n. We suppose that $f \in C'(\overline{\Omega})$, i.e., that the differential $Df(x; h)$ of f exists and is continuous for all x in some open set containing $\overline{\Omega}$. We also assume that the condition (2) is satisfied.

Step I. We consider a linear map $l: E^n \to E^n$ which is not singular. Then $\det l \neq 0$ and we define the index $j(l)$ by setting

$$j(l) = \begin{cases} 1 & \text{if } \det l > 0, \\ -1 & \text{if } \det l < 0, \end{cases} \tag{3}$$

where det stands for determinant.

Step II. Suppose y_0 is a regular value for f, i.e., either (i) the equation (1) has no roots in Ω, or (ii) for every root $x \in \Omega$ of (1) the differential $l_x(h) = Df(x; h)$ is nonsingular. We then set

$$d(f, \Omega, y_0) = 0 \tag{4}$$

in Case (i). Now the fact that y_0 is a regular value together with (2) implies that (1) has at most a finite number of roots. (This is a special case of Lemma 22 in Chapter 1.) Let, in Case (ii), $x_1, x_2, \ldots, x_r$ be these roots. Then the differentials $l_\rho(h) = Df(x_\rho; h)$, $\rho = 1, 2, \ldots, r$, are not singular and the indices $j_\rho = j(l_\rho)$ are defined. We then set

$$d(f, \Omega, y_0) = \sum_{\rho=1}^{r} j_\rho = p - q, \tag{5}$$

where p and q are the numbers of those ρ for which $\det l_\rho > 0$ and $\det l_\rho < 0$ resp.

Step III. Suppose y_0 is not a regular value for f. In this case the definition of the degree is obviously made possible by Theorems III$_1$ and III$_2$ below.

THEOREM III$_1$. *Every neighborhood $N(y_0)$ of y_0 contains a point y which is a regular value for f.*

THEOREM III$_2$. *A neighborhood $N(y_0)$ of y_0 can be chosen in such a way that*

$$d(f, \Omega, y') = d(f, \Omega, y'')$$

for any couple y', y'' of regular values in $N(y_0)$.

We note that Theorem III$_1$ is a special case of a well-known theorem of Sard [49] which will be stated in Appendix B.

It takes a fourth step to extend degree theory from maps $f \in C'(\Omega)$ to continuous maps $\overline{\Omega} \to E$. This extension is based on the fact that such continuous f can be uniformly approximated by maps $f_0 \in C'(\overline{\Omega})$.

5. The method sketched in the preceding section will be referred to as the differential method. In the present section we give an intuitive description of the "simplicial" approach to a degree theory in E^n introduced by Brouwer [8]. For each nonnegative integer $r \le n$, simplices σ^r of dimension r in E^n are defined: σ^0 is a point, σ^1 an open interval, σ^2 an open triangle, σ^3 an open tetrahedron and so on. We assume E^n to be oriented. This orientation induces an orientation of each n-simplex σ^n in E^n. This induced orientation is called the positive orientation of σ^n.

Now let $\sigma_1^n, \sigma_2^n, \ldots, \sigma_s^n$ be a finite number of (positively) oriented disjoint n-simplices in E^n. Let $\overline{\Omega}$ be the union of the closures of these simplices, and let Ω be the interior of $\overline{\Omega}$. Then a continuous map $\overline{\Omega} \to E^n$ is called simplicial if for each $\rho = 1, 2, \ldots, s$ the restriction f_ρ of f to σ_ρ^n is affine, i.e., is of the form $f_\rho(x) = c_\rho + l_\rho(x)$ where c_ρ is a constant and l_ρ is linear. We set $\det f_\rho = \det l_\rho$ and assume that l_ρ is nonsingular. Before defining the degree for simplicial f we need the concept of "general position": a point set $S \subset E^n$ is said to be in general position if for $r \le n$ no $r + 1$ points of S lie in an $(r - 1)$-dimensional plane of E^n (e.g., four points of S do not lie in a two-dimensional plane of E^n if $n \ge 3$). The definition of $d(f, \Omega, y_0)$ is then given in the following steps.

Step I. Let $f = a$ be an affine map $E \to E$: $a(x) = \text{const.} + l(x)$ where l is a nonsingular linear map $E \to E$. We then define the index $j(a)$ by

$$j(a) = \begin{cases} 1 & \text{if } \det l > 0, \\ -1 & \text{if } \det l < 0. \end{cases}$$

Step II. Let f be a simplicial map $\overline{\Omega} \to E^n$ and suppose that y_0 is a (simplicially) regular value for f, i.e., that y_0 together with the vertices of the simplices $f(\sigma_\rho^n)$, $\rho = 1, 2, \ldots, s$, form a point set in general position. This implies that y_0 does not lie on the boundary of any of the simplices $f(\sigma_\rho^n) = f_\rho(\sigma_\rho^n)$, $\rho = 1, 2, \ldots, s$. Since the determinant of the linear part of f_ρ is $\neq 0$, it follows that in each σ_ρ^n the equation (1) has either none or exactly one root. Thus the number of roots in Ω is at most finite. We set $d(f, \Omega, y_0) = 0$ if (1) has no root. If there are roots, we set

$$d(f, \Omega, y_0) = \sum_\rho j(f_\rho) = p - q, \tag{7}$$

where the summation is extended over those ρ for which σ_ρ^n contains a root of (1) and where p and q denote the number of those simplices for which $\det f_\rho$ is positive and negative resp.

Step III. Now let y_0 satisfy (2) but not be a regular value for f. Then the definition of $d(f, \Omega, y_0)$ is made possible by Theorems III$_1$ and III$_2$ below.

THEOREM III$_1$. *Every neighborhood $N(y_0)$ of y_0 contains a point y which is a (simplicially) regular value for f.*

THEOREM III$_2$. *A neighborhood $N(y_0)$ of y_0 can be chosen in such a way that $d(f, \Omega, y') = d(f, \Omega, y'')$ for any couple y', y'' of simplicially regular values in $N(y_0)$.*

We note that Theorem III$_1$ is a special case of a well-known theorem on point sets in general position (see Theorem 1.11 in Chapter 6). Thus this theorem plays for the simplicial method a role corresponding to the one that the Sard theorem plays for the differential method.

The possibility of extending the degree theory to arbitrary continuous maps $\overline{\Omega} \to E^n$ is based on the fact that every such map can be uniformly approximated by simplicial maps.

6. The first one to define a degree for continuous maps in finite-dimensional spaces—now known as the Brouwer degree—was Brouwer in his 1911 paper [8]. When he introduced the simplicial method described in §5, degree theory became part of the so-called polyeder—or simplicial—topology and was treated in this way, e.g., in unpublished lecture notes by H. Hopf of 1926/27 [27] and in the topology book of 1934 by P. Alexandroff and H. Hopf [2]. The latter book treats degree theory as a special case of intersection theory and contains earlier important degree-theoretic results of Kronecker [32] and extensions thereof.

7. In their now classical 1934 paper [36] Leray and Schauder defined a degree $d(f, \Omega, y_0)$, where Ω is a bounded open set in a Banach space E and where f is

a map $\overline{\Omega} \to E$ of the form
$$f(x) = x - F(x), \tag{8}$$
where F is completely continuous, i.e., where the closure of $F(\overline{\Omega})$ is compact, and where y_0 satisfies (2). They did this by taking Brouwer's degree theory for the finite-dimensional space for granted and then using a limit process in the dimension.

Calling maps of the form (8) Leray-Schauder (L.-S.) maps, their procedure may be described as consisting of the following two steps:

Step I. They defined a degree for those L.-S. maps for which $F(\Omega)$ lies in a finite-dimensional subspace E^n of E. A map of this kind is called a finite layer map with respect to E^n. The definition of a degree for layer maps is based on the following:

LEMMA (FIRST LERAY-SCHAUDER). *Let the L.-S. map f be a layer map with respect to each of the finite-dimensional subspaces E^n and E^m of E. Suppose that (2) is satisfied and that $y_0 \in E^n \subset E^m$. Then the Brouwer degrees $d(f, \Omega \cap E^n, y_0)$ and $d(f, \Omega \cap E^m, y_0)$ are equal (see subsections 3.2 and 3.3 in Chapter 3).*

This lemma makes it possible to define $d(f, \Omega, y_0) = d(f, \Omega \cap E^n, y_0)$.

Step II. The definition of $d(f, \Omega, y_0)$ for arbitrary L.-S. maps is now made possible on account of the following two Lemmas II$_1$ and II$_2$:

LEMMA II$_1$ (SECOND LERAY-SCHAUDER) (cf. Lemma 4.3 in Chapter 3). *If f is an L.-S. map $\overline{\Omega} \to E$ and if (2) is satisfied, then, with ε being a given positive number, there exists a finite layer map $\lambda: \overline{\Omega} \to E$ such that*
$$\|f(x) - \lambda(x)\| < \varepsilon \quad \text{for all } x \in \overline{\Omega}. \tag{9}$$
Here $\| \cdot \|$ denotes the norm in E.

LEMMA II$_2$. *There exists an $\varepsilon > 0$ of the following property. If λ_1 and λ_2 are two finite layer maps such that (9) holds with $\lambda = \lambda_1$ and $\lambda = \lambda_2$, then*
$$d(\lambda_1, \Omega, y_0) = d(\lambda_2, \Omega, y_0).$$

8. It is natural to ask whether the degree theory in Banach spaces can be established intrinsically, i.e., without assuming the finite-dimensional theory and without the passage to the limit in the dimension. This does not seem possible with the simplicial method of §6 since there is no analogue in an infinite-dimensional Banach space of an n-dimensional simplex in E^n. Since differentials in Banach space are well defined, it is different with the differential method described in §4.

This method introduced in 1951 [**42**] in an important paper by Nagumo made degree theory in E^n part of differential topology. It is clear from §4 what is needed to apply the differential method in the Banach space case: looking at Step I in §4 we note that for defining an index $j(l)$ for a nonsingular linear L.-S. map of a Banach space E into itself we need an analogue of the division of nonsingular linear maps $l: E^n \to E^n$ into those of negative and positive determinant. Now, the simplest map of determinant $+1$ in E^n is the identity map, and a simple map of determinant -1 is the map I^-, which in coordinates $x_1, x_2, \ldots, x_n$ for E^n is given by $x_1' = -x_1, x_2' = x_2, \ldots, x_n' = x_n$. Now the identity map is also defined in the Banach space E, and it is easy to define in E a linear nonsingular map which is the analogue of the map I^- (see subsections 4.4 and 4.5 in Appendix A). If we denote this analogue again by I^-, then the following linear homotopy theorem (proved in 1955 [48] and independently in 1970 [20]) holds:

THEOREM. *The family of linear nonsingular L.-S. maps of a Banach space E into itself consists of two disjoint homotopy classes H^+ and H^- where $I \in H^+$ and $I^- \in H^-$.*

This theorem allows us to define an index $j(l)$ for every nonsingular linear L.-S. map $E \to E$ by setting

$$j(l) = \begin{cases} +1 & \text{if } l \in H^+, \\ -1 & \text{if } l \in H^-. \end{cases} \tag{10}$$

Turning to Step II of §4, we note first of all that the definition of regular value given there for E^n makes sense for an arbitrary Banach space E (see §17 of Chapter 1). It can be proved that also in E the equation (1) has at most a finite number of roots if y_0 is a regular value for f and satisfies (2) (see §22 of Chapter 1). Moreover, according to a lemma of Krasnoselskiĭ [31] (see §18 of Chapter 1), the fact that f is an L.-S. map implies that the linear map $l_x(h) = Df(x; h)$ is also of the L.-S. type. If, in particular, x is a root of (1) with a regular y_0, then $l_x(h)$ is nonsingular by definition. It follows that if $x_1, x_2, \ldots, x_s$ are the roots of (1) with regular y_0 and $l_\rho(h) = Df(x_\rho; h)$, $\rho = 1, 2, \ldots, s$, then the indices $j(l_\rho)$ are well defined and the degree $d(f, \Omega, y_0)$ can be defined by the first part of the equality (5) with j defined by (10). If there are no roots in Ω of equation (1), the definition is given by (4).

To find a definition of the degree if y_0 is not a regular value for f, we recall that the definition given in Step III of §4 in the finite-dimensional case was based on Theorems III$_1$ and III$_2$ of that section whose proof in turn was based on Sard's theorem. What is now needed is a theorem generalizing Sard's theorem to the infinite-dimensional case. Such a theorem was established by Smale in 1965 [53]. Using this theorem, now known as the Sard-Smale theorem, C. Fenske established in 1971 [21] the degree theory "intrinsically" for a mapping $f \in C''(\overline{\Omega})$. (The Sard-Smale theorem had already been used in 1970 [20] to establish a degree theory for "Banach manifolds" by Elworthy and Tromba.) Both of these papers contain a proof for the generalization to Banach spaces of Theorem III$_2$ stated in §4 above.

We finally note that the method described in the last paragraph of §4 cannot be applied for arbitrary Banach spaces E: it is known from investigations of Kurzweil [33] and Bonic-Frampton [5] that not in every Banach space can an

L.-S. map be uniformly approximated by differentiable maps. Thus this "intrinsic" method does not lead to the same generality as the original method of Leray and Schauder described in §7. It supposes in fact that the mappings involved are twice differentiable. (However, according to a paper [**41**] by Moulis one-time differentiability is sufficient for the Hilbert manifold.)

9. It is always instructive to look at a given subject from different points of view even if this sometimes entails repetitions. Therefore, this book will not give preference to any of the various approaches to degree theory sketched above but will follow a "mixed method" which will give an opportunity to make clear the relationship between these approaches.

10. In this section an outline of this book's content will be given.

Chapter 1 contains preliminary material to be consulted if in the later part of the book the reader encounters definitions or theorems with which he is not familiar.

Chapter 2 develops the "intrinsic" degree theory sketched in §8 of this introduction. §2.1 (i.e., §1 of Chapter 2) contains the definition of the degree for linear L.-S. maps, which is based on the linear homotopy theorem whose proof is given in Appendix A following the "spectral method" used in [**48**]. §2.2 contains the definition of the degree $d(f, \Omega, y_0)$ for differentiable L.-S. maps provided y_0 is a regular value for f. Some of the properties of this degree are immediate consequences of the definition and are collected in Lemma 2.4; they are referred to as elementary properties in contradistinction to the continuity properties of the degree. §2.2 also contains the definition of and a later often-used lemma on layer maps, i.e., L.-S. maps f for which the perturbation $x - f(x)$ of the identity lies in a proper subspace of the Banach space E.

Before considering the case where y_0 is not a regular value for f, we treat in §2.3 the one-dimensional case because it is instructive to see the general case reduced to this elementary one.

The latter section starts with the definition of the degree $d(f, \Omega, y_0)$ with y_0 not being a regular value for f provided f is twice differentiable. This definition is based on Theorems 4.1 and 4.2 in Chapter 2 (i.e., on the Banach space version of Theorems III$_1$ and III$_2$ stated in Step III of §4 of this introduction). Theorem 4.1 follows from the Sard-Smale theorem whose proof is given in Appendix B. The proof of Theorem 4.2 uses methods contained in papers by Fenske [**21**] and by Elworthy-Tromba [**20**] in conjunction with a method employed by Nagumo in the finite-dimensional case [**42**, p. 491f.] by which the proof is ultimately reduced to the one-dimensional case, i.e., to Theorem 3.6 in Chapter 2.

The continuity of the degree $d(f\Omega, y_0)$ in y_0 (Theorem 4.14 in Chapter 2) is essentially a consequence of Theorem 4.2. Following again an idea used by Nagumo in the finite-dimensional case [**42**, pp. 492–493], the continuity of the degree in f is proved by applying Theorem 4.14 to a Banach space "one dimension higher" than the given one.

The degree theory developed in Chapter 2 assumes that the L.-S. maps are twice differentiable since this is assumed in Theorem 4.2. The object of Chapter 3 is to develop degree theory without that assumption. A first step in that direction is the extension lemma, Lemma 1.3, which—speaking vaguely—states: let $C = C(\overline{\Omega})$ be the normed space of all completely continuous maps $\overline{\Omega} \to E$ (see subsection 1.1 in Chapter 3), and let C_1 be a linear subspace of C. Suppose a degree having certain specified properties has been defined for all mappings $f_1(x) = x - F_1(x)$ for which F_1 belongs to $C_1 = C_1(\overline{\Omega})$; then the definition of the degree can be extended to all maps $\overline{f}(x) = x - \overline{F}(x)$ for which $\overline{F}$ is an element of the closure $\overline{C}_1$ of C_1 in such a way that the specified properties hold for the extended degree.

This lemma is applied twice. In §3.2 it is applied with

$$C_1(\overline{\Omega}) = C(\overline{\Omega}) \cap C''(\overline{\Omega}), \tag{11}$$

where $C''(\overline{\Omega})$ denotes the set of twice differentiable maps $\overline{\Omega} \to E$. It follows that the degree theory is established for all L.-S. maps in those Banach spaces for which

$$\overline{C_1(\overline{\Omega})} = C(\overline{\Omega}), \tag{12}$$

with $C_1(\overline{\Omega})$ given by (11). As already pointed out in the last paragraph of §8 of this introduction, the equality (12) does not hold in all Banach spaces. It holds, however, for certain types of Banach spaces (see subsections 2.3–3.8 in Chapter 3).

In particular (12) holds for finite-dimensional Banach spaces. Since in such spaces a map is L.-S. if and only if it is continuous, the degree theory is established for continuous maps of a finite-dimensional Banach space into itself.

This allows us to carry out the first step of the Leray-Schauder theory sketched in §7 of this introduction and thus to establish the degree theory for finite layer maps (see §3.3).

We now make a second application of the extension lemma, Lemma 1.3 (see §3.4): we apply it with $C_1(\overline{\Omega})$ being the subset of those $F \in C(\overline{\Omega})$ for which $F(\overline{\Omega})$ lies in some finite-dimensional subspace $E^n = E^n(F)$ of the Banach space E. But for this $C_1(\overline{\Omega})$ the equality (12) holds on account of the second Leray-Schauder Lemma II_1 stated in §7 of this introduction (and proved in §3.4). Thus by the extension lemma the degree theory is established for all L.-S. maps in arbitrary Banach spaces.

In §3.6 the degree is defined for generalized Leray-Schauder (g.L.-S.) maps, which essentially are maps of the form $\lambda(x)f(x)$ where f is an L.-S. map and λ a positive function bounded away from zero. The reason for introducing this slight generalization [47] is that a central projection of an L.-S. map is in general not L.-S., while a g.L.-S. map remains g.L.-S. under central projections.

It is easy to see that the assertion of the continuity theorem, Theorem 6.6 in Chapter 3, remains true if the "homotopy" is a linear convex combination of two g.L.-S. maps f_0 and f_1, i.e., if $f_t(x) = (1-t)f_0(x) + tf_1(x)$, $0 \le t \le 1$. The resulting theorem will be called the Poincaré-Bohl theorem in accordance with terminology customary in the finite-dimensional case. Chapter 4, which consists mainly of an updating of papers [46] and [47], deals with some consequences of this theorem. Among these are (i) the fact that the degree $d(f, \Omega, y_0)$ depends only on the values of f on the boundary $\partial \Omega$ of Ω, which allows one to define the "winding number" $u(f(\partial \Omega), y_0)$ of a g.L.-S. map f defined only on $\partial \Omega$ with respect to the point y_0 (provided that $y_0 \notin f(\partial \Omega)$); (ii) Theorem 1.9 in Chapter 4, which is a generalization to Banach spaces of the classical Rouché theorem; (iii) Theorem 1.12 together with its corollaries and Theorem 1.17 (these are useful for giving existence proofs in analysis); (iv) the fixed point theorem, Theorem 1.18.

§4.2 gives an interpretation of degree and winding number as "intersection" numbers.

Chapter 5 deals with the product theorem, i.e., a formula which expresses the degree $d(gf, \Omega, z_0)$ of a composite map gf in terms of the degrees of g and f and with consequences of that theorem. The results of this chapter are all due to Leray, who in a short 1935 Note ([34], see also [35]) established the product theorem for L.-S. maps in Banach spaces by a limit process from the finite-dimensional case, used this theorem to prove the "invariance of the domain," i.e., the fact that an open set in a Banach space remains open under a one-to-one L.-S. map, and deduced from the product theorem also the validity of the Jordan theorem in its generalized form in Banach spaces (see subsection 3.1 in Chapter 5). This last result will be referred to as the Jordan-Leray theorem.

The proof of the product theorem will be given in §5.1. We first assume that f and g are twice differentiable and that z_0 is a regular value for gf. In this case the theorem is a direct consequence of the chain rule for differentials and of Theorem 6.1 of Appendix A according to which $j(l_1 l_2) = j(l_1) j(l_2)$, where j denotes the index defined by equation (10) of this introduction. Thus, in this case, our theorem is established directly in Banach space without recourse to a limit process in the dimension. For the proof without the special assumptions made above, a complicated approximation procedure is needed for which we refer to the text. The invariance of the domain theorem is proved in §5.2. This proof is followed by some consequences of that theorem (subsections 2.4–2.10 in Chapter 5).

§5.3 contains the proof of the Jordan-Leray theorem.

§§5.1 and 5.2 are influenced by the presentation in Nagumo's paper [42] dealing with the finite-dimensional case, and for part of §5.3 we used the method employed by Lloyd [37, pp. 47–50] for the proof of the finite-dimensional versions of the theorem in question.

Chapter 6 deals with finite-dimensional linear spaces.

§6.1 contains prerequisites about subjects like point sets in general position, simplices, affine maps, and orientation.

If Ω is a bounded open subset in the space E_1^n of finite dimension n, if f is a continuous map $\overline{\Omega} \to E_1^n$ and y_0 a point of E_1^n, then the degree $d(f, \Omega, y_0)$ was already defined in subsection 2.4 of Chapter 3. Since two linear spaces of the same dimension are linearly isomorphic, the degree $d(f, \Omega, y_0)$ can, in an obvious way, be defined if f is a continuous map of $\overline{\Omega}$ into another linear space E_2^n of dimension n and if y_0 is a point of E_2^n. It is easy to verify that this degree has the properties corresponding to those of the previously defined degree. This is done in §6.2.

In §6.3 simplicial maps are introduced. Since simplicial maps are continuous, the degree theory developed in §3.2 applies to them (see in particular subsection 2.4 in Chapter 3). We note that by part (iv) of Lemma 3.7 in Chapter 6 the equality (7) of this introduction holds, an equality which has been a matter of definition in the original simplicial theory sketched in §5 of this introduction.

§§6.5 and 6.6 deal with subdivision and simplicial approximations resp.

Chapter 7 treats mappings between finite-dimensional spheres. Following H. Hopf's presentation in [27], the definition of the degree for such maps is reduced to the "plane" case by means of stereographic projection.

§7.1 deals first with elementary geometric properties of the sphere and with its orientation. Furthermore if S_1^n and S_2^n are two spheres of dimension n, if f is a continuous map $S_1^n \to S_2^n$, and if y_0 is a point of S_2^n, the degree $d(f, S_1^n, y_0)$ is defined and proved to be independent of the particular choice of y_0.

This allows us in §7.2 to denote this number by $d(f, S_1^n, S_2^n)$ and derive the main properties of this degree.

§7.3 starts with the definition of the "order" $v(f(S_1^n), c_1)$ of a continuous map $f \colon S_1^n \to E_2^n - c_1$ where S_1^n is an n-sphere, where E_2^n is a linear space of dimension n, and where c_1 is a point of E_2^n: let B_2^{n+1} be a ball with center c_1 whose closure does not intersect $f(S_1^n)$ and let $S_2^n = \partial B_2^{n+1}$. If π is the projection of $f(S_1^n)$ on S_2^n from c_1 and $f^* = \pi f$, then the degree $d(f^*, S_1^n, S_2^n)$ is defined and can be proved to be independent of the particular choice of B_2^{n+1}. The order $v(f(S_1^n), c_1)$ is then defined as this degree. The main result states that the order equals the winding number $u(f(S_1^n), c_1)$, or what is the same, by Definition 1.5 of the winding number in Chapter 4, that $v(f(S_1^n), c_1) = d(\overline{f}, B_1^{n+1}, c_1)$, where $\overline{f}$ is an arbitrary continuous extension of f to the ball B_1^{n+1} whose boundary is S_1^n. The proof presented here follows the one given by Hopf in [27].

Relations established in this proof allow us in subsection 3.13 of Chapter 7 to interpret the order and the degree as intersection numbers in more detail than in §4.2.

Theorem 3.4 in Chapter 7 will be of use in Chapter 8.

To describe the content of Chapter 8, we note that Definition 1.5 in Chapter 4 of the winding number u together with an elementary property of the degree implies that if f is a g.L.-S. map $\partial\Omega \to E - y_0$ and if $u(f(\partial\Omega), y_0) \neq 0$, then for every g.L.-S. extension $\overline{f}$ of f to $\overline{\Omega}$ the equation $\overline{f}(x) = y_0$ has at least one solution x in Ω. In other words, the following assertion A holds:

(A) for f to have a g.L.-S. extension $\overline{f}$ to $\overline{\Omega}$ which maps $\overline{\Omega}$ into $E - y_0$, the condition

$$u(f(\partial\Omega), y_0) = 0 \qquad (13)$$

is necessary. We note next that Theorem 1.7 in Chapter 4 implies

(B) for the g.L.-S. maps f and g mapping $\partial\Omega$ into $E - y_0$ to be homotopic in $E - y_0$, the equality

$$u(f(\partial\Omega), y_0) = u(g(\partial\Omega), y_0) \qquad (14)$$

is a necessary condition.

The object of §8.1 is the proof of Theorems 1.2 and 1.4 in Chapter 8, which state that conditions (13) and (14) are not only necessary but also sufficient. The finite-dimensional versions of these theorems are both special cases of a classical theorem by H. Hopf (see [2, p. 499ff.] and the papers by Hopf quoted on p. 621 of [2]). We therefore refer to Theorem 1.2 as the Hopf extension theorem. Theorem 1.4 will be called the Hopf-Krasnoselskiĭ homotopy theorem since Krasnoselskiĭ was the first to establish the theorem in the generality stated; see [31, p. 110]. (For "strictly convex" Banach spaces the theorem was established in [47], and for Hilbert spaces with Ω being a ball in [45].)

The proof of the Hopf extension theorem will first be given for the finite-dimensional case following the method used in [2, pp. 500–505]. It is then proved for finite layer maps in a Banach space E and finally for g.L.-S. maps. For the last two steps the "homotopy extension" theorem, Theorem 1.12 in Chapter 8, is used, whose proof for Banach space is due to Granas (see [24, p. 37 and 23]).

Krasnoselskiĭ's proof of Theorem 1.2 assumes Hopf's finite-dimensional result and then uses a limit process in the dimension. The proof given in the present book generalizes to Banach space a method used by Hopf in the proof he gives for the finite-dimensional case in his lectures [27]. In sketching this method by which the homotopy theorem is shown to be a consequence of the extension theorem, we restrict ourselves at this moment for brevity's sake to the case that the convex set V_1 is a ball B_1^{n+1} with center y_0: we extend the map f given on $S_1^n = \partial B_1^{n+1}$ to $\overline{B_1^{n+1}}$, denoting the extension again by f. Due to the fact that by assumption $f \neq y_0$ on S_1^n, we can choose a ball $B_0^{n+1} \subset B_1^{n+1}$ with center y_0 such that $f(x) \neq y_0$ for $x \in \overline{B_1^{n+1}} - B_0^{n+1}$. For $x \in S_0^n = \partial B_0^{n+1}$, we set $g_0(x) = g(x')$ where x' is the point on S_1^n which lies on the same ray from y_0 as x. If $\dot{f} = f$ on S_1^n and $\dot{f} = g_0$ on S_0^n, then assumption (14) together with the definition of g_0 implies that $u(f(\partial\Omega_0), y_0) = 0$ where $\Omega_0 = B_1^{n+1} - \overline{B_0^{n+1}}$. Consequently we can apply the Hopf extension theorem to obtain an extension $\overline{f}$ from $\dot{f}$ to $\overline{\Omega_0}$ with $\overline{f} \neq y_0$. Now if $\dot{x}_0$ and $\dot{x}_1$ are points on S_0^n and S_1^n resp. which lie on the same ray from y_0, then for $t \in [0,1]$ the point $\dot{x}_0(1-t) + t\dot{x}_1$ lies in $\overline{\Omega_0}$ and therefore $h(x,t) = \overline{f}(\dot{x}_0(1-t) + t\dot{x}_i) \neq y_0$. Since $h(x,0) = g_0(\dot{x}_0) = g(\dot{x}_1)$ and $h(x,1) = f(\dot{x}_1)$, we obtain a homotopy of the desired property.

Part of Hopf's finite-dimensional theorem states [**2**, p. 502ff.]: if Ω^{n+1} is an open bounded set of E^{n+1} and if f is a continuous map $\partial\Omega^{n+1} \to E^m - y_0$, where $m > n + 1$ and where y_0 is a point of E^m, then f can always be extended to a continuous map $\overline{\Omega^{n+1}} \to E^m - y_0$. We prove this assertion in §8.2. It will be used in Chapter 9.

§9.1 contains a proof of a theorem by Borsuk [**7**] and its generalization to Banach space: if Ω is a bounded open set in the Banach space E which is symmetric with respect to the zero point θ of E and if f is an L.-S. map $\partial\Omega \to E$ which is odd, i.e., satisfies

$$f(-x) = -f(x), \tag{15}$$

then the winding number $u(f(\partial\Omega), \theta)$ is odd. The proof for the finite-dimensional case uses the extension theorem, Theorem 2.2 in Chapter 8, and follows the proof given by J. T. Schwartz [**51**]. The generalization to Banach space is based on Lemma 1.9 in Chapter 9, which is due to Granas [**24**, p. 41].

§9.2 contains some consequences of the (generalized) Borsuk theorem. Lemma 2.2, for example, states that the conclusion of the Borsuk theorem remains true if the assumption (15) is replaced by the weaker one that for no $x \in \partial\Omega$ do the "vectors" $f(x)$ and $f(-x)$ have the same direction. Another consequence (Theorem 2.5 of Chapter 9) is a generalization to Banach spaces of a theorem conjectured by Ulam, proved by Borsuk, and now known as the Borsuk-Ulam theorem according to which a continuous map of an n-sphere S^n into the space E^n maps at least one pair of antipodal points of S^n into the same point of E^n (cf. [**24**, Chapter IV]).

We finally note that topological tools (like subdivision and simplicial approximations) used in this book are defined and treated not in the degree of abstraction and generality to be found in topology books but only to the degree needed for the purpose at hand.

CHAPTER 1

Function-Analytic Preliminaries

1. The term "Banach space" will always denote a Banach space over the reals except in Appendix A. For definition and elementary properties of a Banach space the reader is referred to [**18**].

2. Let E be a Banach space with norm $\|\cdot\|$ and zero element θ. If S is a subset of E, then its closure and boundary will be denoted by $\overline{S}$ and ∂S resp., and the empty set by $\varnothing$. I denotes the identity map E onto E. By "subspace" of E we always mean "linear subspace." If x_0 is a point of E and ρ a positive number, the set $\{x \in E \mid \|x - x_0\| < \rho\}$ is called the ball with center x_0 and radius ρ. It will be denoted by $B(x_0, \rho)$.

3. DEFINITION. (i) Let E_1 and E_2 be closed subspaces of E and suppose every point $x \in E$ can be represented uniquely as

$$x = x_1 + x_2, \qquad x_1 \in E_1, \; x_2 \in E_2. \tag{1}$$

Then we write

$$E = E_1 \dot{+} E_2 \tag{2}$$

and say: E is the direct sum of E_1 and E_2. Each of the spaces E_1 and E_2 is said to be a direct summand of E, and E_1 and E_2 are said to be complementary to each other.

We note that the uniqueness of the representation (1) is equivalent to saying that

$$E_1 \cap E_2 = \theta, \tag{3}$$

where the symbol "$\cap$" denotes intersection.

(ii) If E_1 and E_2 are Banach spaces, then the "product" $E_1 \times E_2$ is defined as the space of couples (x_1, x_2), $x_1 \in E_1$, $x_2 \in E_2$ with the linear operations

$$(x_1, x_2) + (y_1, y_2) = (x_1 + y_1, x_2 + y_2),$$

$$\lambda(x_1, x_2) = (\lambda x_1, x_2), \qquad \lambda \text{ real}.$$

With a proper norm, $E_1 \times E_2$ becomes a Banach space (see [**18**, p. 89]), and by identifying the points (x_1, θ) and (θ, x_2) of $E_1 \times E_2$ with the points x_1 of E_1

and x_2 of E_2 resp., the spaces E_1 and E_2 become subspaces of $E_1 \times E_2$ and $E_1 \times E_2 = E_1 \dotplus E_2$.

REMARK. If the direct summand $E_1 \subset E$ is given, then E_2 is not uniquely determined by E. However, if E_2 is of finite dimension n and $E = E_1 \dotplus E_2 = E_1 \dotplus E_2'$, then E_2' is also of dimension n. Indeed for $x \in E$ we have the direct decompositions $x = x_1 + x_2 = x_1' + x_2'$ where $x_2 \in E_2$, $x_2' \in E_2'$ while x_1 and x_1' are elements of E_1. Thus $x_2 = \bar{x}_1 + x_2'$ with $\bar{x}_1 = x_1' - x_1 \in E_1$. This direct decomposition of x_2 shows that x_2' is uniquely determined by x_2. We thus obtain a linear map l of E_2 into E_2'. Thus $n \leq n'$ where n' denotes the dimension of E_2'. Since $n' \leq n$ can be proved the same way, we see that $n' = n$. The common dimension n is called the codimension of E_1.

4. LEMMA. *A finite dimensional subspace E_1 of E is a direct summand of E.*

PROOF. Let $b_1, b_2, \ldots, b_n$ be a base for E_1. Then every $x \in E_1$ can be written uniquely as

$$x = \sum_{i=1}^{n} \alpha_i b_i, \tag{4}$$

where the $\alpha_i = \alpha_i(x)$ are real-valued functions with domain E_1 which are obviously linear. But it is well known that they are also continuous (see e.g. [**18**, pp. 244–245]). Therefore, by the Hahn-Banach theorem [**18**, II.3.10] they can be extended to continuous linear functionals $A_i(x)$ on E. We now set

$$x_1 = \sum_{i=1}^{n} A_i(x) b_i, \qquad x_2 = x - x_1. \tag{5}$$

As x varies over E, then obviously x_1 varies over E_1, and x_2 over a linear subspace E_2 of E. Now (1) is obviously true. It remains to prove (3). Let $x \in E_1 \cap E_2$. From $x \in E_2$ we see that

$$x = \bar{x} - \sum_{i=1}^{n} A_i(\bar{x}) b_i$$

for some $\bar{x} \in E$. But this equality implies that $\bar{x} \in E_1$ since $x \in E_1$. Thus

$$\bar{x} = \sum_{i=1}^{n} A_i(\bar{x}) b_i \quad \text{and} \quad A_i(\bar{x}) = \alpha_i(\bar{x}).$$

It follows that $x = \theta$.

5. DEFINITION. A continuous linear map $P \colon E \to E$ is called a projection if

$$P^2(x) = P(x) \quad \text{for all } x \in E. \tag{6}$$

6. LEMMA. *Let x_1, x_2 be as in (5); then the maps $P_1 \colon x \to x_1$ and $P_2 \colon x \to x_2$ are projections.*

PROOF. By (5)

$$P_1^2(x) = P_1(x_1) = \sum_{i=1}^{n} A_i(P_1(x))b_i$$

$$= \sum_{i=1}^{n} A_i \left(\sum_{j=1}^{n} A_j(x)b_j \right) b_i = \sum_{i,j=1}^{n} A_j(x)A_i(b_j)b_i.$$

But obviously

$$A_i(b_j) = \begin{cases} 1, & i = j, \\ 0, & i \neq j. \end{cases}$$

Therefore $P_1^2(x) = \sum_{j=1}^{n} A_j(x)b_j = P_1(x)$. To prove (6) with $P = P_2$, we apply P_2 to the second of the equations (5) and obtain

$$P_2^2(x) = P_2(x_2) = P_2 x - P_2 P_1(x). \tag{6a}$$

But applying P_2 to the first equation (5) we obtain

$$P_2 P_1 x = P_2 x_1 = \theta \tag{7}$$

since $P_2 b_i = \theta$. (6a) together with (7) proves our assertion, since the boundedness of P_1 and P_2 is clear from (5), the A_i being bounded.

7. DEFINITION. A subset S of E is called compact if every sequence of points in S contains a subsequence which converges to a point of S.

8. DEFINITION. A continuous map F of a subset S of E into E is called completely continuous if for every bounded subset S_1 of S the set $\overline{F(S_1)}$ is compact.

9. DEFINITION. A map $f: S \to E$ is called a Leray-Schauder, or L.-S., map if $F = I - f$ is completely continuous. Here as elsewhere the symbol "$f: S \to E$" means: f is a map mapping S into E.

10. LEMMA. *Let S be a bounded subset of E, and let f be an L.-S. map $S \to E$. Then*

(i) *$f(S)$ is bounded;*

(ii) *if S is closed, then $f(S)$ is closed;*

(iii) *if Γ is a compact set in E, then $f^{-1}(\Gamma)$ is compact;*

(iv) *let S be a bounded closed set in E, and let $f: S \to E$ be an L.-S. map. Suppose f maps S onto $f(S)$ in a one-to-one fashion. Then the $g = f^{-1}$ is an L.-S. map.*

PROOF. Assertion (i) follows trivially from Definition 9 and the fact that a compact set is bounded.

Proof of (ii). Let $y_1, y_2, \ldots$ be a convergent sequence of points in $f(S)$ and let

$$y_0 = \lim_{j \to \infty} y_j. \tag{8}$$

We have to show that there exists an $x_0 \in S$ such that

$$y_0 = f(x_0). \tag{9}$$

Now since $y_j \in f(S)$ there exist $x_j \in S$ such that

$$y_j = f(x_j) = x_j - F(x_j), \qquad j = 1, 2, \dots, \tag{10}$$

where F is completely continuous. Consequently there exists a subsequence $\{x_{j_i}\}$ of the sequence $\{x_j\}$ such that the $F(x_{j_i})$ converge. Since by (8) the y_{j_i} converge, it follows from (10) that the x_{j_i} converge. If x_0 is the limit of that sequence, then $x_0 \in S$ since S is closed, and assertion (9) follows from (8) and (10).

Proof of (iii). If $f^{-1}(\Gamma)$ is empty or finite, then the assertion is trivially true. Suppose then that $f^{-1}(\Gamma)$ contains an infinite sequence $x_1, x_2, \dots$. We have to prove that this sequence contains a convergent subsequence for whose limit x_0

$$f(x_0) \in \Gamma. \tag{11}$$

Now let the sequence $y_1, y_2, \dots$ of points in Γ be defined by (10). Since Γ is compact and F completely continuous, there exists a subsequence $\{j_i\}$ of the integers such that the y_{j_i} converge to a point $y_0 \in \Gamma$ and such that the $F(x_{j_i})$ converge. It then follows from (10) that the x_{j_i} converge to some point x_0 and that $y_0 = f(x_0)$, i.e., that $x_0 \in f^{-1}(\Gamma)$. This proves assertion (iii).

Proof of (iv). Let $G(y) = y - g(y)$. We have to prove that $G(y)$ is completely continuous, i.e., that

(a) every sequence $y_1, y_2, \dots$ of points in $f(\overline{\Omega})$ contains a subsequence $\{y_{j_i}\}$ for which $G(y_{j_i})$ converges, and

(b) $G(y)$ is continuous.

Now from $f(x) = x - F(x)$, and from $y = f(x)$, $x = g(y)$ we see that

$$G(y) = y - g(y) = f(x) - x = -F(x).$$

Now if $\{y_j\}$ is a sequence in $f(S)$, then $\{x_j = g(y_j)\}$ is a sequence in S. Since F is completely continuous and $G(y) = -F(x)$, assertion (a) follows.

The proof of assertion (b) is based on the following remark. If the sequence $\{z_j\}$ in E satisfies the following two conditions:

(α) every subsequence of the $\{z_j\}$ has a convergent subsequence;

(β) all convergent subsequences of the sequence $\{z_j\}$ converge to the same point z_0,

then the sequence $\{z_j\}$ converges to z_0.

Indeed, if this were false, there would exist an $\varepsilon > 0$ and a subsequence $\{z_{j_i}\}$ of the $\{z_j\}$ such that $\|z_0 - z_{j_i}\| \geq \varepsilon$. This inequality holds for every subsequence of the $\{z_{j_i}\}$. This leads to a contradiction since by (α) we can choose a convergent subsequence of the sequence $\{z_{j_i}\}$, and by (β) such a subsequence converges to z_0.

This proves the remark and we return to the proof of (b). We have to show: if $y_1, y_2, \dots$ is a sequence of points in $f(S)$ which converges to a point y_0, then the sequence $\{z_i = G(y_i)\}$ converges to $G(y_0)$. Now by assertion (a) the sequence $\{z_i\}$ satisfies assumption (α) of the remark. It remains to show that it also satisfies assumption (β). Suppose $\{y_{n_i}\}$ is a subsequence of the $\{y_i\}$ for which

$z_{n_i} = G(y_{n_i})$ converges. Then we have to show that the limit of the z_{n_i} is $G(y_0)$. Now $G(y_{n_i}) = -F(x_{n_i})$ with $x_{n_i} = g(y_{n_i})$. Thus the $F(x_{n_i})$ converge. But $y_{n_i} = f(x_{n_i}) = x_{n_i} - F(x_{n_i})$. Since $y_{n_i} \to y_0$, it follows that the x_{n_i} converge to some point x_0. Then $y_0 = x_0 - F(x_0) = f(x_0)$, and $G(y_0) = -F(x_0) = -\lim F(x_{n_i}) = \lim G(y_{n_i})$ as we wanted to prove.

REMARK. It is easily seen that the assertions (ii) and (iii) of Lemma 10 remain valid if the assumption that S is bounded is replaced by the assumption that $f^{-1}(B)$ is bounded for any bounded set B.

11. DEFINITION. Let l be a linear continuous map $E \to E$. Then l is called regular or nonsingular if it maps E onto E. l is called singular if it is not regular. The kernel $K = K(l)$ of l is defined by $K(l) = \{x \in E \mid l(x) = 0\}$. $K(l)$ is obviously a subspace of E.

12. LEMMA (FREDHOLM ALTERNATIVE). *Let l be a linear L.-S. map $E \to E$, and let the range $P = P(l)$ be defined by*

$$P(l) = \{y \in E \mid y = l(x) \quad \text{for some } x \in E\}.$$

Then

(i) *$\dim K(l)$ is finite;*

(ii) *there exists a subspace $K^*(l)$ such that*

$$E = P(l) \dotplus K^*(l). \tag{12}$$

K^* is of course not uniquely determined. However

(iii)

$$\dim K^*(l) = \dim K(l) \tag{13}$$

for every $K^*(l)$ satisfying (12). Each such $K^*(l)$ is called a co-kernel of l.

(iv) *l is nonsingular if and only if it is one-to-one.*

(v) *If l is a nonsingular linear L.-S. map $E \to E$ then l^{-1} is a linear L.-S. map $E \to E$.*

PROOF. A proof for assertions (i), (ii) and (iii) may be found in [**4**, Chapter X, §§2, 4]. Assertion (iv) is an immediate consequence of the first three assertions.

Proof of assertion (v). l^{-1} is a one-valued map E onto E which is obviously linear. By a classical theorem of Banach, it is also continuous (see [**4**, Chapter III, Theorem 5, p. 41] or [**18**, Theorem II.2.2, p. 57]. Now $l = I - L$ with completely continuous L. Let

$$k = l(h) = h - L(h) = h - L(l^{-1}(k))$$

or

$$l^{-1}(k) = h = k + L(l^{-1}(k)).$$

Since L is completely continuous and l^{-1} is continuous, it follows that the map $L(k) = -L(l^{-1}(k))$ is completely continuous.

13. The differential. Let x_0 be a point of E, and $\Omega \subset E$ an open set containing x_0. Let f be a continuous map of Ω into a Banach space $\tilde{E}$. If there exists a continuous linear map $l_{x_0} : \Omega \to \tilde{E}$ such that

$$\lim_{h \to \theta} \frac{f(x_0 + h) - f(x_0) - l_{x_0}(h)}{\|h\|} = \theta, \tag{14}$$

then it is easily seen that l_{x_0} is uniquely determined by (14) (see e.g. [**16**, VIII.1]), and l_{x_0} is called the differential of f at x_0, while f is called differentiable at x_0. We use the notation

$$l_{x_0}(h) = Df(x_0; h). \tag{15}$$

If f is differentiable at every $x_0 \in \Omega$, then it is called differentiable in Ω.

14. The chain rule. Let E, F, G be Banach spaces. Let x_0 be a point of E and U an open neighborhood of x_0. Let f be a continuous map $U \to F$. Let V be an open neighborhood of the point $y_0 = f(x_0)$, and let g be a continuous map $V \to G$. Let f and g be differentiable at x_0 and y_0 resp. Then the composite map $gf : U \to G$ is differentiable at x_0, and

$$Dgf(x_0; h) = Dg(y_0; Df(x_0; h)). \tag{16}$$

For a proof see [**16**, p. 145].

15. The second differential. As is well known, a continuous linear map of E into a Banach space E is bounded, i.e., for all $h \in E$

$$\|l(h)\| \le M\|h\| \tag{17}$$

for some positive M. It is also well known that with the obvious definitions of addition and multiplication by reals the set of linear maps $l : E \to \tilde{E}$ becomes a Banach space if the norm $\|l\|$ of l is defined as the smallest number M which satisfies (17) (see [**18**, p. 61]). We denote this Banach space by $\Lambda(E, \tilde{E})$. With this notation $Df(x_0; \cdot)$ is an element of $\Lambda(E, \tilde{E})$ if f is differentiable at x_0, and if f is differentiable in the open set Ω containing x_0, then Df is a map $\Omega \to \Lambda(E, \tilde{E})$ mapping the point $x \in \Omega$ into the element $Df(x; \cdot)$ of $\Lambda(E, \tilde{E})$. If this map is continuous, we say: f is continuously differentiable, or Df is continuous. It may be that at some point $x_0 \in \Omega$ this map of Ω into the Banach space $\tilde{\tilde{E}} = \Lambda(E, \tilde{E})$ has a differential. Then we denote this differential by $D^2 f(x_0; \cdot)$ and call it the second differential of f at x_0. $D^2 f(x_0; \cdot)$ is an element of the Banach space $\Lambda(E, \tilde{\tilde{E}}) = \Lambda(E, \Lambda(E, \tilde{E}))$. If $D^2 f(x_0; \cdot)$ exists for all $x_0 \in \Omega$, we say f is twice differentiable in Ω. Then D^2 is a map $\Omega \to \Lambda(E, \Lambda(E, \tilde{E}))$; if this map is continuous, we say f is twice continuously differentiable.

15a. The second differential as a bilinear map. Let E and $\tilde{E}$ be Banach spaces. A bilinear map $E \times E \to \tilde{E}$ is a map $l_2 = l_2(h_1, h_2)$, $h_1, h_2 \in E$, into $\tilde{E}$ which is linear in h_1 and in h_2. If l_2 is continuous, then

$$\|l_2(h_1, h_2)\| \le M\|h_1\| \cdot \|h_2\| \tag{18}$$

for some number M. The set of all continuous bilinear maps $l_2 : E \times E \to \tilde{E}$ can be made a Banach space if the norm of l_2 is defined as the smallest number

M satisfying (18) (see e.g. [**16**, V.7]). We denote this Banach space by $\Lambda_2 = \Lambda(E \times E, \tilde{E})$. We define a map of this space into the space $\Lambda(E, \Lambda(E, \tilde{E}))$ defined in the preceding section as follows: given $l_2 \in \Lambda_2$ and $h_1 \in E$ we define a map $l_{h_1} \colon E \to \tilde{E}$ by setting

$$l_{h_1}(h_2) = l_2(h_1, h_2). \tag{19}$$

Then $l_{h_1} \in \Lambda(E, \tilde{E})$ and the map

$$h_1 \to l_{h_1} \tag{20}$$

is an element of $\Lambda(E, \Lambda(E, \tilde{E}))$. Thus we assign to the element l_2 of Λ_2 the element (20) of $\Lambda(E, \Lambda(E, \tilde{E}))$. It can be shown that the map thus defined is a linear isometry onto (see [**16**, V.7.8]). We identify the two spaces in question by identifying elements corresponding under this linear isometry. In particular the element $D^2 f(x_0; \cdot)$ of $\Lambda(E, \Lambda(E, \tilde{E}))$ may then be considered as an element of Λ_2 which we denote by $D^2 f(x_0, h_1, h_2)$. If f is twice differentiable in an open neighborhood Ω of x_0, then $D^2 f$ is said to be continuous at x_0 if the map $D^2 f \colon \Omega \to \Lambda_2$ is continuous. It can be proved that D^2 is symmetric, i.e., that

$$D^2 f(x_0, h_1, h_2) = D^2 f(x_0, h_2, h_1)$$

(see [**16**, VIII.12]). Differentials D^r of order $r > 2$ are defined correspondingly.

16. Let Ω be an open set in E and let f be a map of Ω into the Banach space $\tilde{E}$. We say "$f \in C'(\Omega)$" if f is differentiable in Ω and if the differential is continuous. We say "$f \in C''(\Omega)$" if $D^2 f(x; \cdot)$ exists for all $x \in \Omega$ and is continuous. If S is an arbitrary subset of E and f a map $S \to \tilde{E}$, then $f \in C'(S)$ or $f \in C''(S)$ means: f is defined in some open subset Ω of E containing S and $f \in C'(\Omega)$ or $C''(\Omega)$ resp. $C^r(S)$, $r \geq 2$, is defined correspondingly (cf. §15).

17. DEFINITION. Let Ω be a bounded open set in E. Let $f \in C'(\bar{\Omega})$. Then a point $x_0 \in \bar{\Omega}$ is called a regular point (with respect to f) if the linear map $Df(x_0; h)$ is not singular. The point x_0 is called singular or critical if it is not regular. A point y_0 in E is called a regular value for f if no solution $x \in \bar{\Omega}$ of the equation

$$f(x) = y_0 \tag{21}$$

is a singular point. y_0 is called a singular or critical value if it is not a regular value.

Note that in particular every $y_0 \in E$ for which (21) has no solution in $\bar{\Omega}$ is a regular value for f.

18. KRASNOSELSKIĬ'S LEMMA. *Let F be a completely continuous map $\bar{\Omega} \to E$ which is differentiable in $\bar{\Omega}$. Then for each $x \in \bar{\Omega}$ the linear map $E \to E$ given by $h \to DF(x; h)$ is completely continuous.*

For a proof see [**31**, p. 136]. Obviously the lemma is equivalent to saying that the differential $h - DF(x; h)$ of the differentiable L.-S. map $f(x) = x - F(x)$ is, for each $x \in \bar{\Omega}$, a linear L.-S. map.

19. LEMMA. *Let Ω be a bounded open subset of E and $f \in C'(\overline{\Omega})$. In addition we assume that f is L.-S. Then*

(i) *the set of regular points in Ω is open, and*

(ii) *the union of $\partial\Omega$ with the set $S(\Omega)$ of singular points in Ω is closed.*

PROOF. Let $x_0 \in \Omega$ be a regular point for f. By Lemma 18, $l_0 = Df(x_0; \cdot)$ is an L.-S. map. By assumption l_0 is not singular. Therefore by assertion (v) of Lemma 12, l_0^{-1} exists as a linear L.-S. map and thus is bounded. But by the continuity of $Df(x; \cdot)$ (as a map $\overline{\Omega} \to \Lambda(E, E)$) there exists an open neighborhood $N(x_0)$ of x_0 such that for $x \in N(x_0)$

$$\|Df(x; \cdot) - Df(x_0; \cdot)\| < \|l_0^{-1}\|^{-1}.$$

From this inequality and from Lemma 5.3 of Appendix A, it follows that every $x \in N(x_0)$ is a regular point for f.

Assertion (ii) follows from (i) since $(\partial\Omega) \cup S(\Omega)$ is the complement in $\overline{\Omega}$ of the set of regular points in Ω.

20. LEMMA. *Under the assumptions of Lemma 19 the set R of those regular values y which satisfy*

$$y \notin f(\partial\Omega) \tag{22}$$

is open.

PROOF. The complement of R is the set

$$R' = f(S(\Omega) \cup \partial\Omega).$$

But $S(\Omega) \cup \partial\Omega$ is obviously bounded and, by assertion (ii) of Lemma 19, also closed. Therefore by assertion (ii) of Lemma 10 the set R' is closed. Thus R is open.

21. INVERSION THEOREM. *Let $x_0 \in E$, let Ω be a bounded open neighborhood of x_0, and let $f \in C'(\Omega)$ be a map $\Omega \to E$. Suppose $Df(x_0; \cdot)$ is nonsingular and one-to-one. Then there exists an open neighborhood $\Omega_1 \subset \Omega$ of x_0 and an open neighborhood $N(y_0)$ of $y_0 = f(x_0)$ such that the map f maps Ω_1 one-to-one onto $N(y_0)$ and such that the map $f^{-1}: N(y_0)$ onto Ω_1 is in $C'(N(y_0))$. In addition if $f \in C^s(\Omega)$, $s \geq 1$, then $f^{-1} \in C^s(N(y_0))$ for some neighborhood $N(y_0)$ of y_0.*

For a proof see, e.g., [**16**, X.2].

Note: if f is an L.-S. map, then, by part (iv) of Lemma 12 and by Lemma 18, the assumption that $Df(x_0; \cdot)$ is one-to-one follows from the assumption that this differential is nonsingular.

22. LEMMA. *Let Ω be a bounded open set in E and let $f \in C'(\overline{\Omega})$ be an L.-S. map $\overline{\Omega} \to E$. Suppose y_0 is a regular value for f, and*

$$y_0 \notin f(\partial\Omega). \tag{23}$$

Then (a) *the equation*

$$f(x) = y_0 \tag{24}$$

has at most a finite number of solutions.

(b) *If in addition* $y_0 = \theta \in \Omega$, *if* Ω *is symmetric with respect to* θ *(i.e., contains with* x *also* $-x$*), and if* f *is odd (i.e.,* $f(-x) = -f(x)$*), then* $Df(x; \eta)$ *is even in* x, *i.e.,*

$$Df(x, \eta) = Df(-x, \eta). \tag{25}$$

PROOF. (a) Suppose there exists an infinite sequence $x_1, x_2, \ldots$ of points in Ω which are solutions of (24). Since the set $\{y_0\}$ consisting of the single point y_0 is compact, the set $f^{-1}(y_0)$ is compact by part (iii) of Lemma 10. Therefore there exists a subsequence $x_{i_1}, x_{i_2}, \ldots$ of the sequence $x_1, x_2, \ldots$ which converges to some point $x_0 \in \overline{\Omega}$. Then by continuity of f, x_0 is a solution of (24), and by (23) the point x_0 lies in Ω. Now by Theorem 21 there exists a neighborhood $\Omega_1 \subset \Omega$ of x_0 such that x_0 is the only root of (24) which lies in Ω_1. This, however, contradicts the fact that the solutions $x_{i_1}, x_{i_2}, \ldots$ converge to x_0.

Proof of (b). From the assumption that $f(-x) = -f(x)$ and from the linearity of $Df(x; \eta)$ in η, we see that with $\eta = -h$

$$\frac{f(x+h) - f(x) - Df(x; h)}{\|h\|} = -\frac{f(-x+\eta) - f(-x) - Df(x; \eta)}{\|\eta\|}.$$

By definition of the differential the left member tends to zero as $h \to \theta$. Therefore

$$\lim_{\eta \to 0} \frac{f(-x+\eta) - f(-x) - Df(x; \eta)}{\|\eta\|} = \theta.$$

But, by §13, this shows that $Df(x; \eta)$ is the differential of f at $-x$. This proves (25).

23. The intervals $0 \leq t \leq 1$ and $0 < t < 1$ will be denoted by $[0, 1]$ and $(0, 1)$ resp. If x_0 and x_1 are points in the Banach space E, then the segment $\overline{x_0 x_1}$ is defined as the set $\{x \in E \mid x = (1-t)x_0 + tx_1, \ t \in [0, 1]\}$, and the ray $\overrightarrow{x_0 x_1}$ as the set $\{x \in E \mid x = (1-t)x_0 + tx_1, \ t \geq 0\}$. We also say $\overrightarrow{x_0 x_1}$ is a ray issuing from x_0. A set $C \subset E$ is called convex if for any couple x_0, x_1 of points in C the segment $\overline{x_0 x_1}$ is contained in C.

24. Let $p_0, p_1, \ldots, p_N$ be $N + 1$ distinct points in E. Then the union P of the segments $s_i = \overline{p_i p_{i+1}}$, $i = 0, 1, 2, \ldots, N - 1$, is called a polygon with vertices p_i and edges s_i. P is said to connect p_0 and p_N. Obviously $s_i \cap s_{i+1} = p_{i+1}$, $i = 0, 1, \ldots, N - 1$. If these are the only intersections of two edges s_i and s_j, $i \neq j$, then the polygon P is called simple.

25. We recall that a set $S \subset E$ is called connected if it is not the union of two nonempty disjoint subsets which are open (with respect to S). With this definition the following theorem holds: if S is an open connected set in E, then any two points q_0 and q_1 in S can be connected by a simple polygon $P \subset S$. (See, e.g., [2, Theorem XX, p. 50]; the theorem is stated there only for finite-dimensional spaces, but the proof holds for any (linear) metric space and thus for our Banach space E.)

26. DEFINITION. Let S be a subset of E, let E_1 be a Banach space (which may be equal to E), let $f: S \to E_1$ be a C'-map and let $y_0 \neq y_1$ be points of E_1. Then the segment $\sigma = \overline{y_0 y_1}$ is said to be transversal to f if for every $x \in S$ for which $f(x) \in \sigma$ the differential $Df(x; h)$ together with σ spans E_1.

27. THEOREM. *Let the subset Y of E be precompact, i.e., have the property that its closure is compact. Then if ε is a given positive number, there exist points $y_1, y_2, \ldots, y_s$ in Y such that for every point $y \in \overline{Y}$*

$$\|y - y_\sigma\| < \varepsilon \tag{26}$$

for at least one positive integer $\sigma \leq s$.

PROOF. As y varies over $\overline{Y}$, the balls $B(y, \varepsilon/2)$ form an open covering of $\overline{Y}$. Consequently, there exist a finite number of these balls, say $B(\overline{y}_1, \varepsilon/2), \ldots,$ $B(\overline{y}_s, \varepsilon/2)$ which form an open covering of $\overline{Y}$, $\overline{y}_0, \ldots, \overline{y}_s$ being points of $\overline{Y}$. Now for every $\overline{y}_\sigma$ there exists a $y_\sigma \in Y$ such that $\|y_\sigma - \overline{y}_\sigma\| < \varepsilon/2$. Now if y is an arbitrary point in $\overline{Y}$, then $y \in B(\overline{y}_{\overline{\sigma}}, \varepsilon/2)$ for some $\overline{\sigma} \leq s$. For such $\overline{\sigma}$

$$\|y - y_{\overline{\sigma}}\| \leq \|y - \overline{y}_{\overline{\sigma}}\| + |\overline{y}_{\overline{\sigma}} - y_{\overline{\sigma}}| < \varepsilon.$$

REMARK. This lemma states that "up to ε" a precompact set lies in a finite-dimensional plane, namely, the plane spanned the points $y_1, y_2, \ldots, y_s$ satisfying (26).

28. DEFINITION. Let S be a subset of E. Then the convex hull $\operatorname{co} S$ of S is defined as the smallest convex set containing S. Since the intersection of convex sets is convex, $\operatorname{co} S$ is the intersection of all convex sets containing S. The closed convex hull $\overline{\operatorname{co}} S$ is defined as the intersection of all closed convex sets containing S. Obviously $\overline{\operatorname{co}} S$ is closed. Moreover

$$\overline{\operatorname{co}} S = \overline{\operatorname{co} S} \tag{27}$$

(see [**18**, p. 415]).

29. MAZUR'S LEMMA. *If S is compact, then $\overline{\operatorname{co}} S$ is compact (see, e.g., [**18**, p. 416]).*

30. DUGUNDJI'S EXTENSION THEOREM. *(For a proof see [**17** or **28**, p. 57].) Let S_0 be a closed subset of the Banach space E and let ϕ_0 be a continuous map of S_0 into a Banach space E_1. Then there exists a continuous map $\phi: E \to E_1$ which is an extension of ϕ_0 (i.e., whose restriction to S_0 is ϕ_0) and which has the following property:*

$$\phi(E) \subset \operatorname{co} \phi(S_0). \tag{28}$$

31. LEMMA. *Let S_0 be a bounded closed subset of the Banach space E, and let ϕ_0 be a completely continuous map of S_0 into a Banach space E_1. Then there exists a completely continuous extension ϕ of ϕ_0 to E which satisfies (28).*

PROOF. By Dugundji's theorem in §30, there exists a continuous extension ϕ of ϕ_0 satisfying (28). It remains to verify that such ϕ is completely continuous, i.e., that for every bounded subset Ω_0 of E the set $\overline{\phi(\Omega_0)}$ is compact. Let then $\Omega_0 \subset E$ be bounded. By (28) and (27)

$$\phi(\Omega_0) \subset \phi(E) \subset \operatorname{co}\phi(S_0) \subset \operatorname{co}\overline{\phi(S_0)},$$

$$\overline{\phi(\Omega_0)} \subset \overline{\phi(E)} \subset \overline{\operatorname{co}\overline{\phi(S_0)}} = \overline{\operatorname{co}}\,\overline{\phi(S_0)}. \tag{29}$$

But $\overline{\phi(S_0)}$ is compact by assumption. Therefore, by Mazur's lemma in §29, the set $\overline{\operatorname{co}}\,\overline{\phi(S_0)}$ is compact. Thus (29) shows that the closed set $\overline{\phi(\Omega_0)}$ is contained in a compact set and therefore compact. (For different proofs see [**31**, p. 116 and **24**, p. 29].)

32. Let Ω be an open rectangle in the plane E^2 with coordinates x_1, x_2. Let f be a C'-map $\overline{\Omega} \to E$. Then to each $\varepsilon > 0$ there is a polynomial $p(x_1, x_2)$ such that

$$\left.\begin{array}{c} |f(x_1, x_2) - p(x_1, x_2)| \\[2mm] \left|\dfrac{\partial f}{\partial x_i}(x_1, x_2) - \dfrac{\partial p}{\partial x_i}(x_1, x_2)\right| \end{array}\right\} < \varepsilon$$

for $(x_1, x_2) \in \Omega$ and $i = 1, 2$. (See, e.g., [**13**, p. 68].) If f is just continuous, then the first of the above inequalities holds for some polynomial p. Corresponding statements hold for E^n with $n > 2$.

33. DEFINITION. Let H be an open subset of the Banach space E (which may be equal to E). Let K be a closed bounded subset of H. Then a component of $H - K$ is a maximal open and connected subset of $H - K$.

34. We will need the following properties of components: (i) every point $x_0 \in E - K$ determines uniquely a component $C(x_0)$ of $E - K$ which contains x_0; (ii) among the components of $E - K$ there is exactly one unbounded one; (iii) if B is a ball for which

$$K \subset B, \tag{30}$$

then all bounded components of $E - K$ are subsets of B; (iv) let C_0 be the unbounded component of $E - K$ and let $\{C_\alpha\}_{\alpha \in A}$ be the set of the bounded components of $E - K$, where A is an index set (not containing 0). Suppose that K has a positive distance from ∂B. Then the components of $B - K$ are the sets $C_0 \cap B$ and C_α, $\alpha \in A$.

Assertions (i) through (iii) follow easily from Definition 33 and the fact that $E - \overline{B}$ is an open connected unbounded subset of $E - K$ and therefore a subset of an unbounded component C_0 of $E - K$, while $C(x_0) = C_0$ for every $x_0 \in E - B$.

As to assertion (iv) we note that $E - K = C_0 + \bigcup_\alpha C_\alpha$. If we intersect both members of this inequality by B and note that $E \cap B = B$ and, by (iii), $C_\alpha \cap B = C_\alpha$, we obtain the equality $B - K = C_0 \cap B + \bigcup_{\alpha \in A} C_\alpha$ which is easily seen to imply (iv).

35. THEOREM. *Let X and Y be two arbitrary sets and let A and B be maps $X \to Y$ and $Y \to X$ resp. Suppose that*

$$\text{(a)} \quad AB = I_Y, \qquad \text{(b)} \quad BA = I_X, \tag{31}$$

where I_X and I_Y denote the identity maps on X and Y resp. Then A is an isomorphism (i.e., a one-to-one and onto map of X onto Y) and B is an isomorphism of Y onto X.

PROOF. Let x_1 and x_2 be points of X for which $A(x_1) = A(x_2)$ and therefore $BA(x_1) = BA(x_2)$. But then by (31b) $x_1 = BA(x_1) = BA(x_2) = x_2$. This shows that A is one-to-one. To show that A is onto, let y be an arbitrary point of B. Then by (31a) $AB(y) = y$. This shows that A maps the point $x = B(y)$ onto the point y. This proves the assertion concerning A. The assertion concerning B is proved correspondingly.

36. Let S be an arbitrary set. Then a "vector space V generated by S" can be defined as follows: we consider all maps f of S into the reals R for which $f(x) \neq 0$ for at most a finite number of elements $x \in S$. We define addition $f_1 + f_2$ of two such functions by setting

$$(f_1 + f_2)(x) = f_1(x) + f_2(x), \tag{32}$$

and define (rf) for $r \in R$ by

$$(rf)(x) = rf(x). \tag{33}$$

Finally for $x_0 \in S$ we identify the function

$$f_{x_0}(x) = \begin{cases} 1, & \text{for } x = x_0, \\ 0, & \text{for } x \neq x_0 \end{cases}$$

with the point x_0 of S. With this identification we see from the rules (32) and (33) that the function which maps the points $x_1, x_2, \ldots, x_n$ of S into the real numbers $r_1, r_2, \ldots, r_n$ resp. and which equals 0 for $x \neq x_i$ $(i = 1, 2, \ldots, n)$ may be written as the "linear combination"

$$f(x) = \sum_{i=1}^{n} r_i x_i.$$

Obviously the space V thus defined is a vector space (over the reals).

37. Let V be the vector space defined in §36. A subset G of V is said to generate V if every element x of V may be written as

$$x = \sum_{i=1}^{n} g_i r_i, \qquad g_i \in G, \ r_i \in R,$$

with the integer n depending on x.

A subset Σ of V is said to be linearly independent if for any finite set $\sigma_1, \ldots, \sigma_n$ of elements of Σ the relation

$$\sum_{i=1}^{n} r_i \sigma_i = 0, \qquad r_i \in R,$$

implies that $r_1 = r_2 = \cdots = r_n = 0$. A base of V is a subset B of V which generates V and is linearly independent.

Obviously the set S is a basis for V.

38. Let B^1 and B^2 be two bases of V. Then B^1 and B^2 have the same cardinal number. For a proof see e.g. [**30**, IX, 1 and 2].

39.   THEOREM.   *Let S and T be two arbitrary sets, and let V and W be the vector spaces generated by S and T resp. We assume that there is a linear one-to-one correspondence l between the elements of V and W. Then for any bases B_V and B_W of V and W resp.*

$$c(B_V) = c(B_W), \tag{34}$$

where $c(\ )$ denotes "cardinal number of."

PROOF.   Let $\tilde{B}_W = l(B_V)$. By the properties of l there is a one-to-one correspondence between the elements of $\tilde{B}_W$ and those of B_V. Therefore

$$c(B_V) = c(\tilde{B}_W). \tag{35}$$

Morever the elements of B_W are linearly independent, since a relation $\sum_{i=1}^{n} \tilde{w}_i r_i = 0$, $\tilde{w}_i \in \tilde{B}_w$, $r_i \in R$, implies that

$$0 = \sum l(v_i) r_i = l\left(\sum v_i r_i\right), \qquad l(v_i) = \tilde{w}_i \in B_V, \ r_i \in R, \ v_i \in B_V,$$

and therefore $\sum v_i r_i = 0$. Since the v_i are linearly independent, we see that $r_1 = r_2 = \cdots = r_n = 0$.

On the other hand $\tilde{B}_w$ generates W. For if y is an element of W, there is a unique point $x \in V$ such that $y = l(x)$. But since B_V is a base for V we see that

$$x = \sum_{i=1}^{n} b_v^i r_i, \qquad b_v^i \in B_V, \ r_i \in R.$$

Therefore $y = \sum l(b_v^i) r_i$ with $l(b_v^i) \in \tilde{B}_W$. Thus $\tilde{B}_W$ is linearly independent and generates W, i.e., it is a base for W. Since B_W is also a base of W, we see from Lemma 38 that $c(\tilde{B}_W) = c(B_W)$. This together with (35) implies the assertion (34).

CHAPTER 2

The Leray-Schauder Degree for Differentiable Maps

§1. The degree for linear maps

1.1. Let l be a nonsingular linear Leray-Schauder (L.-S.) map of the Banach space E into itself. Let Ω be a bounded open subset of E and y_0 a point of E. (See §§1, 9, 11 in Chapter 1.) We want to define the degree

$$d = d(l, \Omega, y_0). \tag{1.1}$$

Now l maps E one-to-one on itself (assertion (iv) in §1.12). Since the degree is supposed to be an algebraic count of the number of coverings of y_0 by $l(\Omega)$, we want to set

$$d(l, \Omega, y_0) = \begin{cases} 0, & \text{if } y_0 \notin l(\overline{\Omega}), \\ \pm 1, & \text{if } y_0 \in l(\Omega). \end{cases} \tag{1.2}$$

As pointed out in §3 of the introduction, we assume

$$y_0 \notin l(\partial\Omega). \tag{1.3}$$

With this restriction it still remains to determine when to use the plus sign and when the minus sign in (1.2). It is natural to use the first one if l is the identity map. In the general case the decision will be based on Theorem 1.3 below. To state this theorem we need a definition.

1.2. DEFINITION. Let $l_0(x) = x - L_0(x)$ and $l_1(x) = x - L_1(x)$ be two linear nonsingular L.-S. maps $E \to E$. Then l_0 and l_1 are called linearly L.-S. homotopic if there exists a continuous map $L(x, t) \colon E \times [0, 1] \to E$ such that

(a) for each $t \in [0, 1]$ the map $l(x, t) = x - L(x, t)$ is a linear nonsingular L.-S. map $E \to E$, and

(b) $l(x, 0) = l_0(x)$, $l(x, 1) = l_1(x)$.

A homotopy class H of nonsingular linear L.-S. maps is a family of such maps of the property: if $l_0 \in H$ then all linear nonsingular L.-S. maps l which are L.-S. homotopic to l_0 are in H.

1.3. THE LINEAR HOMOTOPY THEOREM. *The family of all nonsingular linear L.-S. maps $E \to E$ consists of exactly two homotopy classes H^+ and H^-,* where H^+ denotes the class containing the identity map $l = I$. For an example of an $l \in H^-$, see subsection 4.4 in Appendix A.

1.4. DEFINITION.[1] For each linear nonsingular L.-S. map $E \to E$ the index $j(l)$ is defined by

$$j(l) = \begin{cases} +1, & \text{if } l \in H^+, \\ -1, & \text{if } l \in H^-. \end{cases} \tag{1.4}$$

1.5. REMARK. The reader will note: if E is finite dimensional with coordinates $x_1, x_2, \ldots, x_n$, then

$$l \in \begin{cases} H^+, & \text{if determinant } l > 0, \\ H^-, & \text{if determinant } l < 0, \end{cases}$$

and the map $(x_1, x_2, \ldots, x_n) \to (-x_1, x_2, \ldots, x_n)$ belongs to H^-.

1.6. Definition of the degree for linear L.-S. maps. Let l and Ω be as in subsection 1.1, and let y_0 be a point of E satisfying (1.3). We then define

$$d(l, \Omega, y_0) = \begin{cases} 0 & \text{if } y_0 \notin l(\overline{\Omega}), \\ j(l) & \text{if } y_0 \in l(\overline{\Omega}). \end{cases}$$

§2. The degree if y_0 is a regular value for the map f

2.1. Let Ω be as in Definition 1.6, and let f be an L.-S. map $\overline{\Omega} \to E$. Let the point $y_0 \in E$ satisfy

$$y_0 \notin f(\partial \Omega). \tag{2.1}$$

Suppose moreover that $f \in C'(\overline{\Omega})$ and that y_0 is a regular value for f (Definition 17 in Chapter 1). Then by Lemma 22 in Chapter 1 the equation

$$f(x) = y_0 \tag{2.2}$$

has either no solution in Ω or a finite number of solutions, say $x_1, x_2, \ldots, x_r$, and the differentials $l_\rho(h) = Df(x_\rho; h)$, $\rho = 1, 2, \ldots, r$, are nonsingular linear maps which by Lemma 18 of Chapter 1 are L.-S. Consequently their indices are defined (see 1.4). This makes the following definition possible.

2.2. DEFINITION. Under the assumptions and with the notations of subsection 2.1, we set $d(f, \Omega, y_0) = 0$ if the equation (2.2) has no roots in Ω. Otherwise we set

$$d(f, \Omega, y_0) = \sum_{\rho=1}^{r} j(Df(x_\rho; h)). \tag{2.3}$$

2.3. The reader will notice: if E is finite dimensional, then $j(Df(x; h))$ is the sign of the Jacobian of f at the point x.

In the following lemma we list a number of elementary but basic properties of the degree. The proof will be given in the Notes to Chapter 2.[2]

2.4. LEMMA. *The assumptions are the same as in Definition 2.2. In partic-ular y_0 is a regular value for f satisfying* (2.1). *I denotes the identity map. It is asserted:*

(i)

$$d(I, \Omega, y_0) = \begin{cases} 1, & \text{if } y_0 \in \Omega, \\ 0, & \text{if } y_0 \notin \bar{\Omega}. \end{cases}$$

(ii) *If $d(f, \Omega, y_0) \neq 0$, then the equation* (2.2) *has at least one root in Ω.*

This assertion is the base of existence proofs for solutions of (2.2). *We note, however, that* (2.2) *may have solutions even if the degree equals zero. This is, for instance, the case if Ω is the interval $(-1, +1)$ of the real axis, if $f(x) = x^2$, and if $0 < y_0 < 1$ (cf. §3 of the Introduction).*

(iii) *("Sum Theorem"). Let $\Omega_1, \Omega_2, \ldots, \Omega_s$ be disjoint open subsets of Ω for which*

$$\bigcup_{\sigma=1}^{s} \bar{\Omega_\sigma} = \bar{\Omega} \quad and \quad y_0 \notin \bigcup_{\sigma=1}^{s} f(\partial\Omega_\sigma).$$

Then

$$d(f, \Omega, y_0) = \sum_{\sigma=1}^{s} d(f, \Omega_\sigma, y_0). \tag{2.4}$$

(iv) *Suppose all roots $x \in \Omega$ of* (2.2) *lie in the open subset Ω_1 of Ω. Then*

$$d(f, \Omega, y_0) = d(f, \Omega_1, y_0). \tag{2.5}$$

(v) *For any point $b \in E$*

$$d(f, \Omega, y_0) = d(f_b, \Omega, y_0 - b), \tag{2.6}$$

where $f_b(x) = f(x) - b$.

(vi) *Let x_0 be a point of E, let $\xi = x - x_0$, let $\Omega_0 = \{\xi \in E \mid x = \xi + x_0 \in \Omega\}$, and let $g(\xi) = f(x)$. Then*

$$d(f, \Omega, y_0) = d(g, \Omega_0, y_0). \tag{2.7}$$

(vii) *With the notation of* (vi) *let $h(\xi) = f(x) - y_0 = g(\xi) - y_0$. Then*

$$d(f, \Omega, y_0 + b) = d(h, \Omega_0, b) \quad \text{for any } b \in E. \tag{2.8}$$

Before stating a further property of the degree we need a definition.

2.5. DEFINITION. *If there exists a subspace $E_1 \neq E$ of E such that $F(x) = x - f(x) \in E_1$, then f is called a layer map, more precisely a layer map with respect to E_1. Obviously f is then also a layer map with respect to any sub-space $E_2 \neq E$ which contains E_1. If f is a layer map with respect to a finite-dimensional subspace we say: f is a finite layer map.*

2.6. LEMMA. *If the L.-S. map $f(x) = x - F(x): \bar{\Omega} \to E$ is a layer map with respect to a subspace E_1 which is a direct summand of E (Definition 3 in Chapter 1), then for $y_0 \in E_1$*

$$d(f, \Omega, y_0) = d(f, \Omega \cap E_1, y_0), \tag{2.9}$$

provided that $f \in C'$, that y_0 satisfies (2.1), *and that y_0 is a regular value for f.*

PROOF. We first have to show that the right member of (2.9) is defined, i.e., that

$$y_0 \notin f(\partial_1(\Omega \cap E_1)), \tag{2.10}$$

where ∂_1 denotes the boundary of $\Omega \cap E_1$ (with respect to E_1). But if (2.10) were not true, then (2.1) would not be true as follows from the inclusion

$$\partial_1(\Omega \cap E_1) \subset \partial \Omega, \tag{2.11}$$

which the reader will easily verify.

We now turn to the proof of (2.9). Since y_0 and $F(x)$ both lie in E_1, it follows that every root x of (2.2) lies in E_1 and is therefore also a root of

$$f_1(x) = y_0, \tag{2.12}$$

where f_1 denotes the restriction of f to $\overline{\Omega} \cap E_1$. The converse is obvious. Thus either both equations have no roots, or they have the same roots, say $x_1, x_2, \ldots, x_r$. In the first case (2.9) is true, both members of that equality being equal to zero by assertion (ii) of Lemma 2.4. In the second case

$$d(f, \Omega, y_0) = \sum_{\rho=1}^{r} j(Df(x_\rho; h)), \qquad h \in E,$$

$$d(f_1, \Omega \cap E_1, y_0) = \sum_{\rho=1}^{r} j(Df_1(x_\rho; h_1)), \qquad h_1 \in E_1,$$

and our assertion (2.9) will be proved once it is shown that

$$j(Df(x_\rho; h)) = j(Df_1(x_\rho; h_1)), \qquad \rho = 1, 2, \ldots, r. \tag{2.13}$$

Now, by assumption, $F(x) \in E_1$ for all $x \in \overline{\Omega}$. Therefore

$$DF(x, h) \in E_1 \tag{2.14}$$

for all $h \in E$.[3]

Also, by assumption, E_1 is a direct summand in E, i.e., there exists a subspace E_2 of E such that E is the direct sum of E_1 and E_2, and every h in E has the unique representation $h = h_1 + h_2$, $h_1 \in E_1$, $h_2 \in E_2$. Thus

$$Df(x; h) = h_1 + h_2 - DF(x; h_1) - DF(x; h_2), \tag{2.14a}$$

where, by (2.14), $DF(x; h_1)$ and $DF(x; h_2)$ lie in E_1. Therefore, (2.14a) reads in matrix form

$$Df(x; h) = \begin{pmatrix} h_1 - DF(x; h_1) & -DF(x; h_2) \\ 0 & h_2 \end{pmatrix},$$

and since $x_\rho \in E_1$,

$$Df(x_\rho; h) = \begin{pmatrix} Df_1(x_\rho; h_1) & -DF(x_\rho; h_2) \\ 0 & h_2 \end{pmatrix}. \tag{2.15}$$

Putting a parameter t here in front of the term in the upper right corner, we see that the linear map (2.15) is L.-S. homotopic to the map

$$l_\rho(h) = \begin{pmatrix} Df_1(x_\rho; h_1) & 0 \\ 0 & h_2 \end{pmatrix}, \qquad h = h_1 + h_2, \tag{2.16}$$

and, by the linear homotopy theorem, we conclude that

$$j(Df(x_\rho; h)) = j(l_\rho(h)), \qquad \rho = 1, 2, \ldots, r. \tag{2.17}$$

Finally it follows from (2.16) and Theorem 6.2 in Appendix A that $j(l_\rho(h)) = j(Df_1(x_\rho; h_1))$. This equality together with (2.17) proves the assertion (2.13).

We finish this section by proving some continuity statements which will be used in Chapter 4.

2.7. THEOREM. *Let Ω, f and y_0 be as in subsection 2.1. Then there exists a positive ρ_0 with the following property: if $0 < \rho < \rho_0$ then for each $y \in B(y_0, \rho)$ (see §2 in Chapter 1) the following assertions hold.*

(i) $y \notin f(\partial\Omega)$,

(ii) y *is a regular value for f, and*

(iii) *if the equation (2.2) has solutions in Ω, then the equation*

$$f(x) = y \tag{2.18}$$

has solutions in Ω.

(iv)

$$d(f, \Omega, y) = d(f, \Omega, y_0). \tag{2.19}$$

PROOF. $\partial\Omega$ and therefore, by Lemma 10 in Chapter 1, $f(\partial\Omega)$ is closed. Consequently (2.1) implies that y_0 has a positive distance from $f(\partial\Omega)$. This together with Lemma 20 in Chapter 1 shows that ρ_0 can be chosen in such a way that assertions (i) and (ii) are satisfied.

To prove the remaining assertions, we consider first the case that (2.2) has no solutions in Ω and therefore by (2.1) has no solutions in $\overline{\Omega}$. Then $y_0 \notin f(\overline{\Omega})$, and since by Lemma 10 in Chapter 1 $f(\overline{\Omega})$ is closed, y_0 has a positive distance ρ_1 from $f(\overline{\Omega})$, and (2.18) has no solutions for $y \in B(y_0, \rho_1)$. Thus (iii) is vacuously satisfied, and (iv) holds, both members of (2.19) being zero by part (ii) of Lemma 2.4.

Now suppose (2.2) has roots in Ω. Since y_0 is a regular value, there are only a finite number, say $x_1, x_2, \ldots, x_s$. Now let $\delta_1 > 0$ be such that the closures of the balls $B(x_\sigma, \delta_1)$, $\sigma = 1, 2, \ldots, s$, are disjoint from each other and from $\partial\Omega$. By Lemma 20 in Chapter 1 we can also require that the differentials $Df(x; h)$ are nonsingular for $x \in B(x_\sigma, \delta_1)$, $\sigma = 1, 2, \ldots, s$. Then by the inversion theorem in §1.21 there exists a positive $\delta_2 < \delta_1$ of the following property: there is a $\rho_2 = \rho_2(\delta_2)$ such that for

$$y \in B(y_0, \rho), \qquad 0 < \rho < \rho_2, \tag{2.20}$$

the equation (2.18) has for each $\sigma = 1, 2, \ldots, s$ one and only one solution $x = x_\sigma(y)$ satisfying

$$x_\sigma(y) \in B(x_\sigma, \rho), \quad x_\sigma(y_0) = x_\sigma, \quad 0 < \rho < \rho_2. \tag{2.21}$$

This proves assertion (iii). It remains to prove assertion (iv). Now, since by Theorem 4.6 in Appendix A the index $j(l)$ of a nonsingular linear L.-S. map is invariant under a continuous change of l, it follows from the continuity of $Df(x; h)$ in x (cf. also the proof of Lemma 20 in Chapter 1) that a positive δ_2 can be chosen in such a way that $j(Df(x_\sigma; h)) = j(Df(x; h))$ for $x \in B(x_\sigma, \delta_2)$. Thus for such x

$$\sum_{\sigma=1}^{s} j(Df(x_\sigma(y); h)) = \sum_{\sigma=1}^{s} j(Df(x_\sigma; h)). \tag{2.22}$$

Here, by Definition 2.2. of the degree, the right member equals the right member of the asserted equality (2.10) by (2.19). We now show also that the left member of (2.22) equals the left member of (2.19) for small enough ρ. Again by definition of the degree this amounts to showing that for ρ small enough $x_1(y), x_2(y), \ldots, x_s(y)$ are the only roots in Ω of (2.18) for y satisfying (2.20). In other words, we have to show that for such y equation (2.18) has no roots in the set

$$\Omega' = \overline{\Omega} - \bigcup_{\sigma=1}^{s} B(x_\sigma, \delta_2).$$

To see this we have only to recall that $x_1, x_2, \ldots, x_s$ are the only roots of (2.2) in Ω and that $x_\sigma \in B(x_\sigma, \delta_2)$. Consequently y_0 does not lie in the closed set $f(\Omega')$ and has therefore a positive distance $2\delta_3$ from this set, and the ball $B(y_0, \delta_3)$ does not intersect $f(\Omega')$. In other words, for $y \in B(y_0, \delta_3)$ all roots of (2.18) lie in $\Omega - \Omega'$, as we wanted to show, i.e., not in Ω'.

2.8. COROLLARY TO THEOREM 2.7. *Let $y = y_0(t)$, $0 \leq t \leq 1$, be a continuous curve consisting of regular values for f satisfying (2.1). Then $d(f, \Omega, y_0(t))$ is independent of t.*

Indeed by assertion (iv) of Theorem 2.7 the degree is constant in "the small," and the corollary follows by applying the Heine-Borel theorem.

§3. The one-dimensional case
and the case of polynomial maps

3.1. It is instructive to consider the above special cases. In the present section E^1 denotes the Banach space of real numbers x with the usual norm $\|x\| = |x|$. Ω will be the open interval (α, β), where $\alpha < \beta$ are two fixed points in E^1 such that $\partial\Omega$ is the union of the points α and β. We start by stating some auxiliary lemmas whose elementary proofs will be given in the Notes.

3.2. LEMMA. *Let f be a continuous map $\overline{\Omega} \to E^1$ and let a be a point of E^1. It is asserted: (i) if the equation*

$$f(x) = a \qquad (3.1)$$

has no roots $x \in \overline{\Omega}$, then there exists an $\varepsilon_1 > 0$ of the following property: if g is a continuous map $\overline{\Omega} \to E^1$ and if b is a point of E^1 for which

$$\left. \begin{array}{c} |f(x) - g(x)| \\ |b - a| \end{array} \right\} < \varepsilon_1, \qquad (3.2)$$

then the equation

$$g(x) = b \qquad (3.3)$$

has no roots in $\overline{\Omega}$.

(ii) If (3.1) has roots in $[\alpha, \beta]$ and these roots are contained in an open subset U of $[\alpha, \beta]$, then there exists a positive ε_1 such that all roots of (3.3) lie in U if (3.2) is satisfied.[4]

3.3. LEMMA. *If f is a C'-map $[\alpha, \beta] \to E^1$, then the set C_f of critical points in $[\alpha, \beta]$ is the set of roots of the equation*

$$f'(x) = 0, \qquad x \in [\alpha, \beta], \qquad (3.4)$$

and the set S_f of critical values of f is the set $f(C_f)$. If f is a polynomial on $[\alpha, \beta]$ (i.e., the restriction of a polynomial to $[\alpha, \beta]$), then the sets C_f and S_f are finite.[5]

3.4. LEMMA. *Let f be a polynomial in $[\alpha, \beta]$ for which α and β are regular points. Let V be an open set containing the set S_f of singular values for f. Then there exists an $\varepsilon > 0$ of the following property: if g is a polynomial in $[\alpha, \beta]$ such that*

$$\left. \begin{array}{c} |f(x) - g(x)| \\ |f'(x) - g'(x)| \end{array} \right\} < \varepsilon \quad \text{for } x \in [\alpha, \beta], \qquad (3.5)$$

then

$$S_g \subset V, \qquad (3.6)$$

where S_g is the set of singular values for g.[6]

3.5. LEMMA. *Let f be a polynomial in $[\alpha, \beta]$. Let Γ be a closed bounded set in E^1 whose points are regular values for f. Then there exists an $\varepsilon > 0$ such that all points of Γ are regular values for the polynomial g in $[\alpha, \beta]$ if (3.5) is satisfied.*[7]

3.6. THEOREM. *Let $f(x, t)$ be a continuous map $[\alpha, \beta] \times [0, 1] \to E^1$. For each $t \in [0, 1]$ the map $f_t(x) = f(x, t)$ is supposed to be C'. Let $a_0 < a_1$ be two points in E^1, and let*

$$a(t) = (1 - t)a_0 + ta_1, \qquad t \in [0, 1]. \qquad (3.7)$$

It is assumed that for $t \in [0, 1]$

$$a(t) \cap f_t(\partial\Omega) = \varnothing, \tag{3.8}$$

and that the points $a_0 = a(0)$ and $a_1 = a(1)$ are regular values for f_0 and f_1 resp. It is asserted that then

$$d(f_0, \Omega, a_0) = d(f_1, \Omega, a_1) \tag{3.9}$$

(Note: for some t in the open interval $(0, 1)$ the point $a(t)$ may be a singular value for f_t. Thus our theorem is not a consequence of Corollary 2.8 even if f does not depend on t.)

3.7. We will first show that for the proof of this theorem it will be sufficient to assume that $f(x, t)$ is a polynomial in x and t. To this end we prove the following lemma.

3.8. LEMMA. *Let f be a C'-map $\overline{\Omega} = [\alpha, \beta] \to E^1$. Let a be a point of E^1 which is a regular value for f and satisfies the relation*

$$a \notin f(\partial\Omega). \tag{3.10}$$

Then there exists a positive ε of the following property: if g is a C'-map $\overline{\Omega} \to E^1$ satisfying the inequalities

$$\left. \begin{array}{r} |g(x) - f(x)| \\ |g'(x) - f'(x)| \end{array} \right\} < \varepsilon \quad \text{for } x \in [\alpha, \beta], \tag{3.11}$$

then a is a regular value for g, and

$$d(g, \Omega, a) = d(f, \Omega, a). \tag{3.12}$$

PROOF. We consider first the case that the equation (3.1) has no roots in (α, β). Then $d(f, \Omega, a) = 0$. On the other hand, it follows from part (i) of Lemma 3.2 with $b = a$ that there is a ε such that the equation $g(x) = a$ also has no roots if (3.11) is satisfied. Consequently $d(g, \Omega, a) = 0$ and the assertion (3.12) follows.

Suppose now that (3.1) has roots on (α, β). Since a is a regular value for f, there are only a finite number, and if $x_1, x_2, \ldots, x_r$ are these roots, then $f'(x_\rho) \neq 0$ (cf. Lemma 3.3). Therefore we can choose a δ such that

$$0 < \delta < 1 \tag{3.13}$$

and such that the closures of the open intervals $i_\rho = (x_\rho - \delta, x_\rho + \delta)$ are disjoint from each other, lie in (α, β), and are such that $|f'(x)| > 0$ for $x \in \bar{i}_\rho$, and consequently

$$|f'(x)| \geq 2m_1 \quad \text{for } x \in \bar{i}_\rho \tag{3.14}$$

for some positive number m_1. We also note that by (3.10) the point a has a positive distance $2m_2$ from $\partial\Omega$, and consequently

$$a \notin g(\partial\Omega) \tag{3.15}$$

if

$$|f(x) - g(x)| < m_2 \quad \text{for } x \in [\alpha, \beta]. \tag{3.16}$$

Finally to define an ε of the asserted property we note that by part (ii) of Lemma 3.2 with $b = a$ there exists an $\varepsilon_1 > 0$ such that all roots in (α, β) of

$$g(x) = a \tag{3.17}$$

lie in the open set

$$U = \bigcup_\rho i_\rho \tag{3.18}$$

if g satisfies (3.2). We now claim that

$$\varepsilon = \min(\varepsilon_1, m_1\delta, m_2) \tag{3.19}$$

has the desired property. Then let g satisfy the assumption (3.11) with that ε. It is clear from the sum theorem (part iii of Lemma 2.4) that

$$d(f, \Omega, a) = \sum_{\rho=1}^{r} d(f, i_\rho, a). \tag{3.20}$$

But since all roots in (α, β) of (3.17) lie in the set U given by (3.18), we also have by the sum theorem

$$d(g, \Omega, a) = \sum_{\rho=1}^{r} d(g, i_\rho, a). \tag{3.21}$$

Therefore to prove the asserted equality (3.12) it will be sufficient to show that

$$d(g, i_\rho, a) = d(f, i_\rho, a), \qquad \rho = 1, 2, \ldots, r. \tag{3.22}$$

Now $x_\rho \in i_\rho$ is by definition a root of (3.1), and by (3.14) this is the only one in i_ρ. Consequently,

$$d(f, i_\rho, a) = j(Df(x_\rho; h)) = \operatorname{sign} f'(x_\rho). \tag{3.23}$$

But also (3.17) has one and only one root in i_ρ. Indeed there is at most one since by (3.14), (3.11), and (3.19)

$$|g'(x)| \geq |f'(x)| - |f'(x) - g'(x)| \geq 2m_1 - \varepsilon \geq m_1 > 0 \quad \text{for } x \in i_\rho.$$

To show that (3.17) has a root in i_ρ, we note that by (3.14) $f'(x)$ has a constant sign in i_ρ. It will be enough to consider the case that $f'(x) > 0$ in i_ρ since the case $f'(x) < 0$ in i_ρ can be treated in a similar way. Then since $a = f(x_\rho)$ we see from the mean value theorem that for some $\xi_\rho \in i_\rho$

$$g(x_\rho + \delta) - a = f(x_\rho + \delta) - f(x_\rho) + g(x_\rho + \delta) - f(x_\rho + \delta)$$
$$= f'(\xi_\rho)\delta + g(x_\rho + \delta) - f(x_\rho + \delta).$$

Therefore by (3.14), (3.11) and (3.19)

$$g(x_\rho + \delta) - a \geq f'(\xi_\rho) - |g(x_\rho + \delta) - f(x_\rho + \delta)|$$
$$\geq 2m_1\delta - \varepsilon \geq m_1\delta > 0.$$

Similarly we see that $a - g(x_\rho - \delta) > 0$. Thus $g(x_\rho - \delta) < a < g(x_\rho + \delta)$, and we see that (3.17) has a unique root $\tilde{x}_\rho \in i_\rho$. Therefore

$$d(g, i_\rho, a) = j(Dg(\tilde{x}_\rho; h)) = \operatorname{sign} g'(\tilde{x}_\rho). \tag{3.24}$$

But by (3.14), (3.13), (3.11), and (3.19), $\operatorname{sign} g'(\tilde{x}_\rho) = \operatorname{sign} f'(\tilde{x}_\rho)$, and since as already noticed $\operatorname{sign} f'(x)$ is constant for $x \in i_\rho$, the assertion (3.22) follows from (3.23) and (3.24).

3.9. It is now easily seen that the assertion of subsection 3.7 is true: by Lemma 3.8 there exists an ε such that with $f_t(x)$ and $a(t)$ as in Theorem 3.6

$$d(g_i, \Omega, a(i)) = d(f_i, \Omega, a(i)), \qquad i = 0, 1, \tag{3.25}$$

if g_0 and g_1 are C'-maps $\overline{\Omega} \to E^1$ for which

$$\left. \begin{array}{c} |g_i(x) - f_i(x)| \\ |g_i'(x) - f_i'(x)| \end{array} \right\} < \varepsilon, \qquad i = 0, 1. \tag{3.26}$$

Now by §1.32 there exists a polynomial $g(x, t)$ such that

$$\left. \begin{array}{c} |g(x, t) - f(x, t)| \\ |g'(x, t) - f'(x, t)| \end{array} \right\} < \varepsilon \quad \text{for } x \in [\alpha, \beta], \ t \in [0, 1],$$

where the prime denotes differentiation with respect to x. Then (3.26) is satisfied with $g_0(x) = g(x, 0)$ and $g_1(x) = g(x, 1)$, and therefore (3.25) is satisfied for these g_0 and g_1. Now suppose Theorem 3.6 is true for polynomials. Then (3.9) is true with f_0 and f_1 replaced by g_0 and g_1:

$$d(g_0, \Omega, a_0) = d(g_1, \Omega, a_1).$$

But this equality together with (3.25) proves (3.9).

From now on we will suppose that $f(x, t)$ is a polynomial. We will prove Theorem 3.6 first for the special case that f does not depend on t, i.e., the following lemma.

3.10. LEMMA. Let $f(x)$ be a polynomial on E^1. Let α, β, Ω be as in subsection 3.1 and let $a(t)$ be defined by (3.7). It is assumed that for $0 \le t \le 1$

$$a(t) \cap f(\partial\Omega) = \varnothing \tag{3.27}$$

and that the points $a_0 = a(0)$ and $a_1 = a(1)$ are regular values for f. We assert that then

$$d(f, \Omega, a_0) = d(f, \Omega, a_1). \tag{3.28}$$

PROOF. Since a_0 and a_1 are by assumption regular values for f, it is clear from Borel's covering theorem that it will be sufficient to prove that to every $t_0 \in [0, 1]$ there corresponds a $\tau = \tau(t_0)$ such that the interval $T(t_0) = (t_0 - \tau, t_0 + \tau)$ has the following properties:

(i) for every $t \in \overline{T(t_0)}$ the point $a(t)$ is a regular value for f with the possible exception of $t = t_0$;

(ii) the degree
$$d(f, \Omega, a(t)) \tag{3.29}$$
is constant in $\overline{T(t_0)}$ if $a(t_0)$ is a regular value for f, and in $\overline{T(t_0)} - \{t_0\}$ otherwise.

Assertion (i) is obvious from Lemma 3.3. Moreover, since by Lemma 3.3 the set S_f of singular values of f is finite, the first part of assertion (ii) follows from part (iv) of Theorem 2.7.

Now suppose $a(t_0)$ is a singular value for f. Then the equation
$$f(x) = a(t_0) \tag{3.30}$$
has roots in (α, β). Since f is a polynomial, there are only a finite number of them. Call them $x_1, x_2, \ldots, x_r$. Consequently there exists a $\delta > 0$ of the following property: the closures of the intervals $i_\rho = (x_\rho - \delta, x_\rho + \delta)$ lie in (α, β) and are disjoint from each other. Moreover, by Lemma 3.3 we can choose δ such that also
$$f'(x) \neq 0 \quad \text{for } x \in \bar{i}_\rho - \{x_\rho\}. \tag{3.31}$$
On the other hand, since $a(t)$ is continuous, we see from part (ii) of Lemma 3.2 (with $g(x) = f(x)$) that there exists a positive τ_1 such that all roots $x \in (\alpha, \beta)$ of the equation
$$f(x) = a(t) \tag{3.32}$$
lie in the union U of the i_ρ if
$$t \in T_1(t_0) = (t_0 - \tau_1, t_0 + \tau_1). \tag{3.33}$$
From this together with (3.31) we see that for $t \in T_1(t_0) - \{t_0\}$ the point $a(t)$ is a regular value for f. Therefore on account of (3.27) we see that for such t the degree $d(f, \Omega, a(t))$ is defined and that by the sum theorem
$$d(f, \Omega, a(t)) = \sum_{\rho=1}^{r} d(f, i_\rho, a(t)) \quad \text{for } t \in T_1(t_0) - \{t_0\}. \tag{3.34}$$

Now to define a subinterval $T(t_0)$ of $T_1(t_0)$ containing t_0 such that the degree (3.34) is constant for $t \in T(t_0) - \{t_0\}$, we note that by (3.31) f is strictly monotone in each of the open intervals.
$$i_\rho^- = (x_\rho - \delta, x_\rho), \qquad i_\rho^+ = (x_\rho, x_\rho + \delta). \tag{3.35}$$
Since $f(x_\rho) = a(t_0)$, it follows that
$$m = \min_\rho(|f(x_\rho + \delta) - a(t_0)|, |f(x_\rho - \delta) - a(t_0)|) > 0. \tag{3.36}$$
Consequently there exists a $\tau_2 > 0$ such that
$$|a(t) - a(t_0)| < m \tag{3.37}$$
if $|t - t_0| < \tau_2$. Now let $\tau = \min(\tau_1, \tau_2)$ and $T(t_0) = (t_0 - \tau, t_0 + \tau)$.

To show that the degree (3.29) is constant for $t \in \overline{T(t_0)} - \{t_0\}$ it is by (3.34) sufficient to prove that for these t
$$d(f, i_\rho, a(t)) = c_\rho, \qquad \rho = 1, 2, \ldots, r, \tag{3.38}$$

where c_ρ is independent of t. But for fixed ρ the sign of $f'(x)$ is constant in each of the intervals (3.35). We distinguish two cases:

(I) sign $f'(x^+) = -$sign $f'(x^-)$;

(II) sign $f'(x^+) = $ sign $f'(x^-)$,

where x^+ and x^- denote arbitrary points in i_ρ^+ and i_ρ^- resp.

To prove (3.38) in Case I it will be enough to consider the case that

$$f'(x^+) > 0, \qquad f'(x^-) < 0, \tag{3.39}$$

the proof for the case $f'(x^+) < 0$, $f'(x^-) > 0$ being quite similar.

Now, by (3.39), the restriction of f to i_ρ takes its absolute minimum at $x = x_\rho$. Therefore the equation (3.32) has no roots in i_ρ if

$$a(t) < f(x_\rho) = a(t_0), \tag{3.40}$$

and therefore by part (ii) of Lemma 2.4 for such t

$$d(f, i_\rho, a(t)) = 0. \tag{3.41}$$

We now claim that (3.41) is still true if

$$a(t_0) < a(t) < a(t_0) + m. \tag{3.42}$$

Indeed by (3.39) f decreases monotonically as x increases from $x_\rho - \delta$ to x_ρ. Consequently, by definition of m the equation (3.32) has exactly one root $x^- \in i_\rho^-$ if $a(t)$ satisfies (3.42), and therefore if

$$t_0 < t < t_0 + \tau. \tag{3.43}$$

Similarly one sees that (3.32) has exactly one root $x^+ \in i_\rho^+$ if t satisfies (3.43). Consequently, for such t, by the sum theorem and by (3.39)

$$d(f, i_\rho, a(t)) = d(f, i_\rho^+, a(t)) + d(f, i_\rho^-, a(t))$$
$$= \text{sign } f'(x^+) + \text{sign } f'(x^-) = 0,$$

as we wanted to show.

It remains to consider the simpler Case II where sign $f'(x)$ is constant in the whole interval i_ρ except possibly at $x = x_\rho$ at which point f' may be zero. In any case f is strictly monotone in all of i_ρ, and (3.32) has exactly one solution on $x \in i_\rho$ if $t_0 - \tau < t < t_0 + \tau$. Consequently $d(f, i_\rho, a(t)) = $ sign $f'(x)$, and therefore is constant in $i_\rho - \{x_\rho\}$ by definition of Case II.

3.11. Conclusion of the proof of Theorem 3.6. As we saw it is sufficient to assume that $f(x, t)$ is a polynomial. We note first of all that there exists a positive constant m_1 such that

$$\text{dist}(a(t), f_t(\partial\Omega)) > 3m_1 \tag{3.44}$$

for all $t \in [0, 1]$. Indeed otherwise there would exist a sequence of points $x_\gamma \in \partial\Omega$ and $t_\gamma \in [0, 1]$ such that for at least one of the points α, β of $\partial\Omega$, say α, $\lim[a(t_\gamma) - f_{t_\gamma}(\alpha)] = 0$. We may assume that the t_γ converge to a point

$\bar{t} \in [0, 1]$. But then $a(\bar{t}) = f_t(\alpha)$, and this equality contradicts the assumption (3.8).

Next we note that if $R(f_{t_0})$ denotes the set of all $y \in E^1$ which are regular values for f_{t_0} (as polynomials in $[\alpha, \beta]$), if J is an interval whose closure does not intersect $f(\partial\Omega)$, and if

$$y \in R(f_{t_0}) \cap J, \tag{3.45}$$

then we may define

$$d(f_{t_0}, \Omega, R(f_{t_0}) \cap J) = d(f_{t_0}, \Omega, y), \tag{3.46}$$

since by Lemma 3.10 the right member of (3.46) has the same value for all y satisfying (3.45).

In particular $d(f_{t_0}, \Omega, R(f_{t_0}) \cap B(a(t_0), m_1))$ is well defined (cf. (3.44)). Corresponding to each $t_0 \in [0, 1]$ we will now define a positive number $\tau(t_0)$ such that the intervals $T(t_0) = (t_0 - \tau(t_0), t_0 + \tau(t_0))$ have the following property: if t_1 and t_2 are a couple of points in $[0, 1]$ for which the intersection $T(t_1) \cap T(t_2)$ is not empty, then

$$d(f_{t_1}, \Omega, R(f_{t_1}) \cap B(a(t_1), m_1)) = d(f_{t_2}, \Omega, R(f_{t_2}) \cap B(a(t_2), m_1)). \tag{3.47}$$

Before constructing a $\tau(t_0)$ of the asserted property, we remark that (3.47) implies the assertion (3.9) of our Theorem 3.6. Indeed using Borel's theorem we conclude from (3.47) that

$$d(f_0, \Omega, R(f_0) \cap B(a_0, m_1)) = d(f_1, \Omega, R(f_1) \cap B(a_1, m_1)). \tag{3.48}$$

But by assumption $a_0 = a(0)$ is a regular value for f_0. Therefore by (3.46) the left member of (3.48) equals the left member of (3.9). In the same way one sees that the right members of (3.48) and (3.9) are equal to each other.

We now turn to the definition of $\tau(t_0)$. Since for fixed $t_0 \in [0, 1]$ there are at most a finite number of points in E^1 which are singular values for f_{t_0}, there exists a positive $\overline{m}(t_0)$ such that

$$\overline{m}(t_0) < m_1 \tag{3.49}$$

and such that all points of the closure of the interval

$$J(t_0, \overline{m}(t_0)) = B(a(t_0), \overline{m}(t_0)) \tag{3.50}$$

are regular values for f_{t_0} with the possible exception of the point $a(t_0)$. In particular the point

$$\overline{y}(t_0) = a(t_0) + \overline{m}(t_0)/2 \tag{3.51}$$

is a regular value for f_{t_0}. Consequently by Lemma 3.8 and the continuity properties of the polynomial $f_t(x) = f(x, t)$, there exists a positive number $\overline{\tau}(t_0)$ of the following property: if

$$|t - t_0| < \overline{\tau}(t_0), \tag{3.52}$$

then $\overline{y}(t_0)$ is a regular value for f_t and

$$d(f_t, \Omega, \overline{y}(t_0)) = d(f_{t_0}, \Omega, \overline{y}(t_0)). \tag{3.53}$$

We now define $\tau(t_0)$ as a positive number satisfying

$$2\tau(t_0) = \min(\bar{\tau}(t_0), m_1|a_1 - a_0|^{-1}). \tag{3.54}$$

Then

$$|\bar{y}(t) - \bar{y}(t_0)| < 2m_1 \tag{3.55}$$

if

$$|t - t_0| < 2\tau(t_0). \tag{3.56}$$

Indeed by (3.51), (3.7), (3.49), (3.56), and (3.54)

$$\begin{aligned} |\bar{y}(t) - \bar{y}(t_0)| &= |a(t) + \bar{m}(t)/2 - (a(t_0) + \bar{m}(t_0)/2)| \\ &\leq |a(t) - a(t_0)| + (\bar{m}(t) + \bar{m}(t_0))/2 \\ &\leq |t - t_0| |a_1 - a_0| + m_1 < \tau(t_0) + m_1 < 2m_1. \end{aligned}$$

This proves (3.55). Now $\bar{y}(t)$ and $\bar{y}(t_0)$ are regular values for f_t. Therefore the inequality (3.55) in conjunction with (3.44) and Lemma 3.10 shows that

$$d(f_t, \Omega, \bar{y}(t)) = d(f_t, \Omega, \bar{y}(t_0)) \tag{3.57}$$

if (3.56) is satisfied.

The intervals

$$T(t_0) = B(a(t_0), \tau(t_0)) \tag{3.58}$$

form an open covering of $[0, 1]$ as t_0 varies over that interval. We are now ready to prove that (3.47) holds for any couple t_1, t_2 of points in $[0, 1]$ for which $T(t_1) \cap T(t_2)$ is not empty. Then if t_{21} is a point of this intersection, we see that $|t_2 - t_1| \leq |t_2 - t_{21}| + |t_{21} - t_1| \leq \tau(t_1) + \tau(t_2)$. We may assume that $\tau(t_1) \geq \tau(t_2)$. Then

$$|t_2 - t_1| \leq 2\tau(t_1), \tag{3.59}$$

i.e., (3.56) is satisfied with $t_0 = t_1$ and $t = t_2$. Therefore by (3.54) and (3.53)

$$d(f_{t_2}, \Omega, \bar{y}(t_1)) = d(f_{t_1}, \Omega, \bar{y}(t_1)). \tag{3.60}$$

But also (3.57) holds with $t = t_2$ and $t_0 = t_1$. Therefore by (3.60)

$$d(f_{t_2}, \Omega, \bar{y}(t_2)) = d(f_{t_1}, \Omega, \bar{y}(t_1)).$$

By the definition (3.46) this equality is equivalent to (3.47).

§4. The degree for a not necessarily regular value y_0

The definition of the degree in this case is based on Theorems 4.1 and 4.2 below.

4.1. THEOREM. *Let $f \in C'(\overline{\Omega})$ be an L.-S. map $\overline{\Omega} \to E$, and let y_0 be a point of E satisfying* (2.1). *Then every neighborhood of y_0 contains a point a which is a regular value for f.*

4.2. THEOREM. *Let $f \in C''(\overline{\Omega})$ be an L.-S. map, and let Γ be a connected open set satisfying*

$$\Gamma \cap f(\partial\Omega) = \varnothing. \tag{4.1}$$

Then $d(f, \Omega, y)$ is constant for all points $y \in \Gamma$ which are regular values for f.

Obviously on account of these theorems the following definition is legitimate.

4.3. DEFINITION. *Let $f \in C''(\overline{\Omega})$ be an L.-S. map $\overline{\Omega} \to E$, and let y_0 be a point of E satisfying* (2.1). *Let U be an open connected neighborhood of y_0 not intersecting $f(\partial\Omega)$, and let y be a point of U which is a regular value for f. We then define the degree $d(f, \Omega, y_0)$ by the equality*

$$d(f, \Omega, y_0) = d(f, \Omega, y). \tag{4.2}$$

The proofs of Theorems 4.1 and 4.2 are based on the following theorem, which we refer to as the Sard-Smale theorem and which is a special case of Smale's generalization of Sard's theorem.

4.4. THEOREM OF SARD-SMALE. *Let Π be a Banach space, and let E be a direct summand of Π which is of finite codimension n (see the remark to §1.3). Let Z be a bounded open subset of Π, and let $\phi \in C^r(\overline{Z})$, where r is a positive integer, be an L.-S. map $\overline{Z} \to E$.*

We assume that

$$n < r \tag{4.3}$$

and assert that then the regular values of ϕ are dense in $E - \phi(\partial Z)$.

The proof of this theorem will be given in Appendix B.

4.5. *Proof of Theorem* **4.1.** We apply Theorem 4.4 with $\Pi = E$, $Z = \Omega$, and $\phi = f$. Then $n = 0$ and the assumption (4.3) is satisfied for $r = 1$. Therefore our theorem is a consequence of the Sard-Smale theorem.

The proof of Theorem 4.2 is considerably harder.

4.6. *Start of the proof of Theorem* **4.2.** We have to show that under the assumptions made

$$d(f, \Omega, y_0) = d(f, \Omega, y_1) \tag{4.4}$$

for any couple of points y_0, y_1 in Γ which are regular values for f. The first step for the proof of (4.4) consists in showing that it is sufficient to consider the special case that the segment $\overline{y_0 y_1}$ (see §1.23) lies in Γ.

Since the points y_0 and y_1 lie in the open connected set Γ, there exists a simple polygon $P \subset \Gamma$ connecting y_0 and y_1 (see §1.25). Let $p_0 = y_0$, $p_1, \ldots, p_n = y_1$ be the vertices of P. Except for the vertices p_0 and p_n these vertices are not necessarily regular values for f. But we may as follows replace P by a simple polygon $Q \subset \Gamma$ whose vertices are regular values for f: since $\partial\Gamma$ is closed and P is compact, these two sets have a positive distance. For any positive number ρ less than this distance, the balls $B(p_i, \rho)$, $i = 0, 1, \ldots, n$, lie in Γ. Now by Theorem 4.1 we can choose for $i = 1, 2, \ldots, n-1$ a point $q_i \in B(p_i, \rho)$ which is a regular value for f. Then the simple polygon Q whose vertices are $q_0 = y_0$, $q_1, \ldots, q_{n-1}$, $q_n = y_1$ lies in Γ. Indeed the edge $\beta_i = \overline{q_i q_{i+1}}$ of Q is given by

$$\beta_i(t) = (1 - t)q_i + tq_{i+1}, \qquad 0 \le t \le 1,$$

while the edge $\overline{p_i p_{i+1}}$ of P is given by

$$\alpha_i(t) = p_i(1 - t) + p_{i+1}t, \qquad 0 \le t \le 1.$$

Therefore

$$\|\beta_i(t) - \alpha_i(t)\| \le \|q_i - p_i\|(1 - t) + \|q_{i+1} - p_{i+1}\|t < \rho,$$

which obviously proves the assertion that $Q \subset \Gamma$. We thus see that (4.4) will be proved once we can show that

$$d(f, \Omega, q_i) = d(f, \Omega, q_{i+1}) \quad \text{for } i = 0, 1, \ldots, n-1.$$

4.7. It is now clear that for the proof of (4.4) we may assume that the segment

$$\overline{y_0 y_1} = (1 - t)y_0 + ty_1, \qquad 0 \le t \le 1$$

is contained in Γ.

The assertion (4.4) is trivial if neither the equation (2.2) nor the equation $f(x) = y_1$ has a root in Ω, since then by Lemma 2.4 both members of (4.4) equal zero. We therefore will always suppose that at least one of these equations has a root in Ω, and we may then assume that (2.2) has at least one such root.

Our next step consists in changing y_0 and y_1 into points $\overline{y}_0$ and $\overline{y}_1$ resp. which have the following property: if

$$\overline{\sigma} = (1 - t)\overline{y}_0 + t\overline{y}_1, \qquad 0 \le t \le 1, \tag{4.5}$$

then $\overline{\sigma}$ is transversal to f (see §1.26).

4.8. We first show that if ρ is a positive number which is not only smaller than the distance of the segment $\overline{y_0 y_1}$ from $\partial\Gamma$ but which also has the properties asserted in Theorem 2.7 and those obtained by replacing y_0 by y_1 in that theorem, then the asserted equality (4.4) will be implied by the equality

$$d(f, \Omega, \overline{y}_0) = d(f, \Omega, \overline{y}_1), \tag{4.6}$$

provided that

$$\left. \begin{array}{l} \|y_0 - \overline{y}_0\| \\ \|y_1 - \overline{y}_1\| \end{array} \right\} < \rho. \tag{4.7}$$

Indeed, by Theorem 2.7, every

$$\overline{y}_0 \in B(y_0, \rho) \tag{4.8}$$

is a regular value for f and satisfies

$$d(f, \Omega, \overline{y}_0) = d(f, \Omega, y_0), \tag{4.9}$$

and the equation

$$f(x) = \overline{y}_0 \tag{4.10}$$

has a solution. Moreover every $\overline{y}_1$ satisfying (4.7) is a regular value for f and satisfies

$$d(f, \Omega, \overline{y}_1) = d(f, \Omega, y_1). \tag{4.11}$$

(4.9) and (4.11) show that (4.6) implies (4.4). If $\overline{y}_0 \in B(y_0, \rho)$ is given, we will always choose

$$\overline{y}_1 = \overline{y}_0 + y_1 - y_0. \tag{4.12}$$

This $\overline{y}_1$ obviously satisfies (4.7).

4.9. To find a $\overline{y}_0$ satisfying (4.7) such that the segment $\overline{\sigma} \subset \Gamma$ given by (4.5) (with $\overline{y}_1$ defined by (4.12)) is transversal to f, we introduce the Banach space Π which is the direct sum of E with the real line R considered as a one-dimensional Banach space (see part (ii) of §1.3). If r_1 is a unit element in R, the elements of Π are of the form $z = x + tr_1$ with $x \in E$, t a real number, and the norm $\|z\| = \|x\| + |t|$. Let Z be the open bounded subset of Π defined by

$$Z = \{z = x + tr_1 \mid x \in \Omega, \ -\eta < t < 1 + \eta\}, \tag{4.13}$$

where the positive number η is chosen in such a way that the extension of the interval (4.5) to the interval $-\eta \leq t \leq 1 + \eta$ still lies in Γ. Finally we define a map $g \colon Z \to E$ by

$$g(x + tr_1) = f(x) - t(\overline{y}_1 - \overline{y}_0). \tag{4.14}$$

4.10. LEMMA. *If $\overline{y}_0$ is a regular value for g, then $\overline{\sigma}$ is transversal to f.*

PROOF. By (4.5) and (4.14) the equation

$$g(x + tr_1) = \overline{y}_0 \tag{4.15}$$

is equivalent to

$$f(x) = \overline{y}_0 + t(\overline{y}_1 - \overline{y}_0) \in \overline{\sigma}. \tag{4.16}$$

But if $\overline{y}_0$ is a regular value for g, then for x, t satisfying (4.16) and therefore (4.15) the differential

$$Dg(x + tr_1; h + \tau r_1) = Df(x; h) - \tau(\overline{y}_1 - \overline{y}_0), \tag{4.17}$$

where $h \in E$ and τ, a real number, is a map Π onto E. This proves the transversality of $\overline{\sigma}$.

4.11. LEMMA. *There exists a $\overline{y}_0 \in B(y_0, \rho)$ which is a regular value for g.*

PROOF. The subspace E of Π is of codimension $n = 1$. By assumption $f \in C''(\overline{\Omega})$. Therefore, by (4.14), $g \in C''(\overline{Z})$. Thus the assumption (4.3) of the Sard-Smale theorem 4.4 is satisfied, and our lemma follows from that theorem.

4.12. To finish the proof of Theorem 4.2 we have to show: if $\overline{y}_0$ and $\overline{y}_1$ are chosen as above, then (4.6) holds. We will follow the lines of the proof given by Elworthy and Tromba [**20**, pp. 74 ff.] in the case of Banach *manifolds* with considerable simplifications made possible by the facts that we deal with a "plane" space and that the "transverse" set $\bar{\sigma}$ is a straight line segment.

We recall that the sets $f^{-1}(\bar{y}_0)$ and $f^{-1}(\bar{y}_1)$ are finite and that (see subsection 4.7) we may assume that $f^{-1}(\bar{y}_0)$ is not empty, while $f^{-1}(\bar{y}_1)$ may or may not be empty. Denoting the elements of $f^{-1}(\bar{y}_0)$ by $x_1, x_2, \ldots, x_r$ and those of $f^{-1}(\bar{y}_1)$, if this set is not empty, by $w_1, w_2, \ldots, w_s$, we see from Definition 2.2 that our assertion (4.6) is equivalent to

$$\sum_{j=1}^{r} j(Df(x_\rho; \cdot)) = \begin{cases} \sum_{j=1}^{s} j(Df(w_\sigma; \cdot)) & \text{if } f^{-1}(\bar{y}_1) \neq \varnothing, \\ 0 & \text{if } f^{-1}(\bar{y}_1) = \varnothing. \end{cases} \tag{4.18}$$

For the proof of (4.18) we will need the following properties (a)–(d) of the set $f^{-1}(\bar{\sigma})$:

(a) $f^{-1}(\bar{\sigma})$ is the union of maximal connected disjoint subsets called path components each of which is determined by any of its points;

(b) if x_0 is one of the points

$$x_1, x_2, \ldots, x_r, w_1, w_2, \ldots, w_s, \tag{4.19}$$

then the path component C_0 determined by x_0 may be written in the form

$$x = x_0(s), \qquad 0 \leq s \leq 1, \ x_0(0) = x_0 \tag{4.20}$$

with $x_0(s)$ being continuously differentiable. x_0 will be called the starting point of C_0;

(c) with C_0 as in (b) the point $x_0(1)$ is again one of the points (4.19);

(d) with the above notations

$$x_0(1) \neq x_0(0) = x_0. \tag{4.21}$$

A sketch for a proof of these properties will be given after assertion (4.18) has been established.

For the proof of (4.18) we establish certain relations between the indices j occurring in the sums of that equality. We first consider the case that $f^{-1}(\bar{y}_1)$ is not empty. Let $C_\rho : x = x_\rho(s)$ be the path component with $x_\rho(0) = x_\rho$ for some integer $\rho \leq r$. By properties (c) and (d) there are two possible cases:

$$\begin{cases} \text{Case I.} & x_\rho(1) = x_\sigma, \quad 1 \leq \sigma \leq r, \ \sigma \neq \rho; \\ \text{Case II.} & x_\rho(1) = w_\sigma, \quad 1 \leq \sigma \leq s. \end{cases} \tag{4.22}$$

We will prove that

$$j(Df(x_\rho; \cdot)) = \begin{cases} -j(Df(x_\sigma; \cdot)) & \text{in Case I,} \\ j(Df(w_\sigma; \cdot)) & \text{in Case II.} \end{cases} \tag{4.23}$$

It is easy to see that (4.23) implies our assertion (4.18): if for some $\rho \leq r$ Case I takes place, then by (4.23) the ρth and σth terms in the sum of the left member of (4.18) cancel each other. Thus the "reduced sum" obtained by keeping only those terms for which Case II holds has the same value as the original sum. Let

r' be the number of terms in the reduced sum, and let s' be the number of the correspondingly reduced sum of the right member of (4.18). Then

$$r' \leq s'. \tag{4.24}$$

This is obvious if r' equals 0 or 1. But if $r' \geq 2$ and if ρ_1 and ρ_2 are two different ρ-values $\leq r'$, then the path component $x_{\rho_1}(s)$ and $x_{\rho_2}(s)$ starting at x_{ρ_1} and x_{ρ_2} resp. are disjoint, and therefore $x_{\rho_1}(1) \neq x_{\rho_2}(1)$; and since we are in Case II, $w_{\sigma_1} = x_{\rho_1}(1)$ and $w_{\sigma_2} = x_{\rho_2}$ are different members of the reduced sum of the right member of (4.18). This implies (4.24). But since $s' \leq r'$ is proved the same way, we see that $r' = s'$, and the equality of the reduced sums (and therefore the original sums) in (4.18) follows from the part of (4.23) referring to Case II.

This proves (4.18) in case $f^{-1}(\bar{y}_2)$ is not empty. But if this set is empty, there are no roots w_σ of $f(x) = \bar{y}_1$, and therefore Case I in (4.22) holds for all terms in the sum of the left of (4.18), and by the previous argument this shows that this sum is zero as was to be proved.

It remains to prove (4.23). It is clear that the indices appearing in that equation are defined, i.e., that the differentials at $x_\rho = x_\rho(0)$ and x_σ ($= x_\rho(1)$ in Case I) and w_σ ($= x_\rho(1)$ in Case II) are nonsingular since the points $\bar{y}_0$ and $\bar{y}_1$ are regular values for f and $f(x_\rho) = f(x_\sigma) = \bar{y}_0$ and $f(w_\sigma) = \bar{y}_1$. However the differentials of f at $x = x_\rho(s)$ are not necessarily nonsingular for all $s \in [0, 1]$ since $f(x_\rho(s)) \in \bar{\sigma}$, i.e.,

$$f(x_\rho(s)) = \bar{y}_0 + t(\bar{y}_1 - \bar{y}_0), \qquad 0 \leq t = t(s) \leq 1, \tag{4.25}$$

and since not all points $y \in \bar{\sigma}$ are necessarily regular values for f. We therefore will define for each $s \in [0, 1]$ an auxiliary L.-S. map $\phi_s(x)$ which has the following properties:

(A) the differential of ϕ_s at $x_\rho(s)$ is nonsingular;

(B) the index of the differential referred to in (A) is independent of s;

(C) for those $s \in [0, 1]$ for which the differential of f at $x = x_\rho(s)$ is nonsingular where $t = t(s)$ is given by (4.25)

$$j(Df(x_\rho(s); h)) = -\operatorname{sign} \frac{dt}{ds} j(D\phi(x_\rho(s); h)). \tag{4.26}$$

The definition of ϕ_s is as follows: for a fixed $s = s_0 \in [0, 1]$ we denote by X_1 the (one-dimensional) tangent space to $x_\rho(s)$ at $s = s_0$, and by X_2 a complementary space to X_1. If moreover ξ_1 is the unit vector in X_1 directed in the direction of increasing s, then every $x \in E$ has the unique representation

$$x = \alpha \xi_1 + x_2, \qquad \alpha \text{ real}, \ x_2 \in X_2. \tag{4.27}$$

On the other hand we denote by Y_1 the one-dimensional subspace of E generated by $\bar{y}_1 - \bar{y}_0$, and by Y_2 a complementary subspace to Y_1 (independent of s_0). If $e_1 = (\bar{y}_1 - \bar{y}_0)\|\bar{y}_1 - \bar{y}_0\|^{-1}$, then every $y \in E$ has the unique representation

$$y = \beta e_1 + y_2, \qquad \beta \text{ real}, \ y_2 \in Y_2. \tag{4.28}$$

The projections $y \to \beta e_1$ and $y \to y_2$ will be denoted by π_1 and π_2 resp. Finally if α_0 is the unique real number in the representation (4.27), for $x = x(s_0)$ we define

$$\phi_{s_0}(x) = f(x_\rho(s_0)) + (\alpha - \alpha_0)e_1 + \pi_2(f(x) - f(x_\rho(s_0))). \tag{4.29}$$

To prove property (A) we note first of all: if

$$h = \xi_1 h_1 + h_2, \qquad h_1 \text{ real, } h_2 \in X_2, \tag{4.30}$$

then $Df(x_\rho(s_0); h_1)$ maps the tangent space X_1 to $x(s)$ at $s = s_0$ into the tangent space to $f(x_\rho(s))$ at $s = s_0$. But $f(x_\rho(s)) \in \bar{\sigma} \subset Y_1$ for all s. Thus this tangent space is Y_1, and we see that

$$Df(x_\rho(s_0), \xi_1 h_1) \subset Y_1 \tag{4.31}$$

and therefore

$$\pi_2 Df(x_\rho(s_0); \xi_1 h_1) = 0. \tag{4.32}$$

Since $\pi_2 e_1 = 0$ we see from (4.29) that also

$$\pi_2 D\phi_{s_0}(x_\rho(s_0); \xi_1 h_1) = 0. \tag{4.33}$$

Taking this into account we see from (4.29) and (4.32) that

$$D\phi_{s_0}(x_\rho(s_0); h) = \begin{pmatrix} h_1 e_1 & 0 \\ 0 & \pi_2 Df(x_\rho(s_0); h_2) \end{pmatrix}. \tag{4.34}$$

To prove that this map is not singular, it will be sufficient to show that $\pi_2 Df(x_\rho(s_0); h_2)$ maps X_2 onto Y_2. Then let y_2 be an arbitrary point of Y_2. Since $\bar{\sigma}$ is transversal to f, the differential $Df(x_\rho(s_0); h)$ together with Y_1 spans E. Thus there is an $\bar{h} \in E$ and a real $\bar{\alpha}$ such that $y_2 = Df(x_\rho(s_0); \bar{h}) + \bar{\alpha} e_1$. But $y_2 = \pi_2 y_2$ since $y_2 \in Y_2$. Thus we see that $y_2 = \pi_2 Df(x_\rho(s_0); \bar{h})$. Now if $\bar{h} = \bar{h}_1 \xi_1 + \bar{h}_2$, $\bar{h}_1$ real, $\bar{h}_2 \in E_2$, we see that $y_2 = Df(x_\rho(s_0); \xi_1 \bar{h}_1) + \pi_2 Df(x_\rho(s_0); \bar{h}_2)$. But here, by (4.32), the first term at the right equals zero. Thus y_2 is the image of the point $\bar{h}_2 \in X_2$ as we wanted to show.

To prove property (B) of ϕ_s we note that the map (4.34) depends continuously on s_0 as follows from the continuous differentiability of f and $x_\rho(s)$ since π_2 is by definition independent of s_0. The constancy of the index now follows from Theorems 5.2 and 4.12 of Appendix A.

To prove property (C) we consider an s_0 for which the differential of f at $x_\rho(s_0)$ is not singular. From (4.31) and (4.30) we see that

$$Df(x_\rho(\sigma_0); h) = \begin{pmatrix} Df(x_\rho(s_0); h_1 \xi_1) & \pi_1 Df(x_\rho(s_0); h_2) \\ 0 & \pi_2 Df(x_\rho(s_0); h_2) \end{pmatrix}. \tag{4.35}$$

Denoting for a moment the right member of (4.25) by $l_0(t)$ we see from (4.25) and definition of e_1 that

$$Df(x_\rho(s_0); \xi_1 h) = \frac{dl_0}{dt}\left(\frac{dt}{ds}\right)_{s=s_0} \qquad h_1 = e_1 h_1 \|\bar{y}_1 - \bar{y}_0\| \left(\frac{dt}{ds}\right)_{s=s_0}. \tag{4.36}$$

Substituting this in (4.35) one sees easily that the map (4.35) is linearly L.-S. homotopic to the map

$$m_0(h) = \begin{pmatrix} e_1 h_1 (dt/ds)_{s=s_0} & 0 \\ 0 & \pi_2 Df(x_\rho(s_0); h_2) \end{pmatrix}$$

and that therefore the index of the differential (4.35) equals $j(m_0)$. But using the linear homotopy theorem, the multiplication theorem 6.1 of Appendix A for indices, and the fact that the index of an I^- map equals -1 (see Definition 4.4 and Lemma 4.5 of that Appendix) one sees easily that $j(m_0)$ equals $\text{sign}(dt/ds)_{s=s_0}$ times the index of the map $D\phi_{s_0}$ given by (4.34). (Note that $dt/ds \neq 0$ for $s = s_0$ since otherwise by (4.35) and (4.36) the differential (4.35) would be singular.)

The assertion (4.23) follows now easily from the properties (A), (B), and (C) of the auxiliary map ϕ_s. Consider Case I. $x_\rho = x_\rho(0)$ and $f(x_\rho) = \bar{y}_o$. Thus as s moves increasingly from $s = 0$ to positive s the point $f(x_\rho(s))$ moves from $\bar{y}_0$ to interior points of $\bar{\sigma}$, i.e., from $t = 0$ to positive t (cf. 4.5). Therefore $dt/ds > 0$ and by property (C),

$$j(Df(x_\rho; h)) = j(D\phi_0(x_\rho(0); h)). \tag{4.37}$$

But as s tends to 1 increasingly the point $f(x_\rho(1)) = \bar{y}_0$ is approached through interior points of $\bar{\sigma}$, i.e., t approaches $t = 0$ through positive t. Thus $dt/ds < 0$ and, by property (C)

$$-j(Df(x_\sigma; h)) = j(D\phi_1(x(1); h)). \tag{4.37a}$$

But by property (B) the right members of (4.37) and (4.37a) are equal. This proves (4.23) in Case I. Case II is treated correspondingly.

This finishes the proof of (4.18), and we start outlining a proof of properties (a)–(d) of the set $f^{-1}(\bar{\sigma})$ (stated shortly after equation (4.18)). We note first of all that by §1.10 this set is compact since f is an L.-S. map and $\bar{\sigma}$ is compact. Now let y_0 be a point of $\bar{\sigma}$ and let

$$y_0 = f(x_0). \tag{4.38}$$

We then assert the following statements (α) and (β):

(α) If y_0 is an interior point of $\bar{\sigma}$, then there is an open interval i which is diffeomorphic to a connected open piece of $f^{-1}(\bar{\sigma})$ containing x_0. (Two sets are called diffeomorphic if there is a one-to-one relation between them which is onto and continuously differentiable both ways.)

(β) If $y_0 = \bar{y}_0$, then there exists an $\varepsilon > 0$ such that the half-open interval $0 \leq \tau < \varepsilon$ is diffeomorphic to a connected piece of $f^{-1}(\bar{\sigma})$ with $\tau = 0$ corresponding to x_0 under the diffeomorphism. The corresponding statement holds if $y_0 = \bar{y}_1$.

Now assertion (α) is obvious if y_0 is a regular value for f. For then there exists a ball $B(\bar{y}_0, r)$ which is diffeomorphic to an open neighborhood of x_0 and whose radius r satisfies the inequality

$$r < \min(\|y_0 - \bar{y}_0\|, \|\bar{y}_1 - y_0\|). \tag{4.39}$$

Then obviously the restriction of this diffeomorphism to the interval $B(y_0, r) \cap \bar{\sigma}$ has the required properties.

The corresponding argument proves assertion (β) since $\bar{y}_0$ and $\bar{y}_1$ are regular values for f.

It remains to prove (α) for the case that y_0 is not a regular value of f. Without loss of generality we may assume

$$x_0 = y_0 = f(x_0) = \theta \tag{4.39}$$

(as one sees by using a translation $\xi = x - x_0$, $\eta = y - y_0$). Now by the transversality of $\bar{\sigma}$ with respect to f the range R of $Df(x_0; h)$ together with the 1-space generated by $\bar{y}_1 - \bar{y}_0$ spans E, i.e., the latter subspace is a co-kernel K^* of the differential $l_0(h) - Df(x_0; h)$:

$$E = R \dotplus K^* \tag{4.40}$$

and by (4.39)

$$y_0 = \theta \subset \bar{\sigma} \subset K^*. \tag{4.41}$$

K^* is one dimensional and since, by the Krasnoselskiĭ Lemma 18 of Chapter 1, the differential l_0 is L.-S., it follows from Lemma 12 of that chapter that its kernel K is also one dimensional. Then there is a subspace X_2 of E complementary to K such that

$$E = K \dotplus X_2. \tag{4.42}$$

It follows from (4.40) and (4.42) that the assumptions of Lemma 5 of Appendix B are satisfied with

$$\begin{cases} \Pi = E, \quad p = 0, \quad Z = \Omega, \quad z = x, \\ z_0 = x_0 = \theta, \quad \Pi_1 = X_2, \quad \phi = f. \end{cases} \tag{4.43}$$

This lemma states: there are "coordinate transformations" $y = h$, $\varsigma = h_1(x)$ given by equations (24) and (25) of Appendix B resp. which have the following property: there are open neighborhoods U and V of the zero points for x and ς resp. such that h_1 is a diffeomorphism of U on V under which the zero points correspond to each other. Moreover if ψ is the map f in the "new coordinates," i.e., if $\psi(\varsigma) = hfh_1^{-1}(\varsigma)$, then, for $\varsigma \in V$, $\psi(\varsigma)$ is of the form given in equation (5) of Appendix B, where ς_1 and ς_2 are determined by

$$\varsigma = \varsigma_1 + \varsigma_2, \qquad \varsigma_1 \in X_2, \quad \varsigma_2 \in K \tag{4.44}$$

(cf. 4.42), and where

$$\chi(\varsigma) \subset K^*. \tag{4.45}$$

Since V is an open neighborhood of θ, we may choose $\varepsilon > 0$ such that the interval i_ε defined by $\varsigma_1 = 0$, $|\varsigma_2| < \varepsilon$ lies in V. By (5) in Appendix B and (4.45), we see that $\psi(i_\varepsilon) \subset K^*$. But from (4.41) we see that for small enough ε

$$\psi(i_\varepsilon) \subset \bar{\sigma}, \tag{4.46}$$

i.e., $i_\varepsilon \in \psi^{-1}(\bar{\sigma})$.

"Transforming back" it is now easy to see from the properties of h and h_1 that

$$h_1^{-1}(i_\varepsilon) \subset f^{-1}(\overline{\sigma})$$

and that this map of i_ε into $f^{-1}(\overline{\sigma})$ has the desired properties.

From properties (α) and (β) just proved we conclude that a path component C_0 of $f^{-1}(\overline{\alpha})$ is a one-dimensional C^1-manifold (see, e.g., [1]). Moreover C_0 is closed, and as a subset of the compact set is compact. Now it is known that a compact one-dimensional C^1-manifold is diffeomorphic either to a circle or a line segment.

We finally show that the first cannot happen if x_0 is one of the points (4.19). Then $f(x_0)$ equals either $\overline{y}_0$ or $\overline{y}_1$; each of these points is a regular value of f. It will be sufficient to consider the case $f(x_0) = \overline{y}_0$. Then there exists an $r > 0$ such that the ball $B(x_0; r)$ and the open neighborhood $V_0 = f^{-1}(B(\overline{y}_0, r))$ of x_0 are diffeomorphic. Now if the path component C_0 containing x_0 is diffeomorphic to a circle, it may be represented by $x = x_0(s)$ where

$$\begin{cases} 0 \le s \le 1, \quad x_0(0) = x_0(1) = x_0, \\ \quad x_0(s_1) \ne x(s_2) \end{cases} \tag{4.47}$$

if at least one of the disjoint points s_1, s_2 is different from zero or 1. Moreover $f(x_0(s)) \in \overline{\sigma}$, i.e., by (4.5)

$$f(x_0(s)) = \overline{y}_0 + t(\overline{y}_1 - \overline{y}_0) = y_t \tag{4.48}$$

for some $t \in [0, 1]$. By (4.47) there exist points s_1, s_2 in the open interval $(0, \frac{1}{2})$ such that

$$f(x_0(s)) \subset \overline{\sigma} \quad \text{for } 0 \le s \le s_1 < \tfrac{1}{2},$$
$$f(x_0(1 - s)) \subset \overline{\sigma} \quad \text{for } 0 \le s \le s_2 < \tfrac{1}{2},$$

and, by (4.48), points t_1, t_2 in $[0, 1]$ such that

$$f(x_0(s_1)) = y_{t_1}, \qquad f(x_0(1 - s_2)) = y_{t_2}.$$

Now let $t_0 = \min(t_1, t_2)$. Then there exist s', s'' such that

$$0 < s' \le s_1 < \tfrac{1}{2}, \qquad 0 < s'' \le s_2 < \tfrac{1}{2}, \tag{4.49}$$

and such that $f(x_0(s')) = f(x_0(1 - s'')) = y(t_0)$. But by the diffeomorphism between $B(\overline{y}_0, r)$ and V_0, this implies that $x_0(s') = x_0(1 - s'')$. This equality in turn implies that $s' = (1 - s'')$ as seen from (4.47). Thus $s' + s'' = 1$. This however leads to a contradiction since $s' + s'' < 1$ by (4.49).

4.13. LEMMA. *The assertions of Lemmas 2.4 and 2.6 are true without the assumption that y_0 is a regular value for f provided that $f \in C''$.*

The proof will be given in the Notes.[8],[9]

We finish this section by proving the important continuity properties of the degree: under proper assumptions $d(f, \Omega, y)$ is continuous in y (Theorem 4.14) as well as in f (Theorems 4.15 and 4.16). Note that continuity is equivalent to constancy since the degree is integer-valued.

4.14. THEOREM. *Let f be an L.-S. map $\overline{\Omega} \to E$ which lies in $C''(\overline{\Omega})$. Let $y = y(t)$, $0 \leq t \leq 1$, be a continuous curve in E satisfying*

$$y(t) \notin f(\partial \Omega) \quad \text{for } 0 \leq t \leq 1. \tag{4.50}$$

Then $d(f, \Omega, y(t))$ is independent of t (cf. Definition 4.3).

PROOF. Assumption (4.50) says that the curve $y(t)$ lies in one component of $E - f(\partial \Omega)$. The assertion follows therefore immediately from Theorem 4.2.

4.15. THEOREM. *Let*

$$f_t(x) = f(x, t) = x - F(x, t), \tag{4.51}$$

where $F(x, t)$ is a completely continuous twice-differentiable map $\overline{\Omega} \times [0, 1] \to E$. Let $y_0 \in E$ satisfy

$$y_0 \neq f(x, t) \quad \text{for } (x, t) \in \partial \Omega \times [0, 1]. \tag{4.52}$$

Then

$$d_t = d(f_t, \Omega, y_0) \tag{4.53}$$

is independent of t.

PROOF. As already mentioned in the introduction, the proof follows an idea used by Nagumo [**42**, pp. 492, 493] in the finite-dimensional case: namely, to apply Theorem 4.14 to a certain mapping defined on a subset of the Banach space Π (cf. subsection 4.9) which is the direct sum of E with the axis of the reals.

Since d_t is an integer-valued function of t, it will be sufficient to prove that d_t is continuous at every point $t_0 \in [0, 1]$, i.e., to show that there exists a positive $\delta = \delta(t_0)$ such that

$$d(f_{t_1}, \Omega, y_0) = d(f_{t_0}, \Omega, y_0) \tag{4.54}$$

if

$$|t_1 - t_0| < \delta, \quad t_1, t_0 \in [0, 1]. \tag{4.55}$$

Now for arbitrary t_0, t_1 in $[0, 1]$ we set

$$f^s(x) = (1 - s)f_{t_0}(x) + sf_{t_1}(x), \quad 0 \leq s \leq 1. \tag{4.56}$$

It is clear from (4.51) that $f^s(x)$ is an L.-S. mapping. In order to see that the degree $d(f^s, \Omega, y_0)$ exists we will prove: there exist positive numbers ε and δ such that

$$\|f^s(x) - y_0\| \geq \varepsilon \tag{4.57}$$

for all $x \in \partial \Omega$ and $t_0, t_1 \in [0, 1]$ satisfying (4.55). If there were no such couple ε, δ, there would exist sequences of points t_0^ν, t_1^ν, s^ν, $\nu = 1, 2, \ldots$, in $[0, 1]$, a sequence of positive numbers ε_ν, and a sequence of points $x^\nu \in \partial \Omega$ such that

$$|t_1^\nu - t_0^\nu| \to 0, \quad \varepsilon_\nu \to 0 \tag{4.58}$$

and

$$y_\nu = (1 - s^\nu)f(x^\nu, t_0^\nu) + s^\nu f(x^\nu, t_1^\nu) \to y_0 \tag{4.59}$$

as $\nu \to \infty$ (see 4.56) and (4.57)). Obviously we may assume that the sequences t_0^ν, t_1^ν, s^ν are convergent:

$$\lim_{\nu \to \infty} t_0^\nu = \lim_{\nu \to \infty} t_1^\nu = \bar{t}_1, \qquad \lim_{\nu \to \infty} s^\nu = \bar{s}. \tag{4.60}$$

But by (4.59) and by (4.51)

$$y_\nu = x_\nu - [(1 - s^\nu)F(x^\nu, t_0^\nu) + s^\nu F(x_\nu, t_1^\nu)] \to y_0, \tag{4.61}$$

and by the complete continuity of F the terms in the bracket are convergent for some subsequence ν_i, $i = 1, 2, \ldots$, of the integers ν. It now follows from (4.61) and (4.60) that the x_{ν_i} converge to a point $\bar{x}$, and we see that

$$y_0 = (1 - \bar{s})f(\bar{x}, \bar{t}_1) + \bar{s}f(\bar{x}, \bar{t}_1) = f(\bar{x}, \bar{t}_1). \tag{4.62}$$

But $\bar{x} \in \partial\Omega$ since $x_{\nu_i} \in \partial\Omega$ and $\partial\Omega$ is closed. Thus (4.62) contradicts the assumption (4.52).

Then let t_0 be a given point in $[0, 1]$, and let ε, δ be chosen such that (4.57) holds for t_1 satisfying (4.55). With such a fixed t_1, our assertion (4.54) is, by (4.56), equivalent to the equality $d(f^0, \Omega, y_0) = d(f^1, \Omega, y_0)$ and will therefore be implied by the assertion

$$d(f^s, \Omega, y_0) = \text{const.} \quad \text{for } 0 \leq s \leq 1. \tag{4.63}$$

To prove this assertion we introduce the Banach space Π defined in subsection 4.9 and use the notation employed there. We define the cylinder $Z \subset \Pi$ by

$$Z = \{z = x + tr_1 \in \Pi \mid x \in \Omega, \; -\eta < t < 1 + \eta\}, \tag{4.64}$$

where

$$0 < \eta < \varepsilon \tag{4.65}$$

and define a map $g: \bar{Z} \to E$ by

$$g(z) = sr_1 + f^s(x), \qquad z = sr_1 + x. \tag{4.66}$$

Then by (4.56) and (4.51)

$$\begin{aligned} g(z) &= sr_1 + x + (1 - s)F(x, t_0) + sF(x, t_1) \\ &= z + (1 - s)F(x, t_0) + sF(x, t_1). \end{aligned} \tag{4.67}$$

This shows that g is an L.-S. map $\bar{Z} \to \Pi$. If we set

$$z_t = tr_1 + y_0, \qquad 0 \leq t \leq 1, \tag{4.68}$$

then as will be verified later

$$z_t \notin g(\partial Z). \tag{4.69}$$

But by Theorem 4.14 with E, Ω, f replaced by Π, Z, g resp., it follows from (4.69) that

$$\bar{d}_t = d(g, Z, z_t) = \text{const.} \quad \text{for } t \in [0, 1). \tag{4.70}$$

This shows that our assertion (4.63) will be proved once it is established that, for $0 \leq t \leq 1$,

$$d(f^t, \Omega, y_0) = d(g, Z, z_t). \tag{4.71}$$

The proof of this equality will be based on Lemma 4.13, on part (vii) of Lemma 2.4, and on Lemma 2.6. We set, for z defined in (4.66),

$$\varsigma = z - r_1 t = r_1(s - t) + x, \tag{4.72}$$

$$h(\varsigma) = g(z) - r_1 t. \tag{4.73}$$

It then follows from the lemmas quoted above and from (4.68) that

$$d(g, Z, z_t) = d(g, Z, tr_1 + y_0) = d(h, Z - tr_1, y_0). \tag{4.74}$$

Now by (4.73), (4.67), and (4.72)

$$h(\varsigma) = \varsigma + (1 - s)F(x, t_0) + sF(x, t_1). \tag{4.75}$$

This shows that h is a layer map with respect to E. Therefore by the lemmas quoted above

$$d(h, Z - tr_1, y_0) = d(h_E, \Omega, y_0), \tag{4.76}$$

where h_E denotes the restriction of h to $\Omega = (Z - tr_1) \cap E$. But $t = s$ and $\varsigma = x$ on E as is seen from (4.72). Therefore by (4.75), (4.56), and (4.51)

$$h_E(x) = x + (1 - t)F(x, t_0) + tF(x, t_1) = f^t(x) \tag{4.77}$$

The assertion (4.71) follows now from (4.77), (4.76), and (4.74).

It remains to verify (4.69). Now the boundary ∂Z of the cylinder Z (see (4.64)) is the union of three parts: the "lateral surface" $(\partial Z)_1$, the "bottom" $(\partial Z)_2$, and the "top" $(\partial Z)_3$. More precisely,

$$(\partial Z)_1 = \{z \in \Pi \mid x \in \partial\Omega, \ -\eta \le s \le 1 + \eta\},$$
$$(\partial Z)_2 = \{z \in \Pi \mid x \in \overline{\Omega}, \ s = -\eta\},$$
$$(\partial Z)_3 = \{z \in \Pi \mid x \in \overline{\Omega}, \ s = 1 + \eta\},$$

where $z = r_1 s + x$ and where η satisfies (4.65). Now if (4.69) were not true, then, by definition (4.68) of z_t, at least one of the following three relations would hold:

$$tr_1 + y_0 = g(z) \quad \text{for some } z \in (\partial Z)_1 \text{ and some } t \in [0, 1];$$
$$tr_1 + y_0 = g(z) \quad \text{for some } z \in (\partial Z)_2 \text{ and some } t \in [0, 1];$$
$$tr_1 + y_0 = g(z) \quad \text{for some } z \in (\partial Z)_3 \text{ and some } t \in [0, 1].$$

Now by (4.66) the first of these relations implies that

$$tr_1 + y_0 = sr_1 + f^s(x) \tag{4.78}$$

for some $x \in \partial\Omega$. Since $f^s(x) \in E$, it follows that $s = t$, $y_0 = f^s(x)$, $x \in \partial\Omega$. But this contradicts (4.57).

The second relation implies (4.78) with $t \in [0, 1]$ and some $x \in \overline{\Omega}$. It follows that $s = t \in [0, 1)$, and this contradicts the fact that $s = -\eta$.

The impossibility of the third relation is proved correspondingly.

4.16. THEOREM. *The conclusion of Theorem 4.15 remains valid if the assumption that the map $F(x,t)\colon \overline{\Omega}\times[0,1] \to E$ is completely continuous is replaced by the following two:* (a) $F(x,t)$ *is, as a function of t, continuous and uniformly so as x varies over $\overline{\Omega}$;* (b) *for fixed $t \in [0,1]$ the map $F_t(x) = F(x,t)$ is completely continuous in x.*

PROOF. As will be verified in the Notes,[10] the above assumptions (a), (b) together imply the assumption of Theorem 4.15.

4.17. The assumption of Theorem 4.15, i.e., the complete continuity of the map $F\colon \overline{\Omega} \times [0,1] \to E$ does not imply assumption (a) of Theorem 4.16. An example will be given in the Notes[11].

§5. Notes to Chapter 2

Notes to §1.

1. Definition 1.4 of the index is equivalent to the definition given in subsection 4.3 of Appendix A. This follows from Lemma 4.5 and Theorem 4.6, both in Appendix A.

Notes to §2.

2. *Proof of Lemma 2.4.* (i) If $y_0 \in \Omega$, then $x = y_0$ is the only solution of equation (2.2) with $f = I$. Since the differential of I is the identity map, the right member of (2.3) equals $j(I) = 1$. If $y_0 \notin \overline{\Omega}$, then equation (2.2) with $f = I$ has no solution in Ω, and the degree equals zero by Definition 2.2.

(ii) Suppose (2.2) has no roots. Then the degree is zero by Definition 2.2.

(iii) This assertion follows directly from the definitions involved since every root $x \in \Omega$ of (2.2) lies in exactly one of the Ω_σ.

(iv) Let $\Omega_2 = \Omega - \overline{\Omega}_1$. Then Ω_2 contains no roots of (2.2). Therefore $d(f, \Omega_2, y_0) = 0$ and the assertion (2.5) follows from (2.4) with $s = 2$.

(v) A point $x \in \Omega$ is a root of (2.2) if and only if x is a root of the equation

$$f_b(x) = y_0 - b. \tag{1}$$

Therefore if (2.2) has no roots in Ω, then (1) has no roots in Ω and both members of (2.6) are zero. Suppose now that $x_1, x_2, \ldots, x_r$ are the roots of (2.2). Then (2.3) holds. But $Df_b(x_\rho; h) = Df(x_\rho; h)$ since f and f_b differ by a constant. Consequently the right member of (2.3) equals

$$\sum_{\rho=1}^{r} j(Df_b(x_\rho; h)).$$

But this sum equals $d(f_b, \Omega, y_0 - b)$ since the x_ρ are the roots of (1).

(vi) $x \in \Omega$ is a root of (2.2) if and only if $\xi = x - x_0$ is a root of

$$g(\xi) = y_0 \tag{2}$$

in Ω_0. Therefore if (2.2) has no roots in Ω, then (2) has no roots in Ω_0 and both members of (2.7) are zero. Now suppose $x_1, x_2, \ldots, x_r$ are the roots of (2.2). Then $\xi_1 = x_1 - x_0$, $\xi_2 = x_2 - x_0, \ldots$, $\xi_r = x_r - x_0$ are the roots of (2) in Ω_0. Since $Df(x_\rho; h) = Dg(\xi_\rho; h)$, the assertion (2.7) follows directly from Definition 2.2.

(vii) By (v) and (vi)

$$d(f, \Omega, y_0 + b) = d(f - y_0, \Omega, b) = d(g - y_0, \Omega_0, b) = d(h, \Omega_0, b).$$

3. We have to show: if $F(x) \in E_1$ for all $x \in \overline{\Omega}$, then (2.14) holds. Now it follows easily from the definition of the differential that for real t

$$DF(x; h) = \lim_{t \to 0} \left[\frac{F(x + th) - F(x)}{t} \right].$$

By assumption the term in the bracket is an element of E_1. Since E_1 is closed, our assertion follows.

Notes to §3.

4. *Proof of Lemma* 3.2. If (3.1) has no roots in $[\alpha, \beta]$, then for some positive ε_1

$$|f(x) - a| \geq 3\varepsilon_1 \tag{3}$$

for all $x \in [\alpha, \beta]$, and (3.2) implies that for all these x

$$|g(x) - b| \geq |f(x) - a| - |f(x) - g(x)| - |b - a| \geq \varepsilon_1 > 0. \tag{4}$$

This proves assertion (i) of our lemma. If all roots of (3.1) lie in the open subset U of $\overline{\Omega}$, then no root of that equation lies in the closed set $U' = [\alpha, \beta] - U$. Consequently for some positive ε_1 the inequality (3) holds for all $x \in U'$. But then (3.2) implies that the inequality (4) holds for all $x \in U'$, i.e., that every root of (3.3) lies in U.

5. Lemma 3.3 follows immediately from the relevant definitions since $Df(x; h) = f'(x)h$ and since equation (3.4) has at most a finite number of solutions if f is a polynomial.

6. *Proof of Lemma* 3.4. By Lemma 3.3 and by assumption, $f(C_f) = S_f \subset V$. If C_f is not empty, it consists, by Lemma 3.3, of a finite number of points, say $x_1, x_2, \ldots, x_r$, and S_f consists of the points $f(x_1), f(x_2), \ldots, f(x_r)$ which lie in V. Therefore there exists a $\delta > 0$ such that for $\rho = 1, 2, \ldots, r$

$$B(f(x_\rho), 2\delta) \subset V. \tag{5}$$

By the continuity of f there exists a positive ς such that $f(B(x_\rho, \varsigma)) \subset B(f(x_\rho), \delta)$. Therefore if $U = \bigcup_\rho B(x_\rho, \varsigma)$, then

$$f(U) \subset \bigcup_\rho B(f(x_\rho), \delta). \tag{6}$$

Now by part (ii) of Lemma 3.2 with f and g replaced by f' and g' resp. and with $a = b = 0$, there exists a positive ε_1 of the following property: if g is a polynomial in $[\alpha, \beta]$ for which

$$|g'(x) - f'(x)| < \varepsilon_1 \quad \text{for } x \in [\alpha, \beta],\tag{7}$$

then the set C_g of roots in $[\alpha, \beta]$ of the equation $g'(x) = 0$ lies in the neighborhood U of C_f:

$$C_g \subset U.\tag{8}$$

Now suppose assumption (3.5) is satisfied with $\varepsilon = \min(\varepsilon_1, \delta)$. Then (7), and therefore (8), is true. Consequently by (6)

$$f(C_g) \subset f(U) \subset \bigcup_\rho B(f(x_\rho), \delta).\tag{9}$$

But by (3.5), $|f(x) - g(x)| < \varepsilon \leq \delta$. Therefore, by (9) and (5), $S_g = g(C_g) \subset \bigcup_\rho B(f(x_\rho), 2\delta) \subset V$.

This proves our lemma if C_f is not empty. But if C_f is empty, then by the first part of Lemma 3.3 the equation $f'(x) = 0$ has no roots in $[\alpha, \beta]$. Consequently by Lemma 3.2 (with f and g replaced by f' and g' resp., and with $a = b = 0$) there exists a positive ε such that the equation $g'(x) = 0$ has no roots in $[\alpha, \beta]$ if $|f'(x) - g'(x)| < \varepsilon$ for $x \in [\alpha, \beta]$. Thus for such g the set C_g, and therefore the set S_g, is empty.

7. *Proof of Lemma 3.5.* By assumption $S_f \cap \Gamma = \varnothing$. Therefore the set S_f containing at most a finite number of points has a positive distance from the closed set Γ. Thus there exists an open set V such that $S_f \subset V$ and

$$V \cap \Gamma = \varnothing.\tag{10}$$

But by Lemma 3.4 there exists a positive ε such that (3.5) implies (3.6). Therefore for such g, $S_g \cap \Gamma \subset V \cap \Gamma$ and, by (10), $S_g \cap \Gamma = \varnothing$.

Notes to §4.

8. *Proof of the part of Lemma 4.13 relating to Lemma 2.4.* Assertion (i) is vacuously true since every point $y_0 \in E$ is a regular value for the identity map I. For the proof of the remaining assertions of Lemma 2.4, we recall that the assumption (2.1) implies the existence of a ρ_0 such that, for $0 < \rho < \rho_0$,

$$y \notin f(\partial \Omega) \quad \text{for } y \in B(y_0, \rho),\tag{11}$$

and that, by Theorem 4.2, ρ_0 may be chosen in such a way that

$$d(f, \Omega, y) = d(f, \Omega, y_0)\tag{12}$$

for those points $y \in B(y_0, \rho)$ which are regular values for f.

Proof of assertion (ii). It follows from Theorem 4.1 that there exists a sequence $\{y_i\}$ of regular values for f which satisfy

$$y_i \in B(y_0, \rho), \qquad \lim_{i \to \infty} y_i = y_0.\tag{13}$$

By assumption, $d(f, \Omega, y_0) \neq 0$. It therefore follows from (11) and (12) that $d(f, \Omega, y_i) \neq 0$ for $i = 1, 2, \ldots$, and from Lemma 2.4 that the equation $f(x) = y_i$ has a root $x_i \in \Omega$. Thus

$$f(x_i) = x_i - F(x_i) = y_i, \qquad i = 1, 2, \ldots . \tag{14}$$

It now follows from (13), (14), and the complete continuity of F that some subsequence $\{x_{i_\nu}\}$ of the $\{x_i\}$ converges to some point $x_0 \in \overline{\Omega}$, and that $f(x_0) = y_0$. But $x_0 \in \Omega$ by (2.1).

Proof of assertion (iii). Since the sum in (2.4) is finite, we can choose ρ_0 in such a way that not only does (12) hold for $0 < \rho < \rho_0$ but also for $\sigma = 1, 2, \ldots, s$ and $y \in B(y_0, \rho)$

$$d(f, \Omega_\sigma, y) = d(f, \Omega_\sigma, y_0) \tag{15}$$

if y is a regular value for f and for such y

$$y \notin f(\partial \Omega_\sigma). \tag{16}$$

Then by Lemma 2.4, the equation (2.4) holds with y_0 replaced by y. But by (12) and (15) it then holds also for y_0.

Proof of assertion (iv). This assertion follows from assertion (iii), as was shown in the proof of Lemma 2.4.

Proof of assertion (v). Let y be a regular value for f satisfying (11) and (12). It is then easily verified that $y - b$ is a regular value for $f_0 = f - b$ and that

$$d(f_0, \Omega, y - b) - d(f_0, \Omega, y_0 - b). \tag{17}$$

But since y is a regular value for f, we see from Lemma 2.4 that

$$d(f, \Omega, y) = d(f_0, \Omega, y - b), \tag{18}$$

and we see from (12) and (17) that (18) remains true if y is replaced by y_0.

Proof of assertion (vi). Let y be a regular value for f satisfying (11) and (12). It is easily verified that y is a regular value for g and that

$$d(g, \Omega_0, y) = d(g, \Omega_0, y_0). \tag{19}$$

Since y is a regular value for f, we see from Lemma 2.4 that

$$d(f, \Omega, y) = d(g, \Omega_0, y). \tag{20}$$

But by (12) and (19) the relation (20) remains true if y is replaced by y_0.

Finally as shown in the proof of Lemma 2.4 the assertion (vii) follows from assertions (v) and (vi).

9. *Proof of the part of Lemma* 4.13 *referring to Lemma* 2.6. We have to show that (2.9) holds even if y_0 is not a regular value for f. Now by (2.1) there exists a neighborhood N of y_0 whose closure does not intersect $f(\partial \Omega)$. Consequently $N_1 = N \cap E_1$ is a neighborhood (with respect to E_1) of y_0 such that $y \notin f(\partial \Omega)$

for $y \in N_1$. By the Sard-Smale theorem there exists a $\bar{y}_0 \in N_1$ which is a regular value for the restriction of f to $\Omega \cap E_1$. Then

$$d(f, \Omega \cap E_1, y_0) = d(f, \Omega \cap E_1, \bar{y}_0), \qquad d(f, \Omega, y_0) = d(f, \Omega, \bar{y}_0). \qquad (21)$$

But since $F(\overline{\Omega})$ and $\bar{y}_0$ lie in E_1, all roots of $f(x) = x - F(x) = \bar{y}_0$ lie in E_1. Moreover for each such root x we see: if $\theta = Df(x; h) = h - DF(x; h)$, then $h \in E_1$, and therefore $h = \theta$ since $Df(x; h)$ is not singular for the restriction of f to $E_1 \cap \Omega$. This shows that $\bar{y}_0$ is a regular value also for f. Therefore by Lemma 4.6 the right members of (21) are equal.

10. *Note to the proof of Theorem* 4.16. We have to show that assumptions (a) and (b) of our theorem together imply the complete continuity of F. We first prove the continuity. For x and x_0 in $\overline{\Omega}$ and for t and t_0 in $[0, 1]$,

$$\|F(x, t) - F(x_0, t_0)\| \le \|F(x, t) - F(x, t_0)\| + \|F(x, t_0) - F(x_0, t_0)\|. \qquad (22)$$

By assumption (a) there corresponds to a given $\varepsilon > 0$ a $\delta > 0$ independent of x such that the first term in the right member of (22) is less than $\varepsilon/2$ if $|t - t_0| < \delta$. The continuity of $F(x, t)$ at every point $(x_0, t_0) \in \overline{\Omega} \times [0, 1]$ now follows from (22).

For the proof of the complete continuity of F it remains to show: every sequence $\{x_\nu, t_\nu\} \in \overline{\Omega} \times [0, 1]$ contains a subsequence $\{x_{\nu_i}, t_{\nu_i}\}$ for which the sequence $F(x_{\nu_i}, t_{\nu_i})$ converges. We assert first that it will be sufficient to show that to every $\eta > 0$ there corresponds a subsequence $\{x_{\nu_i}, t_{\nu_i}\}$ of the sequence $\{x_\nu, t_\nu\}$ such that

$$\|F(x_{\nu_i}, t_{\nu_i}) - F(x_{\nu_j}, t_{\nu_j})\| < \eta \qquad (23)$$

for all $i, j = 1, 2, \ldots$. Indeed, let $\eta_1, \eta_2, \ldots$ be a sequence of positive numbers converging to zero. Assuming (23), we can select a subsequence $\{x_\nu^1, t_\nu^1\}$ of the sequence $\{x_\nu, t_\nu\}$ such that

$$\|F(x_i^1, t_i^1) - F(x_j^1, t_j^1)\| < \eta_1$$

for all i, j. Going on in this way we obtain for every positive integer ν a sequence $\{x_i^\nu, t_i^\nu\}$ which is a subsequence of the sequence $\{x_i^\mu, t_i^\mu\}$ for $\mu = 1, 2, \ldots, \nu - 1$, and for which

$$\|F(x_i^\nu, t_i^\nu) - F(x_j^\nu, t_j^\nu)\| < \eta_\nu \qquad (24)$$

for $i, j = 1, 2, \ldots$. We claim that the "diagonal" sequence $\{x_i^i, t_i^i\}$ is a Cauchy sequence, i.e., to every $\varepsilon > 0$ there corresponds an integer ν_0 such that

$$\|F(x_i^i, t_i^i) - F(x_j^j, t_j^j)\| < \varepsilon \qquad (25)$$

for $i, j \ge \nu_0$. To see this we choose ν_0 in such a way that $\eta_\nu < \varepsilon$ for $\nu \ge \nu_0$. Then by (24) with $\nu = \nu_0$

$$|F(x_i^{\nu_0}, t_i^{\nu_0}) - F(x_j^{\nu_0}, t_j^{\nu_0})\| < \eta_{\nu_0} < \varepsilon \qquad (26)$$

for all i, j. But for $i \ge \nu_0$ the sequence $\{x_i^i, t_i^i\}$ is a subsequence of the sequence

$\{x_i^{\nu_0}, t_i^{\nu_0}\}$. Thus (26) implies that (25) holds, i.e., that the sequence $F(x_i^i, t_i^i)$ converges.

It remains to show that (23) holds for some subsequence $\{x_{\nu_i}, t_{\nu_i}\}$ of the given sequence $\{x_i, t_i\}$. Obviously we may assume that the t_i converge and therefore form a Cauchy sequence. Consequently, by assumption (a) there exists a positive integer $\bar{i}$ such that

$$\|F(x, t_i) - F(x, t_{\bar{i}})\| < \eta/3 \tag{27}$$

for $i \geq \bar{i}$ and for all $x \in \bar{\Omega}$. But by assumption (b), $F(x, t_{\bar{i}})$ is completely continuous. Therefore there exists a subsequence $\{x_{i_\alpha}\}$ of the sequence $\{x_i\}$ such that the sequence $\{F(x_{i_\alpha}, t_{\bar{i}})\}$ converges. Consequently there exists an integer $\bar{\bar{i}}$ such that

$$\|F(x_{i_\alpha}, t_{\bar{i}}) - F(x_{i_\beta}, t_{\bar{i}})\| < \eta/3 \tag{28}$$

for $i_\alpha, i_\beta \geq \bar{\bar{i}}$. We now omit from the sequence $\{x_{i_\alpha}\}$ all those elements for which $i_\alpha \leq \max(\bar{i}, \bar{\bar{i}})$. Then (28) holds for all elements of the sequence thus obtained. Now for this sequence

$$\|F(x_{i_\alpha}, t_{i_\alpha}) - F(x_{i_\beta}, t_{i_\beta})\|$$
$$\leq \|F(x_{i_\alpha}, t_{i_\alpha}) - F(x_{i_\alpha}, t_{\bar{i}})\| + \|F(x_{i_\alpha}, t_{\bar{i}}) - F(x_{i_\beta}, t_{\bar{i}})\|$$
$$+ \|F(x_{i_\beta}, t_{\bar{i}}) - F(x_{i_\beta}, t_{i_\beta})\|.$$

The first and the last of the three summands in the right member of this inequality are less than $\eta/3$ by (27) and the middle one by (28). This proves (23).

11. *Note to* 4.17. Let E be the Hilbert space of points $x = (x_1, x_2, \ldots)$ where the x_i are real numbers with $\sum x_i^2 < \infty$, with the usual definitions of linear operations, and with the scalar product (x, y) and norm $\|x\| = (x, x)^{1/2}$. Let Ω be the unit ball in E. We define a map $F \colon \bar{\Omega} \times [0, 1]i \to E$ as follows:

$$y = F(x, t) = (y_1(x, t), y_2(x, t), \ldots),$$

where

$$y_1(x, t) = x_1 + \sum_{\nu=2}^{\infty} x_\nu^2 t^\nu, \qquad y_2(x, t) = y_3(x, t) = \cdots = 0. \tag{29}$$

We will show: (i) F is completely continuous; (ii) condition (a) of Theorem 4.16 is not satisfied.

Proof of (i). $F(\bar{\Omega} \times [0, 1])$ is a bounded set in a one-dimensional subspace of E. Therefore it will be sufficient to show that $F(x, t)$ is continuous at every point (a, t_0) of $\bar{\Omega} \times [0, 1]$, i.e., that to every $\varepsilon > 0$ there corresponds a positive $\delta = \delta(x, t_0)$ such that the norm of

$$\Delta^0 = F(a + h, t) - F(a, t_0) \tag{30}$$

is less than ε if $\|h\|$ and $|t - t_0|$ are less than δ (provided that $a + h \in \bar{\Omega}$ and $t \in [0, 1]$). Without loss of generality we assume $\varepsilon < 1$. If $a = (a_1, a_2, \ldots)$, and

$h = (h_1, h_2, \ldots)$, and $\Delta^0 = (\Delta^0_1, \Delta^0_2, \ldots)$, then by (29) and (30)

$$\left.\begin{aligned}
\Delta^0_1 &= h_1 + \sum_{\nu=2}^{\infty}[(a_\nu + h_\nu)^2 t^\nu - a_\nu^2 t_0^\nu], \\
\Delta^0_2 &= \Delta^0_3 = \cdots = 0.
\end{aligned}\right\} \tag{31}$$

We now choose

$$\|\delta\| < \varepsilon/8. \tag{32}$$

Moreover, since $\sum a_\nu^2$ converges, we can choose an integer N such that

$$\sum_{N+1}^{\infty} a_\nu^2 < \varepsilon^2/8. \tag{33}$$

Then using the Schwarz inequality and taking into account the fact that the nonnegative numbers t, t_0, and ε are all ≤ 1, we see from (32) and (33) that

$$\left\|\sum_{N+1}^{\infty}[(a_\nu + h_\nu)^2 t^\nu - a_\nu^2 t_0^\nu]\right\| \leq \sum_{N+1}^{\infty}[2a_\nu^2 + 2|h_\nu a_\nu| + h_\nu^2]$$

$$\leq 2\sum_{N+1}^{\infty} a_\nu^2 + 2\left(\sum_{N+1}^{\infty} h_\nu^2 \cdot \sum_{N+1}^{\infty} a_\nu^2\right)^{1/2} + \sum_{N+1}^{\infty} h_\nu^2$$

$$\leq 2\sum_{N+1}^{\infty} a_\nu^2 + 2\|h\|\left(\sum_{N+1}^{\infty} a_\nu^2\right)^{1/2} + \|h\|^2 < \frac{\varepsilon}{2}.$$

But since $|h_\nu| \leq \|h\|$ it is obvious that there exists a $\delta_1 > 0$ such that the finite sum

$$\left|h_1 + \sum_{2}^{N}[a_\nu + h_\nu^2)t^\nu - a_\nu^2 t_0^\nu]\right| < \frac{\varepsilon}{2}$$

if $|h|$ and $|t - t_\nu|$ are less than δ_1. It is now clear from (31) that $|\Delta^0_1| < \varepsilon$ if $|h|$ and $(t - t_0)$ are less than $\delta = \min(\varepsilon/8, \delta_1)$.

This finishes the proof of assertion (i), and we turn to the proof of assertion (ii). Suppose that property (a) of Theorem 4.16 is satisfied. Then there would exist a $\delta > 0$, independent of x, such that

$$\|F(x, t) - F(x, 1)\| < 1/2 \tag{34}$$

for all $x \in \overline{\Omega}$ and all t satisfying

$$0 < 1 - t < \delta. \tag{35}$$

However, if x^j is the point of $\overline{\Omega}$ whose jth coordinate is 1 and whose other coordinates are zero, then by (29) for $j \geq 2$, $F(x^j, t) - F(x^j, 1) = (t^j - 1, 0, 0, \ldots)$, and therefore for each $t \in [0, 1]$

$$\|F(x^j, t) - F(x^j, 1)\| = 1 - t^j. \tag{36}$$

Now choose a fixed t satisfying (35) and then a fixed j so great that $1 - t^j > 1/2$. For such t and y we conclude from (36) that $\|F(x^j, t) - F(x^j, 1)\| > 1/2$. But this contradicts the fact that (34) holds for all $x \in \overline{\Omega}$ and t satisfying (35).

CHAPTER 3

The Leray-Schauder Degree
for Not Necessarily Differentiable Maps

§1. An extension lemma

1.1. As in Chapter 2, Ω denotes a bounded open set in the Banach space
E. The family $C(\overline{\Omega})$ of completely continuous maps $F: \overline{\Omega} \to E$ is, with the
obvious definitions of addition and multiplication by (real) scalars, a linear space.
Moreover, if we define

$$\|F\| = \sup_{x \in \overline{\Omega}} \|F(x)\|, \tag{1.1}$$

then it is easily seen that $C(\overline{\Omega})$ is a *closed* linear subspace of the space of bounded
maps $\overline{\Omega} \to E$ with norm (1.1) (see e.g. [**31**, pp. 15–16]). In what follows $C(\overline{\Omega})$
will always denote this subspace. If Ω is fixed we will often write C for $C(\overline{\Omega})$.

1.2. In Chapter 2 the Leray-Schauder degree was defined for maps $f: \overline{\Omega} \to E$
for which the elements $F = I - f$ of $C(\overline{\Omega})$ belong to $C''(\overline{\Omega})$. We want to extend
the definition of the degree $d(f, \Omega, y_0)$ to L.-S. maps for which $F = I - f$ lies in
the closure of this subspace of $C(\overline{\Omega})$, always assuming that

$$y_0 \notin f(\partial\Omega). \tag{1.2}$$

The "extension lemma" in subsection 1.3 below serves this purpose. Before
stating it we make the following

REMARK. If $C_1(\overline{\Omega})$ is the set of twice differentiable completely continuous
maps on $\overline{\Omega}$ (cf. §1.16), i.e., if $C_1(\overline{\Omega}) = C(\overline{\Omega}) \cap C''(\overline{\Omega})$ and if $C_1(\overline{U})$ is defined correspondingly for every bounded open set $U \subset E$, then $C_1(\overline{\Omega})$ obviously satisfies
the following conditions (i)–(iii):

(i) $C_1(\overline{\Omega})$ is linear;

(ii) if $F \in C_1(\overline{\Omega})$, then for every element $b \in E$ the map $F_b: \overline{\Omega} \to E$ given by
$F_b(x) = F(x) - b$ belongs to $C_1(\overline{\Omega})$;

(iii) if Ω_{x_0} denotes the translate of Ω by the element $-x_0 \in E$, i.e., if $\Omega_{x_0} = \{\xi \in E \mid x = \xi + x_0 \in \overline{\Omega}\}$, then the map $G: \overline{\Omega}_{x_0} \to E$ belongs to $C_1(\overline{\Omega}_{x_0})$ if and
only if the map $F: \overline{\Omega} \to E$ given by $F(x) = x_0 + G(x - x_0)$ belongs to $C_1(\overline{\Omega})$.

In addition we know from our previous results for C'' maps that the following statements (d_0)–(d_2) are true (under the assumption made and with the notations used in this remark):

(d_0) if $F \in C_1(\overline{\Omega})$, $G \in C_1(\overline{\Omega}_{x_0})$, $f(x) = x - F(x)$, $g(x) = x - G(x)$, $\tilde{\Omega}$ is an arbitrary open subset of Ω, $y_0 \notin f(\partial\Omega)$ and $\tilde{y}_0 \notin f(\partial\tilde{\Omega})$, then degrees $d(f, \Omega, y_0)$, $d(g, \Omega_{x_0}, y_0)$, and $d(f, \tilde{\Omega}, y_0)$ exist and satisfy the following conditions (d_1) and (d_2).

(d_1) The part of Lemma 4.13 in Chapter 2 referring to Lemma 2.4 in Chapter 2 holds.

(d_2) If for each $t \in [0,1]$, the map F_t given by $F_t(x) = F(x,t)$ satisfies conditions (a) and (b) of Theorem 4.16 in Chapter 2 and if (1.2) holds with f replaced by $f_t = I - F_t$, then $d(f_t, \Omega, y_0)$ is independent of t.

1.3. EXTENSION LEMMA. *Let Ω, x_0, and Ω_{x_0} be as in the Remark in subsection 1.2, let $C(\Omega_{x_0})$ be the space of completely continuous maps $\overline{\Omega}_{x_0} \to E$ (such that $C(\overline{\Omega}_\theta) = C(\overline{\Omega})$). For any subfamily $\tilde{C} = \tilde{C}(\overline{\Omega})$ of $C(\overline{\Omega})$ we denote by $\Lambda(\tilde{C})$ the family of L.-S. maps $f(x) = x - F(x)$ where $F \in \tilde{C}$.*

Now let $C_1(\overline{\Omega})$ be an arbitrary subset of $C(\overline{\Omega})$ which satisfies the conditions (i)–(iii) and (d_0)–(d_2) stated in the Remark in subsection 1.2. It is then asserted that the definition $d(f, \Omega, y_0)$ for $f \in \Lambda(C_1)$ can be extended to all $f \in \Lambda(\overline{C}_1)$ in such a way that the properties (d_1) and (d_2) are preserved and that, in addition, for these f the following assertion (d_3) holds: let $y = y(t)$ be a continuous curve in E, $0 \le t \le 1$, and suppose that for each such t the condition (1.2) holds with y_0 replaced by $y(t)$. Then $d(f, \Omega, y(t))$ is independent of t.

PROOF. Let f be an element of $\Lambda(\overline{C}_1)$ for which (1.2) holds. To define a degree for f we note first that, on account of (1.2), y_0 has a positive distance ρ from $f(\partial\Omega)$. Therefore

$$y_0 \notin B(f(x), \rho) \quad \text{for all } x \in \partial\Omega. \tag{1.3}$$

Now $F = I - f \in \overline{C}_1$ by assumption. Consequently there exists an $F_1 \in C_1$ such that

$$\|F_1 - F\| < \rho. \tag{1.4}$$

If $f_1(x) = x - F_1(x)$ then by (1.1)

$$\|f_1(x) - f(x)\| = \|F_1(x) - F(x)\| \le \|F_1 - F\| < \rho$$

for all $x \in \overline{\Omega}$. In particular,

$$f_1(x) \in B(f(x), \rho) \quad \text{for } x \in \partial\Omega. \tag{1.5}$$

From this together with (1.3) we conclude that (1.2) holds with f replaced by f_1. Since $f_1 \in \Lambda(C_1)$ we thus see that the degree $d(f_1, \Omega, y_0)$ is defined. We show next that this degree is independent of the particular choice of $f_1 \in \Lambda(C_1)$ provided that (1.4) holds: let, for $i = 0, 1$, $f_1^i(x) = x - F_1^i(x)$ be maps in $\Lambda(C_1)$ for which (1.4) holds with F replaced by F_1^i. To show that indeed

$$d(f_1^0, \Omega, y_0) = d(f_1^1, \Omega, y_0), \tag{1.6}$$

we recall that (1.4) implies (1.5). Therefore $f_1^i(x) \in B(f(x), \rho)$ for $i = 0, 1$ and $x \in \partial\Omega$. Since a ball is convex, it follows that for $t \in [0, 1]$

$$f_1^t(x) = (1 - t)f_1^0(x) + tf_1^1(x) \in B(f(x), \rho)$$

for $x \in \partial\Omega$. Therefore $y_0 \notin f_1^t(x)$ for $t \in [0, 1]$ and $x \in \partial\Omega$ by (1.3). Since $f_1^t \in \Lambda(C_1)$ the asserted equality (1.6) follows from assumption (d_2) of our lemma.

It is now clear that the following definition is legitimate.

1.4. DEFINITION. Let f, F, and ρ be as above. Let $F_1 \in C_1$ satisfy (1.4) and let $f_1(x) = x - F_1(x)$. Then the degree $d(f, \Omega, y_0)$ is defined by the equality

$$d(f, \Omega, y_0) = d(f_1, \Omega, y_0). \tag{1.7}$$

1.5. It remains to show that the degree thus defined has the properties asserted in the extension lemma in subsection 1.3. The verification of property (d_1) will be given in the Notes,[1] and we turn to the proof of (d_2). By the Borel covering theorem it will be sufficient to show that under the assumption made for (d_1) there corresponds to every $t_0 \in [0, 1]$ a positive $\tau_0 = \tau_0(t_0)$ such that

$$d(f_t, \Omega, y_0) = d(f_{t_0}, \Omega, y_0) \tag{1.8}$$

if

$$|t - t_0| < \tau_0, \qquad t \in [0, 1]. \tag{1.9}$$

Now y_0 has a positive distance $2\rho_0 = 2\rho_0(t_0)$ from $f_{t_0}(\partial\Omega)$. Therefore

$$y_0 \notin B(f_{t_0}(x), \rho_0) \quad \text{for all } x \in \partial\Omega. \tag{1.10}$$

But by assumption (a) we can choose a τ_0 such that

$$\|f_t(x) - f_{t_0}(x)\| < \rho_0/2 \quad \text{for all } x \in \overline{\Omega} \tag{1.11}$$

and all t satisfying (1.9). Now let t_1 be such a fixed t-value. Then (1.11) holds with $t = t_1$, and since f_{t_0} and f_{t_1} are by assumption elements of $\Lambda(\overline{C}_1)$ there exist maps f_1^0 and f_1^1 in $\Lambda(C_1)$ such that

$$\|f_1^i(x) - f_{t_i}(x)\| < \rho_0/2 \tag{1.12}$$

for $i = 0, 1$ and all $x \in \overline{\Omega}$. Then

$$f_1^0(x) \in B(f_{t_0}(x), \rho_0/2) \subset B(f_{t_0}(x), \rho_0). \tag{1.13}$$

But since $\|f_1^1(x) - f_{t_0}(x)\| \leq \|f_1^1(x) - f_{t_1}(x)\| + \|f_{t_1}(x) - f_{t_0}(x)\|$, we see from (1.12) and (1.11) (with $t = t_1$) that also

$$f_1^1(x) \in B(f_{t_0}(x), \rho_0). \tag{1.14}$$

From (1.13), (1.14), and the convexity of a ball, we conclude that $f_1^\theta(x) = (1 - \theta)f_1^0(x) + \theta f_1^1(x) \in B(f_{t_0}(x), \rho_0)$ for $\theta \in [0, 1]$ and therefore, by (1.10), that $y_0 \neq f_1^\theta(x)$ for these θ and for $x \in \partial\Omega$. Since $f_1^\theta \in \Lambda(C_1)$, it follows from our assumptions on C_1 that

$$d(f_1^1, \Omega, y_0) = d(f_1^0, \Omega, y_0). \tag{1.15}$$

Now assertion (d$_2$) will be proved if we prove (1.8) for $t = t_1$ since t_1 is an arbitrary t-value satisfying (1.9). For the proof of (1.8) (with $t = t_1$) it will, because of (1.15), be sufficient to verify that, for $i = 0, 1$,

$$d(f_1^i, \Omega, y_0) = d(f_{t_i}, \Omega, y_0).$$

But this equality follows from Definition 1.4 on account of (1.12) and the following inequalities still to be proved:

$$\text{dist}(y_0, f_{t_i}(\partial\Omega)) \geq \rho_0/2, \qquad i = 0, 1. \tag{1.16}$$

For $i = 0$ this inequality is a consequence of (1.10). For $i = 1$ the inequality (1.16) follows from the inequality

$$\|f_{t_1}(x) - y_0\| \geq \|f_{t_0}(x) - y_0\| - \|f_{t_0}(x) - f_{t_1}(x)\|$$

together with (1.10) and (1.11) (with $t = t_1$).

Proof of (d$_3$). Let $f_t(x) = f(x) - y(t)$. Then $d(f, \Omega, y(t)) = d(f_t, \Omega, \theta)$ by (d$_1$). But the right member of this equality is independent of t by (d$_2$).

§2. An application of the extension lemma

2.1. Let $C_1 = C_1(\overline{\Omega})$ be the subspace of those elements of $C = C(\overline{\Omega})$ which belong to $C''(\overline{\Omega})$ (see §1.16). Then, by the results of Chapter 2, C_1 satisfies the assumptions of the extension lemma in subsection 1.3, and consequently a degree theory is established for those L.-S. maps $f = I - F$ for which F lies in the closure $\overline{C}_1$ of C_1, i.e., those F which belong to $\Lambda(\overline{C}_1)$.

2.2. Especially if the Banach space E has the property that

$$\overline{C}_1 = C \tag{2.1}$$

where C_1 is as above, then it follows that our degree theory is established for all L.-S. maps $f:\Omega \to E$ satisfying (1.2). However (2.1) does not hold in every Banach space.

Indeed the relation (2.1) is equivalent to the following statement: to every completely continuous $F:\overline{\Omega} \to E$ and every $\varepsilon > 0$ there corresponds a completely continuous $F_2 \in C''(\overline{\Omega})$ such that

$$\|F(x) - F_2(x)\| < \varepsilon \tag{2.2}$$

for all $x \in \overline{\Omega}$. An example of a Banach space in which such approximation is not always possible will be given in the Notes.[2]

2.3. In the remainder of this chapter we discuss some important classes of Banach spaces in which the approximation (2.2) is always possible and in which therefore a degree theory is established for arbitrary L.-S. maps.

2.4. Let E be a finite-dimensional Banach space. Then every closed bounded set in E is compact. This implies that every continuous map of a subset of E into E is completely continuous and also an L.-S. map, since $f(x) = x - F(x)$ is continuous if and only if $F(x)$ is continuous. Thus $C = C(\overline{\Omega}) = \Lambda(C)$ is the set of continuous maps $\overline{\Omega} \to E$. $C_1 = C_1(\overline{\Omega})$ defined in subsection 2.1 is identical with the set $C''(\overline{\Omega})$. But it is well known that every continuous map $F: \overline{\Omega} \to E$ can be uniformly approximated by map F_2 in C'' (see, e.g., [13, p. 68]), i.e., that the approximation (2.2) is always possible. Thus (2.1) holds, and by our extension lemma in subsection 1.3, a degree theory with the properties (d_1)–(d_3) stated in that lemma is established for continuous maps. We finally note that conditions (a) and (b) of property (d_2) (see subsection 1.3) may be replaced by the condition that the map $f: \overline{\Omega} \times [0,1] \to E$ is continuous since a continuous map of a closed bounded set in a finite-dimensional space is uniformly continuous.

2.5. In order to obtain a class of infinite-dimensional Banach spaces in which the approximation (2.2) is always possible, we start with the following remark. Let E be a Banach space of the following property: to every open bounded set Ω in E, to every continuous map G of $\overline{\Omega}$ into a Banach space E_1, and to every $\eta > 0$, there exists a C'' map $G_2: \overline{\Omega} \to E_1$ such that

$$\|G(x) - G_2(x)\| < \eta \quad \text{for all } x \in \overline{\Omega}. \tag{2.3}$$

We show that in such space E the approximation (2.2) is always possible. (Note that in (2.2), F and F_2 are both completely continuous.) The proof is based on the "second Leray-Schauder lemma" stated and proved in subsection 4.3. By this lemma there exists to every completely continuous $F: \overline{\Omega} \to E$ and $\varepsilon > 0$ a continuous map G of $\overline{\Omega}$ into a finite-dimensional subspace E_1 of E such that

$$\|F(x) - G(x)\| < \varepsilon/2 \quad \text{for all } x \in \overline{\Omega}. \tag{2.4}$$

But by our assumption on E there exists a C''-map $G_2: \overline{\Omega} \to E$ such that (2.3) is satisfied with $\eta = \varepsilon/2$. Then by (2.4)

$$\|F(x) - G_2(x)\| < \varepsilon \quad \text{for all } x \in \overline{\Omega}. \tag{2.5}$$

Now G_2 is completely continuous. Indeed F is completely continuous and therefore $F(\overline{\Omega})$ is bounded. It follows from (2.4) and (2.5) that the subset $G_2(\overline{\Omega})$ of the finite-dimensional space E_1 is bounded, and this implies the complete continuity of G_2 since G_2 is continuous. Thus (2.5) shows that (2.2) is satisfied with $F_2 = G_2$.

2.6. DEFINITION. A Banach space E is called 2-smooth if there exists a bounded set $U \in E$ and a C'' map of E into the reals which is zero on $E - U$ but not identically zero.

2.7. LEMMA. *Let E be a separable 2-smooth Banach space. Then the approximation (2.3) is always possible.*

This is a special case of a more general theorem by Bonic and Frampton (see [5, Theorem 2]). We refer the reader to their proof.

2.8. It follows from subsections 2.7, 2.5 and 2.2 that a degree theory with the properties (d_1)–(d_3) has been established for all 2-smooth separable Banach spaces. If μ is the Lebesgue measure in a finite-dimensional Euclidean space R^k, if S is a Lebesgue measurable set in R^k, if for any real-valued measurable g defined on S, $[g]$ is the family of those $\tilde{g}$ for which $g(x) = \tilde{g}(x)$ except for a set of measure zero, then it is well known that with proper definitions those $[g]$ for $\int_S |g|^p \, d\mu < \infty$ form a Banach space L_p for $p \geq 1$. It is also well known that such L_p is separable (see, e.g., [18, p. 125]). But for $p \geq 2$ these spaces are also 2-smooth (see [5, pp. 881, 882]).

§3. The degree theory for finite layer maps

3.1. As pointed out in subsection 2.2 the application of the extension lemma of subsection 1.3 to L.-S. maps which are in C'' does not yield a degree theory for all L.-S. maps in all Banach spaces. But in subsection 3.4 of Chapter 2 a degree theory was established for arbitrary continuous maps in finite-dimensional spaces. Based on this fact and on Leray-Schauder's "first lemma" in subsection 3.2, we establish in the present section degree theory for L.-S. maps which are finite layer maps (see Definition 2.5 in Chapter 2). Using Leray-Schauder's "second lemma" we will show in the next section that application of the extension lemma to finite layer maps yields a degree theory for arbitrary L.-S. maps in all Banach spaces (cf. §7 of the Introduction).

3.2. FIRST LERAY-SCHAUDER LEMMA. *Let E_2 be a Banach space of finite dimension n_2, and let E_1 be a subspace of E_2 of dimension $n_1 < n_2$. Let Ω be a bounded open set in E_2 which intersects E_1. Let $f(x) = x - F(x)$ be a continuous map $\overline{\Omega} \to E$ with the property that $F(\overline{\Omega}) \in E_1$. Finally let y_0 be a point in E_1 for which*

$$y_0 \notin f(\partial\Omega). \tag{3.1}$$

Then

$$d(f, \Omega, y_0) = d(f, \Omega \cap E_1, y_0). \tag{3.2}$$

PROOF. The assumption (3.1) implies the existence of ρ such that

$$0 < \rho < \|y_0 - f(x)\| \quad \text{for } x \in \partial\Omega. \tag{3.3}$$

Since F maps the bounded subset $\overline{\Omega}$ of the finite-dimensional space E_2 into the subspace E_1 of E_2, there exists a C'' map $F_1 : \overline{\Omega} \to E_1$ such that

$$\|F_1(x) - F(x)\| < \rho \tag{3.4}$$

for all $x \in \overline{\Omega}$. Now let $f_1(x) = x - F_1(x)$ and

$$f_t(x) = (1-t)f(x) + tf_1(x) = x - [(1-t)F(x) + tF_1(x)] \tag{3.5}$$

for $0 \leq t \leq 1$. Now it follows from (3.4) that $\|f(x) - f_1(x)\| = \|F(x) - F_1(x)\| < \rho$, i.e., $f_1(x) \in B(f(x), \rho)$. Since obviously $f(x) \in B(f(x), \rho)$ we see that $f_t(x) \in B(f(x), \rho)$ for $0 \leq t \leq 1$ and all $x \in \overline{\Omega}$. But for $x \in \partial\Omega$ we see from (3.3) that $y_0 \notin B(f(x), \rho)$. Thus $f_t(x) \neq y_0$ for these x and all $t \in [0, 1]$, and it follows from subsection 2.4 that $d(f, \Omega, y_0) = d(f_1, \Omega, y_0)$, $d(f, \Omega \cap E_1, y_0) = d(f_1, \Omega \cap E_1, y_0)$. But by Lemma 2.6 and (11), both in Chapter 2, the right members of these two equalities are equal. This proves the assertion (3.2).

3.3. LEMMA. *Let E be a Banach space, Ω a bounded open set in E, and let $f:\overline{\Omega} \to E$ be a finite L.-S. layer map. Let E_0 and E_1 be finite-dimensional subspaces of E with respect to which f is a layer map. Let y_0 be a point of E which satisfies (3.1) and which lies in $E_0 \cap E_1$. Then*

$$d(f, \Omega \cap E_1, y_0) = d(f, \Omega \cap E_0, y_0). \tag{3.6}$$

PROOF. Let E_2 the subspace of E spanned by E_0 and E_1. Then by Lemma 3.2 both members of (3.6) equal $d(f, \Omega \cap E_2, y_0)$.

Lemma 3.3 allows us to make the following definition.

3.4. DEFINITION. Let $f:\overline{\Omega} \to E$ be a finite L.-S. layer map. Let y_0 satisfy (3.1) and let E_1 be a finite-dimensional subspace of E with respect to which f is a layer map and which contains y_0. We then define $d(f, \Omega, y_0)$ by the equality $d(f, \Omega, y_0) = d(f, \Omega \cap E_1, y_0)$.

3.5. LEMMA. *The assertions of Lemma 4.13 in Chapter 2 remain true if the assumption that $f \in C''$ is replaced by the assumption that f is a finite L.-S. layer map.*

The proof will be given in the Notes.[3]

3.6. LEMMA. *Let $f_t(x) = f(x,t)$ be a continuous map $\overline{\Omega} \times [0,1]$ into E which is an L.-S. map for each $t \in [0,1]$. Moreover, we assume that f_t is a "uniform finite layer map," i.e., that there exists a finite-dimensional subspace E_1 of E (independent of t) such that*

$$F(x,t) = x - f(x,t) \in E_1 \quad \text{for all } (x,t) \in \overline{\Omega} \times [0,1]. \tag{3.7}$$

Finally it is assumed that

$$y_0 \notin f_t(\partial\Omega) \quad \text{for } t \in [0,1]. \tag{3.8}$$

Then $d(f_t, \Omega, y_0)$ is independent of t.

PROOF. By our assumptions f_t is a finite L.-S. layer map with respect to E_1. Without restriction of generality we may assume that $\Omega \cap E_1$ is not empty. We also may assume that $y_0 \in E_1$. Then by Definition 3.4

$$d(f_t, \Omega, y_0) = d(f_t, \Omega \cap E_1, y_0) \tag{3.9}$$

if $y_0 \in E_1$, and here the right member is independent of t by subsection 2.4.

§4. Another application of the extension lemma

4.1. As in subsection 1.1 let $C = C(\overline{\Omega})$ denote the space of completely continuous maps $F:\overline{\Omega} \to E$. In the present chapter C_1 will denote the subspace of those $F \in C$ which map $\overline{\Omega}$ into a finite-dimensional subspace $E_1 = E_1(F)$ of E.

4.2. LEMMA. *C_1 satisfies the assumptions of the extension lemma in subsection 1.3 and condition (d_3).*

PROOF. That assumptions (i)–(iii) of that lemma are satisfied is obvious. As to conditions (d_1)–(d_3) we note that (d_1) follows from Lemma 3.5 and that, as shown at the end of the proof of the extension lemma, (d_3) is a consequence of (d_2). It thus remains to prove (d_2). Then let $f_t(x) = f(x, t) = x - F(x, t)$, where $F_t(x) = F(x, t) \in C_1$ for every $t \in [0, 1]$. Assume that

$$y_0 \notin f_t(\partial \Omega) \quad \text{for } t \in [0, 1] \tag{4.1}$$

and that $F(x, t)$ is continuous in t uniformly for $x \in \overline{\Omega}$. We have to show that $d(f_t, \Omega, y_0)$ is independent of t. The proof is quite similar to the argument given in subsection 1.5: it will be sufficient to show that to each $t_0 \in [0, 1]$ there corresponds a $\tau_0 = \tau_0(t_0)$ such that (1.9) implies (1.8). Now by assumption (4.1) there exists a $\rho_0 > 0$ such that (1.10) holds, and by our uniformity assumption we can choose a τ_0 such that (1.9) implies (1.11). Again let t_1 be a fixed t-value satisfying (1.11). Then for $x \in \partial \Omega$ the point $f_{t_1}(x)$ lies in the ball $B(f_{t_0}(x), \rho_0/2)$, and since $f_{t_0}(x)$ lies in that ball,

$$f^\theta(x) = (1 - \theta)f_{t_0}(x) + \theta f_{t_1}(x) \in B(f_{t_0}(x), \rho_0/2) \tag{4.2}$$

for $x \in \partial \Omega$ and $0 \leq \theta \leq 1$. Now, by assumption, $F_{t_0}(x)$ and $F_{t_1}(x)$ lie in respective finite-dimensional subspaces E_0 and E_1 of E, and therefore both lie in the subspace E_2 spanned by E_0 and E_1. Thus f^θ is for each $\theta \in [0, 1]$ an L.-S. map which is a layer map with respect to the finite-dimensional space E_2, and the asserted equality $d(f_{t_0}, \Omega, y_0) = d(f_{t_1}, \Omega, y_0)$ will follow from Lemma 3.6 once it is shown that $y_0 \notin f^\theta(\partial \Omega)$ for all $\theta \in [0, 1)$. To prove this we have only to note that by (4.2), (1.10), and (1.11) with $t = t_1$ for $x \in \partial \Omega$ and $t \in [0, 1]$,

$$\|f^\theta(x) - y_0\| = \|f_{t_0}(x) - y_0 + \theta(f_{t_1}(x) - f_{t_0}(x))\|$$
$$\geq \|f_{t_0}(x) - y_0\| - \|f_{t_1}(x) - f_{t_0}(x)\| \geq \rho_0/2 > 0.$$

This finishes the proof of Lemma 4.2, and by the extension lemma a degree theory having properties (d_1)–(d_3) is established for all L.-S. maps $f(x) = x - F(x)$ where $F \in \overline{C}_1$. We will now prove that (2.1) holds, i.e., establish the theory for all L.-S. maps. Now by our definition the equality (2.1) is equivalent to the following lemma.

4.3 SECOND LERAY-SCHAUDER LEMMA. *To every completely continuous map $F: \overline{\Omega} \to E$ and every ε there corresponds a finite-dimensional subspace E_1 of E and a continuous map $F_1: \overline{\Omega} \to E_1$ such that*

$$\|F(x) - F_1(x)\| < \varepsilon \quad \text{for all } x \in \overline{\Omega}. \tag{4.3}$$

PROOF. Let $Y = F(\overline{\Omega})$. Then the set $\overline{Y}$ is compact. Consequently by §1.27 there correspond to given $\varepsilon > 0$ points $y_1, y_2, \ldots, y_s$ such that

$$y_\sigma \in F(\overline{\Omega}), \qquad \sigma = 1, 2, \ldots, s, \tag{4.4}$$

and such that for every $y \in \overline{Y}$

$$\|y - y_\sigma\| < \varepsilon \tag{4.5}$$

for at least one positive integer $\sigma \leq s$. We now define s real-valued continuous functions μ_σ on $\overline{Y}$ by setting, for $\sigma = 1, 2, \ldots, s$,

$$\mu_\sigma(y) = \begin{cases} \varepsilon - \|y - y_\sigma\| & \text{if } \|y - y_\sigma\| < \varepsilon, \\ 0 & \text{if } \|y - y_\sigma\| \geq \varepsilon. \end{cases} \tag{4.6}$$

All μ_σ are nonnegative and for fixed $y \in \overline{Y}$ at least one of the $\mu_\sigma(y)$ is positive by (4.5). Therefore

$$a(y) = \sum_{\sigma=1}^{s} y_\sigma \mu_\sigma(y) \Big/ \sum_{\sigma=1}^{s} \mu_\sigma(y) \tag{4.7}$$

is a continuous map of $\overline{Y}$ into the finite-dimensional space E_1 spanned by the points $y_1, y_2, \ldots, y_s$. Then

$$a(y) - y = \sum_{\sigma=1}^{\sigma} (y_0 - y) \mu_\sigma(y) \Big/ \sum_{\sigma=1}^{\sigma} \mu_\sigma(y). \tag{4.8}$$

Now it follows from (4.6) that the norms of those terms in the numerator of the right member of (4.8) for which $\|y - y_\sigma\| \geq \varepsilon$ are zero while the norms of the remaining ones are not greater than $\varepsilon \mu_\sigma(y)$. Thus we see from (4.8) that

$$\|a(y) - y\| < \varepsilon \tag{4.9}$$

for all $y \in \overline{Y}$. Thus

$$\|a(F(x)) - F(x)\| < \varepsilon \tag{4.10}$$

for all $x \in \overline{\Omega}$. Since a is a continuous map $\overline{Y} \to E_1$ and F is completely continuous, it follows that

$$F_1(x) = a(F(x)) \tag{4.11}$$

is a completely continuous map $\overline{\Omega} \to E_1$. Since F_1 satisfies (4.3) by (4.10) and (4.11) our lemma is proved.

4.4. The degree thus defined for arbitrary L.-S. maps $\overline{\Omega} \to E$ is known as the Leray-Schauder degree. It has properties (d_1)–(d_3) asserted in subsection 1.3.

§5. Two additional properties of the Leray-Schauder degree

5.1. Dealing with twice-differentiable L.-S. maps we proved the continuity theorems in subsections 4.15 and 4.16 of Chapter 2 and showed that the former implies the latter but that the converse implication is not true. Now property (d_2) proved for the Leray-Schauder degree corresponds to Theorem 4.16 in Chapter 2. However, the property corresponding to Theorem 4.15 in Chapter 2 holds also for the Leray-Schauder degree as will be shown in Theorem 5.3 below. We first state a lemma needed for the proof of that theorem.

5.2. LEMMA. *Let* $F(x,t): \overline{\Omega} \times [0,1] \to E$ *be completely continuous. Then to every* $\varepsilon > 0$ *there corresponds a finite-dimensional subspace* E_1 *of* E *and a completely continuous map* $\overline{\Omega} \times [0,1] \to E_1$ *such that for all* (x,t) *in* $\overline{\Omega} \times [0,1]$

$$\|F(x,t) - F_1(x,t)\| < \varepsilon \tag{5.1}$$

PROOF. Let $Y = F(\overline{\Omega} \times [0,1])$. Then $\overline{Y}$ is compact on account of the complete continuity of F. Therefore by the argument used in the proof of Lemma 4.3, the space E_1 and the map $a(y): \overline{Y} \to E_1$ can be defined in such a way that (4.9) holds for all $y \in \overline{Y}$. Then $F_1(x, t) = a(F(x, t))$ satisfies (5.1).

5.3. THEOREM. *Let $f_t(x) = x - F(x, t)$ where $F: \overline{\Omega} \times [0, 1] \to E$ is completely continuous. Suppose that*

$$y_0 \notin f_t(\partial\Omega) \quad \text{for } t \in [0, 1]. \tag{5.2}$$

Then $d(f_t, \Omega, y_0)$ is independent of t.

PROOF. It follows from (5.2) that there exists an $\varepsilon > 0$ which is less than the distance of y_0 from $f_t(\partial\Omega)$. Then let $F_1(x, t)$ satisfy the assertions in Lemma 5.2 with that ε, and let $f_t^1(x) = x - F_1(x, t)$. Then by Definition 1.4

$$d(f_t, \Omega, y_0) = d(f_t^1, \Omega, y_0). \tag{5.3}$$

But f_t^1 is a layer map with respect to the finite-dimensional space E_1 (which is independent of t). Therefore by Lemma 3.6 the right member of (5.3) is independent of t.

5.4. THEOREM. *Let f_t be as in Theorem 5.3. Let $y_0 = y_0(t)$ be a continuous curve in E satisfying*

$$y_0(t) \notin f_t(\partial\Omega) \quad \text{for } t \in [0, 1]. \tag{5.4}$$

Then $d(f_t, \Omega, y_0(t))$ is independent of t.

PROOF. We note first that there exists a positive ε such that

$$\text{dist}(y_0(t), f_t(\partial\Omega)) > \varepsilon \quad \text{for all } t \in [0, 1]. \tag{5.5}$$

Indeed otherwise there would exist a convergent sequence $t_1, t_2, \ldots$ of numbers in $[0, 1]$ and a sequence of points $\dot{x}_1, \dot{x}_2, \ldots$ in $\partial\Omega$ such that

$$\lim_{i \to \infty} \|y_0(t_i) - f(\dot{x}_i)\| = 0. \tag{5.6}$$

Then with $t_0 = \lim t_i$, the $y(t_i)$ converge to the point $y_0(t_0)$ of our continuous curve $y_0 = y_0(t)$, and by (5.6)

$$\lim f_{t_i}(\dot{x}_i) = \lim(\dot{x}_i - F(\dot{x}_i, t_i)) = y_0(t_0).$$

By the complete continuity of $F(x, t)$ we now see that some subsequence of the x_i converges to a point x_0 and that

$$f_{t_0}(x_0) = y_0(t_0). \tag{5.7}$$

Now $\partial\Omega \times [0, 1]$ is a closed set, and since $f(x, t)$ is an L.-S. map $f(\partial\Omega \times [0, 1])$ is also a closed set. Since all $x_i \in \partial\Omega$ it follows that $x_0 \in \partial\Omega$. Thus (5.7) contradicts our assumption (5.4). This contradiction proves the existence of an $\varepsilon > 0$ satisfying (5.5).

To prove our theorem it is sufficient to show that $d(f_t, \Omega, y_0(t))$ is constant in a small enough neighborhood of every $t_0 \in [0, 1]$. With ε as above let $\delta > 0$ be such that

$$\|y_0(t') - y_0(t'')\| < \varepsilon/2 \tag{5.8}$$

for any couple t', t'' of points of $[0, 1]$ satisfying

$$|t' - t''| < \delta. \tag{5.9}$$

Now let $\dot{x} \in \partial\Omega$, and let t_1 be a fixed t-value satisfying $|t_1 - t_0| < \delta$. Then we see from (5.5) and (5.8) that, for any t in the interval bounded by t_0 and t_1,

$$\|f_t(\dot{x}) - y_0(t_1)\| \geq \|f_t(\dot{x}) - y_0(t)\| - \|y_0(t) - y_0(t_1)\| \geq \varepsilon - \varepsilon/2 > 0. \tag{5.10}$$

Therefore, by Theorem 5.3, $d(f_{t_1}, \Omega, y(t_1)) = d(f_{t_0}, \Omega, y(t_1))$, which implies that

$$d(f_{t_1}, \Omega, y(t_1)) - d(f_{t_0}, \Omega, y(t_0)) = d(f_{t_0}, \Omega, y(t_1)) - d(f_{t_0}, \Omega, y(t_0)).$$

Here the right member equals zero by assertion (d$_3$) in subsection 1.3, since for $\dot{x} \in \partial\Omega$ and t in the interval with end points t_0 and t_1, the inequality $\|f_{t_0}(\dot{x}) - y_0(t)\| > 0$ holds, as is seen by a proof quite similar to that of (5.10).

5.5. The computation of the Leray-Schauder degree for an arbitrary L.-S. map can be reduced to that of a map taking place in a finite-dimensional space. More precisely we assert: *let f be an L.-S. map $\overline{\Omega} \to E$ and let y_0 be a point of E satisfying $y_0 \notin f(\partial\Omega)$. Then there exists a subspace E^n of E of finite dimension n containing y_0 and having a nonempty intersection Ω^n with Ω, and a continuous map $f_1^n: \overline{\Omega^n} \to E^n$ such that*

$$d(f, \Omega, y_0) = d(f_1^n, \Omega^n, y_0). \tag{5.10}$$

PROOF. With C_1 as in subsection 4.1 the left member of (5.4) is given by Definition 1.4 where $F_1(\overline{\Omega}) = (I - f_1)(\overline{\Omega})$ satisfies (1.4) and lies in some finite-dimensional space $E_1 \subset E$. But then there exists also a finite-dimensional E^n such that $F_1(\overline{\Omega}) \subset E_1 \subset E^n \subset E$ and such that E^n contains y_0 and points of Ω. Thus, f_1 is a finite layer map with respect to E^n. Therefore by Definition 3.4 of the degree for finite layer maps the right member of (1.7) equals $d(f_1^n, \Omega^n, y_0)$ if f_1^n denotes the restriction of f_1 to $\overline{\Omega^n}$. This proves (5.10).

5.6. REMARK. The proof of (5.10) shows that the following conditions on E^n are sufficient for (5.10) to hold: (i) if $\rho = \text{dist}(y_0, \partial\Omega)$, then there exists a continuous map $F_1: \overline{\Omega} \to E^n$ such that $\|F_1(x) - F(x)\| < \rho$; (ii) $y_0 \in E^n$.

§6. Generalized L.-S. maps

6.1. In this section a slight generalization of L.-S. maps will be introduced. The motivation for this was given in §10 of the introduction.

6.2. DEFINITION. Let S be a closed subset of the Banach space E. A map $f: S \to E$ is called a generalized L.-S. (g.L.-S.) map if it is of the form

$$f(x) = \lambda(x)x - F(x), \qquad x \in S, \tag{6.1}$$

where F is completely continuous and where λ is a real-valued continuous function satisfying

$$0 < m \leq \lambda(x) \leq M, \qquad x \in S, \tag{6.1a}$$

for some constants m and M.

6.3. Before defining a degree for g.L.-S. mappings we note that if f, F, and λ are as in (6.1) and (6.1a), if

$$F_0(x) = y_0(1/\lambda(x) - 1) + F(x)/\lambda(x) \tag{6.2}$$

and

$$f_0(x) = x - F_0(x), \tag{6.3}$$

then direct calculation shows that x is a root of

$$f_0(x) = y_0 \tag{6.4}$$

if and only if x is a root of

$$f(x) = y_0. \tag{6.5}$$

6.4. DEFINITION. Let f be a g.L.-S. map $\overline{\Omega} \to E$, and let the point y_0 satisfy

$$y_0 \notin f(\partial\Omega). \tag{6.6}$$

We then set

$$d(f, \Omega, y_0) = d(f_0, \Omega, y_0). \tag{6.7}$$

Note that by subsection 6.3 the assumption (6.6) implies that $y_0 \notin f_0(\partial\Omega)$ such that the right member of (6.7) is defined, since f_0 is obviously an L.-S. map.

6.5. LEMMA. *Let f and y_0 be as in subsection 6.4. Then f has the elementary properties* (i)–(vii) *stated in Lemma 2.4 of Chapter 2.*

Since $d(f_0, \Omega, y_0)$ has these properties, the lemma follows easily from (6.7) and (6.3).

6.6. THEOREM. *Let the map $F(x, t): \overline{\Omega} \times [0, 1] \to E$ be completely continuous. Let $\lambda(x, t)$ be a continuous map of $\overline{\Omega} \times [0, 1]$ into the reals. Suppose, moreover, the existence of two constants m, M such that for $(x, t) \in \overline{\Omega} \times [0, 1]$*

$$0 < m \leq \lambda(x, t) \leq M. \tag{6.8}$$

Let

$$f_t(x) = f(x, t) = \lambda(x, t)x - F(x, t), \tag{6.9}$$

and let y_0 be a point of E satisfying, for all $t \in [0, 1]$,

$$y_0 \notin f_t(\partial\Omega). \tag{6.10}$$

Then $d(f_t, \Omega, y_0)$ is independent of t.

PROOF. Let

$$F_0(x, t) = y_0(1/\lambda(x, t) - 1) + F(x, t)/\lambda(x, t) \tag{6.11}$$

and
$$f_0^t(x) = f_0(x, t) = x - F_0(x, t).$$
Then by (6.10) and subsection 6.3 for $0 \leq t \leq 1$
$$y_0 \notin f_0^t(\partial\Omega) \tag{6.12}$$
and by Definition 6.4
$$d(f_t, \Omega, y_0) = d(f_0^t, \Omega, y_0). \tag{6.13}$$

Now from the assumed complete continuity of $F(x, t)$ together with (6.8) and (6.11), we see that the map $F_0(x, t): \overline{\Omega} \times [0, 1] \to E$ is completely continuous. Taking into account also (6.12) we see from Theorem 5.3 that the right member of (6.13) is independent of t.

6.7. COROLLARY 1 TO THEOREM 6.6. *The conclusion of Theorem 6.6 remains true if the assumption that $F(x, t)$ is completely continuous is replaced by the assumptions* (a) *and* (b) *of Theorem 4.16 in Chapter 2.*

This is obvious since the latter assumptions imply the complete continuity of $F(x, t)$ as proved in Note 10 to Chapter 2.

6.8. COROLLARY 2 TO THEOREM 6.6. *Let f be a g.L.-S. map $\overline{\Omega} \to E$. Let $y = y(t)$, $0 \leq t \leq 1$, be a continuous curve in E satisfying $y(t) \notin f(\partial\Omega)$. Then $d(f, \Omega, y(t))$ is independent of t.*

This follows by the argument used for the proof of (d₃) in subsection 1.5.

§7. Notes to Chapter 3

Notes to §1.

1. *Proof of assertion* (d₁) *in subsection 1.5, i.e., of the assertions* (i)–(vii) *of Lemma 2.4 in Chapter 2.* Let Λ_1 be the family of L.-S. maps $f(x) = x - F(x)$: $\overline{\Omega} \to E$ for which $F \in C_1$, and $\overline{\Lambda}_1$ the family of those L.-S. maps for which $F \in \overline{C}_1$.

Assertion (i) is trivial since the identity map I belongs to Λ_1.

Proof of assertion (ii). Let $f \in \overline{\Lambda}_1$, and let y_0 and ρ be as in Definition 1.4. Then there exists a sequence $f_n(x) = x - F_n(x)$ of maps in Λ_1 such that $\|f_n - f\| = \|F_n - F\| \to 0$ as $n \to \infty$, i.e., that
$$\lim_{n \to \infty} f_n(x) = f(x) \tag{1}$$
uniformly for $x \in \overline{\Omega}$, and we may suppose that
$$\|f_n(x) - f(x)\| = \|F_n(x) - F(x)\| < \rho \tag{2}$$
for $x \in \overline{\Omega}$ and $n = 1, 2, \ldots$. Then by Definition 1.4
$$d(f_n, \Omega, y_0) = d(f, \Omega, y_0), \qquad n = 1, 2, \ldots. \tag{3}$$
But here, by assumption, the right member is different from zero. Since $f_n \in \Lambda_1$ it follows that there exists an $x_n \in \Omega$ such that
$$f_n(x_n) = y_0. \tag{4}$$

But

$$y_0 = f_n(x_n) = x_n - F_n(x_n) = x_n - F(x_n) + [F(x_n) - F_n(x_n)], \qquad (5)$$

and

$$\|F(x_n) - F_n(x_n)\| \le \|F - F_n\| \to 0. \qquad (6)$$

Since F is completely continuous, there exists a subsequence $\{x_{n_i}\}$ of the sequence $\{x_n\}$ for which the $F(x_{n_i})$ converge. It then follows from (5) and (6) that the x_{n_i} converge to some point $\bar{x} \in \bar{\Omega}$ and that $\bar{x}$ satisfies the equation $y_0 = \bar{x} - F(\bar{x}) = f(\bar{x})$. But $\bar{x} \notin \partial\Omega$ by (1.2). Thus the solution $\bar{x}$ lies in Ω.

Proof of assertion (iii). Let $\Omega_0 = \Omega$ and for $\sigma = 1, 2, \ldots, s$ let Ω_σ be as in part (iii) of Lemma 2.4 in Chapter 2. Let $f \in \bar{\Lambda}_1$. We assume

$$y_0 \notin f(\partial\Omega_\sigma), \qquad \sigma = 0, 1, \ldots, s. \qquad (7)$$

We want to prove

$$d(f, \Omega, y_0) = \sum_{\sigma=1}^{s} d(f, \Omega_\sigma, y_0). \qquad (8)$$

Now from (7) we conclude the existence of a $\rho > 0$ such that

$$\|y_0 - f(x)\| > \rho \quad \text{for } x \in \partial\Omega_\sigma, \ \sigma = 0, 1, \ldots, s. \qquad (9)$$

But there exists an $f_1(x) = x - F_1(x)$ in Λ_1 such that

$$\|f(x) - f_1(x)\| = \|F(x) - F_1(x)\| < \rho \quad \text{for all } x \in \bar{\Omega}. \qquad (10)$$

$$d(f, \Omega_\sigma, y_0) = d(f_1, \Omega_\sigma, y_0), \qquad \sigma = 0, 1, \ldots, s, \qquad (11)$$

for such f_1, by definition. Now by assumption the equality obtained from (8) by replacing f by f_1 holds. Therefore, by (11), the assertion (8) holds.

Assertion (iv) is a consequence of assertions (ii) and (iii).

Proof of assertion (v). Let $f \in \bar{\Lambda}_1$, let ρ be as in (1.3), and let $f_1(x) = x - F_1(x)$ be an element of Λ_1 for which (1.4) holds. Then by Definition 1.4

$$d(f, \Omega, y_0) = d(f_1, \Omega, y_0). \qquad (12)$$

On the other hand, if $g(x) = f(x) - b$ and $g_1(x) = f_1(x) - b$, then $\|g(x) - g_1(x)\| = \|f(x) - f_1(x)\| < \rho$, and therefore

$$d(g, \Omega, y_0) = d(g_1, \Omega, y_0). \qquad (13)$$

Now our assertion is

$$d(f, \Omega, y_0) = d(g, \Omega, y_0 - b). \qquad (14)$$

But by assumption the equality obtained from (14) by replacing f and g by f_1 and g_1 resp. holds. Therefore by (12) and (13) the assertion (14) is true.

Proof of assertion (vi). Let $f(x) = x - F(x)$, where

$$F \in \overline{C_1(\bar{\Omega})}. \qquad (15)$$

Suppose

$$y_0 \notin f(\partial\Omega), \qquad (16)$$

and let

$$g(\xi) = f(x) \tag{17}$$

where

$$\xi = x - x_0. \tag{18}$$

We have to prove that

$$d(g, \Omega_{x_0}, y_0) = d(f, \Omega, y_0). \tag{19}$$

Now by (16) there exists a $\rho > 0$ such that

$$\|y_0 - f(x)\| > \rho \quad \text{for } x \in \partial\Omega. \tag{20}$$

But by (15) there exists an $F_1 \in C_1(\overline{\Omega})$ such that $\|F_1(x) - F(x)\| < \rho$, and therefore

$$\|f_1(x) - f(x)\| < \rho \tag{21}$$

where $f_1(x) = x - F_1(x)$. Therefore by Definition 1.4

$$d(f, \Omega, y_0) = d(f_1, \Omega, y_0). \tag{22}$$

Now if $g_1(\xi) = f_1(x)$ with ξ given by (18), then $g_1 \in C_1(\overline{\Omega}_{x_0})$. Moreover $\|g(\xi) - g_1(\xi)\| = \|f(x) - f_1(x)\| < \rho$ by (21). But since $x \in \partial\Omega$ is equivalent to $\xi \in \partial\Omega_{x_0}$, we see from (20) and (17) that $\|g(\xi) - y_0\| > \rho$ for $\xi \in \partial\Omega_{x_0}$, and it follows from our definition that

$$d(g, \Omega_{x_0}, y_0) = d(g_1, \Omega_{x_0}, y_0). \tag{23}$$

But since (d_1) holds for $F_1 \in C_1(\overline{\Omega})$, the right members of (23) and (22) are equal. This proves the assertion (19).

Assertion (vii) is a consequence of assertions (v) and (vi).

Notes to §2.

2. *An example for the assertion at the end of subsection 2.2.* Let l_1 be the space whose elements are the points $x = (x_1, x_2, \ldots)$, where the x_i are real numbers subject to the condition that

$$\sum_{i=1}^{\infty} |x_i| < \infty. \tag{24}$$

It is well known that with the natural definition of addition and multiplication by real numbers and with the sum (24) taken as norm $\|x\|$ this space is a Banach space (see, e.g., [**18**, Chapter IV.8]). Now let Ω be the unit ball $B(\theta, 1)$ in l_1, let e_1 be a unit element in l_1 (i.e., an element of norm 1), and let E_1 be the one-dimensional subspace of l_1 generated by e_1. Let $F:\overline{\Omega} \to E_1$ be defined by

$$F(x) = e_1 \|x\|. \tag{25}$$

F is obviously completely continuous.

Assertion A. The map F furnishes the desired example, i.e., there is an $\varepsilon > 0$ such that no map $G: \overline{\Omega} \to E$ has the following three properties: G is completely continuous, $G \in C'(\overline{\Omega})$, and

$$\|F(x) - G(x)\| < \varepsilon \tag{26}$$

for all $x \in \overline{\Omega}$. The proof is based on assertions B and C below.

Assertion B. Let $\phi \in C'$ be a completely continuous map of $\overline{\Omega}$ into the reals. Then

$$\phi(\overline{\Omega}) \subset \overline{\phi(\partial \Omega)}. \tag{27}$$

This assertion is a special case of a more general theorem by Bonic and Frampton for which we refer the reader to [**5**, p. 896, Corollary 2].

Assertion C. There exists an $\varepsilon > 0$ such that for no real-valued map $\phi \in C'(\overline{\Omega})$ do we have

$$| \|x\| - \phi(x)| < \varepsilon \quad \text{for all } x \in \overline{\Omega}. \tag{28}$$

Proof of assertion C. Suppose C not to be true. Then there exists a C' map $\phi: \overline{\Omega}$ into the reals such that (28) is true with $\varepsilon = 1/3$. Now

$$|\phi(x) - \phi(\theta)| \geq \|x\| - |\phi(x) - \|x\| | - |\phi(\theta)|.$$

Since $\|x\| = 1$ for $x \in \partial \Omega$ and since $|\phi(\theta)| = |\phi(\theta) - \theta| < 1/3$ by (28) (with $\varepsilon = 1/3$), we see from the above inequality that

$$|\phi(x) - \phi(\theta)| \geq 1/3 \quad \text{for all } x \in \partial \Omega. \tag{29}$$

But $\phi(\theta) \in \overline{\phi(\partial \Omega)}$ by (27). Therefore there exists an $x \in \partial \Omega$ such that $|\phi(x) - \phi(\theta)| < 1/3$ in contradiction to (29).

Proof of assertion A. Suppose A not to be true. Then there exists a sequence $\varepsilon_1, \varepsilon_2, \ldots$ of positive numbers converging to 0 and a sequence $G_1, G_2, \ldots$ of completely continuous C' maps $\overline{\Omega} \to E$ satisfying

$$\|F(x) - G_n(x)\| < \varepsilon_n \tag{30}$$

for all $x \in \overline{\Omega}$. Now let E_2 be a subspace of l_1 complementary to E_1, and let P_1 and P_2 be the corresponding projections (see §1.6). Since P_1 is bounded, we see from (30) that $\|P_1(F(x) - G_n(x))\| < M\varepsilon_n$ for some positive constant M. Now, by (25), $P_1 F(x) = F(x) = e_1 \|x\|$ and $P_1 G_n(x) = e_1 \phi_n(x)$, where ϕ_n is a real-valued C' map. We thus see that $| \|x\| - \phi_n(x)| \leq M\varepsilon_n$. Since $\varepsilon_n \to 0$ this inequality contradicts assertion C.

Notes to §3.

3. *Proof of Lemma* 3.5. First we show that assertions (i)–(vii) of Lemma 2.4 in Chapter 2 hold. Assertion (i) is again obvious. For the proof of assertion (ii)–(vii) we suppose without loss of generality that f is a layer map with respect to a finite-dimensional subspace E_1 of E which contains y_0. We note moreover that by subsections 2.1–2.4 and by subsection 1.5 the assertions (ii)–(vii) are true for continuous maps $\overline{\Omega}_1 \to E_1$ where Ω_1 is a bounded open set in E_1.

Proof of (ii). If $d(f, \Omega, y_0) \neq 0$, then by definition $d(f, \Omega \cap E_1, y_0) \neq 0$, and the equation $f(x) = y_0$ has a solution $x \in \Omega_1 = \Omega \cap E_1 \subset \Omega$.

Proof of (iii). By definition

$$d(f, \Omega_\sigma, y_0) = d(f, \Omega_\sigma \cap E_1, y_0), \qquad \sigma = 0, 1, \ldots, s, \qquad (31)$$

where $\Omega_0 = \Omega$. By assumption $\Omega_\sigma \cap \Omega_\tau = \emptyset$ for $\sigma, \tau = 1, 2, \ldots, s$ and $\sigma \neq \tau$, and $\bigcup_1^s \bar{\Omega}_\sigma = \bar{\Omega}_0 = \bar{\Omega}$. Therefore, for these σ, τ, $(\Omega_\sigma \cap E_1) \cap (\Omega_\tau \cap E_1) = \emptyset$ and $\bigcup_1^s \bar{\Omega}_\sigma \cap E_1 = \bar{\Omega} \cap E_1$. Consequently,

$$d(f, \Omega \cap E_1, y_0) = \sum_{\sigma=1}^s d(f, \Omega_0 \cap E_1, y_0).$$

From this equality together with (31) the assertion

$$d(f, \Omega, y_0) = \sum_{\sigma=1}^s d(f, \Omega_\sigma \cap E_1, y_0)$$

follows.

Assertion (iv) follows from assertion (iii).

Proof of assertion (v). Without restriction of generality we may assume $b \in E_1$. Then $f_b = f - b$ is also a layer map with respect to E_1. Thus by definition

$$d(f_b, \Omega, y_0 - b) = d(f_b, \Omega \cap E_1, y_0 - b), \qquad d(f, \Omega, y_0) = d(f, \Omega \cap E_1, y_0). \quad (32)$$

But the right members of these equalities are equal and the asserted equality of the left members follows.

Proof of assertion (vi). Without restriction of generality we may assume that $x_0 \in E_1$. Then if $\xi = x - x_0$ and $g(\xi) = f(x) = x - F(x) = \xi - [F(\xi + x_0) - x_0]$, we see that, with f also, g is a layer map with respect to E_1. Therefore by definition not only (32) but also the equality

$$d(g, \Omega_{x_0}, y_0) = d(g, \Omega_{x_0} \cap E_1, y_0) \qquad (33)$$

holds. But the right members of (32) and (33) are equal and the asserted equality of the left members of these equalities follows.

Assertion (vii) is a consequence of assertions (v) and (vi).

Finally we have to verify that part of Lemma 4.13 which refers to Lemma 2.6 in Chapter 2. Again let f be an L.-S. layer map with respect to the finite-dimensional subspace E_1 of E which we suppose to contain y_0. Let $\tilde{E}$ be a not necessarily finite-dimensional subspace of E with respect to which f is a layer map. We have to prove

$$d(f, \Omega, y_0) = d(f, \Omega \cap \tilde{E}_1, y_0). \qquad (34)$$

Now by definition

$$d(f, \Omega, y_0) = d(f, \Omega \cap E_1, y_0). \qquad (35)$$

But f is also a layer map with respect to $\tilde{E} \cap E_1$. Therefore we may suppose $E_1 \subset \tilde{E}$. But then by definition

$$d(f, \Omega \cap \tilde{E}, y_0) = d(f, \Omega \cap E_1, y_0), \qquad (36)$$

and the assertion (34) follows from (35) and (36).

CHAPTER 4

The Poincaré-Bohl Theorem
and Some of Its Applications

§1. Poincaré-Bohl theorem and the winding number

1.1. In a Banach space E of finite dimension, the special case of Theorem 6.6 in Chapter 3 that the map f_t (defined by (6.9) of Chapter 3) is a linear convex combination of two continuous maps f_0 and f_1 is known as the Poincaré-Bohl theorem. We will use this terminology also for arbitrary Banach spaces. As we will see, this special case of Theorem 6.6 has a number of important consequences.

1.2. THE POINCARÉ-BOHL THEOREM. *As always let E be a Banach space and Ω a bounded open subset of E. For $i = 0, 1$, let*

$$f_i(x) = \lambda_i(x)x - F_i(x) \tag{1.1}$$

be a g.L.-S. map $\overline{\Omega} \to E$ (see §3.6). Let

$$f_t(x) = (1 - t)f_0(x) + tf_1(x), \qquad t \in [0, 1], \tag{1.2}$$

and suppose

$$y_0 \notin f_t(\partial\Omega) \tag{1.3}$$

for all $t \in [0, 1]$. Then $d(f_t, \Omega, y_0)$ is independent of t.

PROOF. By (1.1) and (1.2)

$$f_t(x) = \lambda_t(x)x - F_t(x), \tag{1.4}$$

where

$$\lambda_t(x) = (1 - t)\lambda_0(x) + t\lambda_1(x), \tag{1.5}$$
$$F_t(x) = (1 - t)F_0(x) + tF_1(x). \tag{1.6}$$

Now by Definition 6.2 in Chapter 3 $\lambda_0(x)$ and $\lambda_1(x)$ both lie in some interval $[m, M]$ with $0 < m < M$ for all $x \in \overline{\Omega}$. By (1.5) we therefore see that the same is true for $\lambda_t(x)$. Thus the assumption (6.8) of Theorem 6.6 in Chapter 3 is satisfied with $\lambda(x, t) = \lambda_t(x)$. Moreover, since F_0 and F_1 are completely continuous, it follows from (1.6) that the map $F(x,t) = F_t(x)$, as a map $\overline{\Omega} \times [0,1] \to E$ is completely continuous. Thus all assumptions of Theorem 6.6 are satisfied, and the Poincaré-Bohl theorem is shown to be a special case of Theorem 6.6.

1.3. THEOREM. *Let f_0 and f_1 be two g.L.-S. maps $\overline{\Omega} \to E$, and suppose that*

$$f_0(x) = f_1(x) \quad \text{for } x \in \partial\Omega. \tag{1.7}$$

If, moreover,

$$y_0 \notin f_0(\partial\Omega), \tag{1.8}$$

then

$$d(f_1, \Omega, y_0) = d(f_0, \Omega, y_0). \tag{1.9}$$

(*In other words: the degree of a map depends only on the boundary values of that map.*)

PROOF. Let f_t be defined as in (1.2). We then see from (1.7) that $f_t(x) = f_0(x)$ for $x \in \partial\Omega$ and all $t \in [0,1]$. Therefore the assumption (1.8) of the present theorem implies the assumption (1.3) of the Poincaré-Bohl theorem. Consequently assertion (1.9) follows from the Poincaré-Bohl theorem.

1.4. THEOREM. *Let f_0 be a g.L.-S. map $\partial\Omega \to E$. Then there exists a g.L.-S. map $\overline{\Omega} \to E$ which is an extension of f_0.*

PROOF. Let

$$f_0(x) = \lambda(x)x - F_0(x), \quad x \in \partial\Omega. \tag{1.10}$$

Then by definition F_0 is completely continuous, and λ_0 satisfies

$$0 < m \leq \lambda_0(x) \leq M, \quad x \in \partial\Omega, \tag{1.11}$$

for some constants m and M. Applying Lemma 31 in Chapter 1 with $S_0 = \partial\Omega$ and $\phi_0 = F_0$, we see that there exists a completely continuous extension F of F_0 to $\overline{\Omega}$. Applying the same lemma with $S_0 = \partial\Omega$ and $\phi_0 = \lambda_0$, we see that there exists a continuous extension λ of λ_0 to $\overline{\Omega}$ satisfying $\lambda(\overline{\Omega}) \subset \text{co}(\lambda_0(\partial\Omega))$. Since the interval $[m, M]$ is convex, it follows that (1.11) remains valid if λ_0 is replaced by λ. It is thus clear that $f(x) = \lambda(x)x - F(x)$ is a g.L.-S. extension of f_0 to $\overline{\Omega}$.

Theorems 1.3 and 1.4 together make the following definition possible.

1.5. DEFINITION. Let f_0 be a g.L.-S. map $\partial\Omega \to E$. Suppose (1.8) holds. Then the winding number $u(f_0(\partial\Omega), y_0)$ of $f_0(\partial\Omega)$ with respect to y_0 is defined by the equality $u(f_0(\partial\Omega), y_0) = d(f, \Omega, y_0)$, where f is a g.L.-S. extension of f_0 to $\overline{\Omega}$.

1.6. EXAMPLE. Let $E = E^2$, the two-dimensional Euclidean plane. Let Ω be the unit ball $B(\theta, 1)$, and $S^1 = \partial B(0,1)$ the one-dimensional unit sphere (unit circle). Let f_0 be a continuous map $S^1 \to E^2$ satisfying

$$\theta \notin f_0(S^1). \tag{1.12}$$

It is then possible to define an integer $n = n(f_0)$ which indicates the number of times "$f(S^1)$ winds around θ" and to prove that

$$n = n(f_0) = u(f_0(S^1), \theta). \tag{1.13}$$

This motivates the use of the term "winding number" in Definition 1.5. For the precise definition of the number $n = n(f_0)$ and the proof of (1.13), see the Notes.[1]

The following theorem shows that the winding number $u(f(\partial\Omega), y_0)$ is "continuous in f."

1.7. THEOREM. *Let*

$$f_t(x) = \lambda(x, t)x - F(x, t) \tag{1.14}$$

be a map $\partial\Omega \times [0, 1] \to E$. *We assume that* (i) $\lambda(x, t)$ *is real valued and continuous and satisfies*

$$0 < m \leq \lambda(x, t) \leq M \quad for \ (x, t) \in \partial\Omega \times [0, 1] \tag{1.15}$$

for some constants m, M; (ii) *the map* $F(x, t): \partial\Omega \times [0, 1]$ *is completely continuous;* (iii)

$$y_0 \notin f_t(\partial\Omega). \tag{1.16}$$

Then $u(f_t(\partial\Omega), y_0)$ *is independent of* t.

PROOF. Let Π be the product space $E \times [0, 1]$. Then $S_0 = \partial\Omega \times [0, 1]$ is a closed subset of Π. If we apply Lemma 31 of Chapter 1 with E replaced by Π, we see by the argument used in the proof of Theorem 1.4 that there exists a completely continuous extension F_1 of F mapping $\overline{\Omega} \times [0, 1]$ into E, and a real-valued continuous extension λ_1 of λ to $\overline{\Omega} \times [0, 1]$ such that (1.15) remains true if λ is replaced by λ_1. Then

$$f_t^1(x) = \lambda_1(x, t)x - F_1(x, t)$$

is a g.L.-S. extension of f_t to $\overline{\Omega} \times [0, 1]$ for which (1.16) remains true if f is replaced by f_t^1. Therefore by Definition 1.5

$$u(f_t(\partial\Omega)) = u(f_t^1(\partial\Omega)) = d(f_t^1, \Omega, y_0),$$

and our assertion follows from Theorem 6.6 of Chapter 3.

We state explicitly the following important special case of the last theorem.

1.8. POINCARÉ-BOHL THEOREM FOR THE WINDING NUMBER. *For* $i = 0, 1$, *let* $f_i(x)$ *be a g.L.-S. map* $\partial\Omega \to E$, *and for* $t \in [0, 1]$ *let* $f_t(x) = (1 - t)f_0(x) + tf_1(x)$. *Suppose that* $y_0 \notin f_t(\partial\Omega)$ *for all* $t \in [0, 1]$. *Then* $u(f_t(\partial\Omega), y_0)$ *is independent of* t.

The next theorem may be considered as a generalization of the classical Rouché theorem in the theory of functions of a complex variable.

1.9. ROUCHÉ THEOREM. *Let f_0 and f_1 be two g.L.-S. maps $\partial\Omega \to E$. We suppose that*

$$y_0 \notin f_0(\partial\Omega) \tag{1.17}$$

and that moreover

$$\|f_1(x) - f_0(x)\| < \|f_0(x) - y_0\| \quad \text{for } x \in \partial\Omega. \tag{1.18}$$

Then $u(f_1(\partial\Omega), y_0)$ exists and equals $u(f_0(\partial\Omega), y_0)$.

PROOF. For $t \in [0,1]$ let $f_t(x) = (1-t)f_0(x) + tf_1(x)$. Then $f_t(x) - y_0 = f_0(x) - y_0 + t(f_1(x) - f_0(x))$. Therefore

$$\|f_t(x) - y_0\| \geq \|f_0(x) - y_0\| - \|f_1(x) - f_0(x)\|.$$

This holds for all $x \in \bar{\Omega}$. But for $x \in \partial\Omega$ the right member is positive by assumption (1.18). Thus the assumption of Theorem 1.8 that $y_0 \notin f_t(\partial\Omega)$ for all $t \in [0,1]$ is satisfied, and that theorem implies the assertions of the present one.

1.10. COROLLARY TO THEOREM 1.9. *In addition to the assumptions of that theorem we suppose that f_0 is the identity map I on $\partial\Omega$ and that $y_0 \in \Omega$. Then*

$$u(f_1(\partial\Omega), y_0) = 1, \tag{1.19}$$

and for every g.L.-S. extension $\bar{f}_1$ of f_1 the equation

$$\bar{f}_1(x) = y_0 \tag{1.20}$$

has at least one solution $x \in \Omega$.

PROOF. By Theorem 1.9

$$u(f_1(\partial\Omega), y_0) = u(f_0(\partial\Omega), y_0). \tag{1.21}$$

But since the identity I on $\bar{\Omega}$ is an extension of f_0, the right member of (1.21) equals, by definition, the degree $d(I, \Omega, y_0)$, and thus equals one since $y_0 \in \Omega$. This proves (1.19). But for any extension $\bar{f}_1$ to $\bar{\Omega}$ of f_1, by definition $d(\bar{f}_1, \Omega, y_0) = u(f_1(\partial\Omega), y_0)$. Thus by (1.19), $d(\bar{f}_1, \Omega, y_0) = 1$ which implies our second assertion.

1.11. In the next theorem we will use the notion of a "star domain." An open bounded set $\Omega \subset E$ is called a star domain if it contains a point y_0 such that every ray issuing from y_0 (see §1.23) intersects $\partial\Omega$ in exactly one point. Ω is then called a star domain with respect to y_0. A bounded open convex domain is a star domain with respect to any of its points. For a proof of this assertion see the Notes.[2]

1.12. THEOREM. *Let the open bounded subset Ω of E be a star domain with respect to the point $y_0 \in \Omega$. Let f_0 and f_1 be g.L.-S. maps $\partial\Omega \to E$ satisfying*

$$y_0 \notin f_i(\partial\Omega), \quad i = 0, 1. \tag{1.22}$$

Suppose that for no $x \in \partial\Omega$ do the vectors $f_0(x) - y_0$ and $f_1(x) - y_0$ have opposite direction, i.e., for no positive λ and $x \in \partial\Omega$ do we have

$$f_1(x) - y_0 = -\lambda(f_0(x) - y_0). \tag{1.23}$$

Then

$$u(f_1(\partial\Omega), y_0) = u(f_0(\partial\Omega), y_0). \tag{1.24}$$

PROOF. For $t \in [0, 1]$ let

$$f_t(x) = (1 - t)f_0(x) + tf_1(x). \tag{1.25}$$

Then our assertion (1.24) will follow from Theorem 1.8 provided we can show that

$$y_0 \notin f_t(\partial\Omega) \quad \text{for } t \in [0, 1]. \tag{1.26}$$

Now we see from (1.25) and assumption (1.22) that (1.26) is true for $t = 0$ and $t = 1$. Now suppose (1.26) were not true for some t in the open interval $(0, 1)$. Then for such t and some $x \in \partial\Omega$, $f_t(x) = y_0$, i.e., by (1.25) $f_1(x) - y_0 = -(1 - t)t^{-1}(f_0(x) - y_0)$. But this contradicts the assumption (1.23) since $(1 - t)/t > 0$.

1.13. COROLLARY 1 TO THEOREM 1.12. *Under the assumption of that theorem,*

$$d(\overline{f}_1, \Omega, y_0) = d(\overline{f}_0, \Omega, y_0) \tag{1.27}$$

for every g.L.-S. extension $\overline{f}_1$ and $\overline{f}_0$ to $\overline{\Omega}$ of f_1 and f_0 resp.

By definition of the winding number this is an obvious consequence of Theorem 1.12.

1.14. COROLLARY 2 TO THEOREM 1.12. *Let f_1 be a g.L.-S. map $\partial\Omega \to E$, where Ω is a star domain with respect to the point $y_0 \in \Omega$. Suppose that for all $x \in \partial\Omega$ and all positive λ*

$$f_1(x) - y_0 \neq -\lambda(x - y_0). \tag{1.28}$$

Then for every g.L.-S. extension $\overline{f}_1$ of f_1 to $\overline{\Omega}$ the equation

$$\overline{f}_1(x) = y_0 \tag{1.29}$$

has at least one solution $x \in \Omega$.

PROOF. If f_0 is the identity map on $\partial\Omega$ and $\overline{f}_0$ the identity map on $\overline{\Omega}$, we see from (1.28) that f_0 and f_1 satisfy the assumptions of Theorem 1.12. It follows, by Corollary 1, that (1.27) holds. But the right member of this equality equals unity since $\overline{f}_0$ is the identity map. Thus $d(\overline{f}_1, \Omega, y_0) = 1$ which implies our assertion.

1.15. COROLLARY 3 TO THEOREM 1.12. *Let Ω and y_0 be as in that theorem. Let the g.L.-S. maps f_0 and $f_1: \partial\Omega \to E$ satisfy (1.22). Finally suppose that*

$$u(f_0(\partial\Omega), y_0) \neq u(f_1(\partial\Omega), y_0). \tag{1.30}$$

Then there exists an $x_0 \in \partial\Omega$ and a positive λ_0 such that

$$f_1(x) - y_0 = -\lambda_0(f_0(x) - y_0). \tag{1.31}$$

Indeed if this assertion were not true, then (1.23) would be true for all $x \in \partial\Omega$ and all $\lambda > 0$. Consequently by Theorem 1.12 the equality (1.24) would hold in contradiction to our assumption (1.30).

1.16. COROLLARY 4 TO THEOREM 1.12. *Let Ω be a bounded open set in a Hilbert space E with scalar product $\langle \cdot, \cdot \rangle$. Let f_0 and f_1 be g.L.-S. maps $\overline{\Omega} \to E$. We assume moreover that Ω is a star domain with respect to the point $y_0 \in \Omega$ and that*

$$\langle f_0(x) - y_0, f_1(x) - y_0 \rangle > 0 \quad \text{for } x \in \partial\Omega. \tag{1.32}$$

Then

$$u(f_0(\partial\Omega), y_0) = u(f_1(\partial\Omega), y_0) \tag{1.33}$$

and

$$d(\overline{f}_0, \Omega, y_0) = d(\overline{f}_1, \Omega, y_0) \tag{1.34}$$

for any g.L.-S. extensions $\overline{f}_0$ and $\overline{f}_1$ to $\overline{\Omega}$ of f_0 and f_1 resp.

PROOF. By definition of the winding number, assertion (1.33) implies assertion (1.34). It thus is sufficient to prove the first of these equalities. Suppose then that (1.33) were wrong. Then by Corollary 3 there would exist a $\lambda_0 > 0$ and an $x_0 \in \partial\Omega$ such that (1.31) was true with $x = x_0$. Scalar multiplication of the latter equality by $f_0(x) - y_0$ yields:

$$\langle f_0(x_0) - y_0, f_1(x_0) - y_0 \rangle = -\lambda_0 \| f_0(x) - y_0 \|^2 < 0$$

in contradiction to assumption (1.32).

1.17. THEOREM. *Let f be a g.L.-S. map of the Hilbert space E into itself and let y_0 be a point of E. We suppose that uniformly*

$$\limsup_{\|x\| \to \infty} \left(f(x) - f(y), \frac{x - y_0}{\|x - y_0\|} \right) = +\infty,$$

i.e., that there exists an increasing sequence of positive real numbers $r_1, r_2, \ldots$ which tends to ∞ and in addition has the following property: to every positive c there corresponds an integer $n(c)$ such that, for $n \geq n(c)$,

$$\left(f(x) - f(y_0), \frac{x - y_0}{\|x - y_0\|} \right) > c \tag{1.35}$$

if $\|x - y_0\| = r_n$. It is asserted that the equation

$$f(x) = y_0 \tag{1.36}$$

has a solution in the ball $B(y_0, r_{n_0})$ with $n_0 = n(c_0)$ where

$$c_0 = \|f(y_0) - y_0\|. \tag{1.37}$$

PROOF. The assertion is trivial if $c_0 = 0$ since then $x = y_0$ is a solution of (1.36). But for $c_0 > 0$ it will be sufficient to show that

$$d(f, B(y_0, r_{n_0}), y_0) = 1. \tag{1.38}$$

Now

$$(f(x) - y_0, x - y_0) = (f(x) - f(y_0), x - y_0) - (y_0 - f(y_0), x - y_0),$$

and therefore by Schwarz's inequality, by (1.37) and (1.35), for $\|x - y_0\| = r_{n_0}$,

$$(f(x) - y_0, x - y_0) \geq (f(x) - f(y_0), x - y_0) - c_0\|x - y_0\|$$
$$> c_0[\|x - y_0\| - \|x - y_0\|] = 0.$$

We thus see that the assumption (1.32) of Corollary 4 to Theorem 1.12 is satisfied with $\Omega = B(y_0, r_{n_0})$ and with $\overline{f}_0 = f$, $\overline{f}_1(x) = x$. Consequently, by that corollary, (1.34) holds with the right member of this equality being unity since $\overline{f}_1$ is the identity map. This proves the assertion (1.38).

So far the applications of the Poincaré-Bohl theorem 1.8 concerned existence proofs. We conclude this section with an application to a fixed point theorem.

1.18. THEOREM. *Let Ω be an open bounded star domain in the Banach space E, and let $F: \overline{\Omega} \to E$ be completely continuous. We assume that*

$$F(\partial\Omega) \subset \overline{\Omega} \tag{1.39}$$

and assert that $\overline{\Omega}$ contains a fixed point for F, i.e., a point x satisfying

$$F(x) = x. \tag{1.40}$$

PROOF. If $\partial\Omega$ contains a fixed point, then our assertion is true. Suppose now

$$F(x) \neq x \quad \text{for } x \in \partial\Omega. \tag{1.41}$$

We will then prove that there is a fixed point in Ω. To this end we consider the L.-S. map

$$f(x) = x - F(x) + y_0, \tag{1.42}$$

where y_0 is a point in Ω with respect to which Ω is a star domain. Then the assertion (1.40) is equivalent to

$$f(x) = y_0, \tag{1.43}$$

and the assumption (1.41) is equivalent to

$$f(x) \neq y_0 \quad \text{for } x \in \partial\Omega. \tag{1.44}$$

Now to prove that (1.43) (and therefore (1.40)) has a solution in Ω, it will, by Corollary 2 to Theorem 1.12, be sufficient to show that the assumption (1.28) of that corollary is satisfied with $f_1 = f$. Suppose (1.28) were not true, i.e.,

$$f(x) - y_0 = -\lambda(x - y_0) \tag{1.45}$$

for some $x \in \partial\Omega$ and some $\lambda > 0$. Then for such x and λ

$$F(x) - y_0 = (x - y_0)(1 + \lambda) \tag{1.46}$$

as seen from (1.42) by elementary calculation. But since $x \in \partial\Omega$, since Ω is a star domain with respect to y_0, and since $1 + \lambda > 1$, we see from (1.46) that $F(x)$ lies in the exterior of Ω in contradiction to our assumption (1.39). Thus (1.45) is shown to be wrong.

§2. The interpretation of degree and winding number as intersection numbers

2.1. As in §4.1 let Ω be a bounded open set in the Banach space E and let $f: \overline{\Omega} \to E$ be a g.L.-S. map. Finally let y_0 be a point of E satisfying

$$y_0 \notin f(\partial\Omega). \tag{2.1}$$

We know that if the equation $f(x) = y_0$ has no solution in Ω, i.e., if $y_0 \notin f(\Omega)$, then $d(f, \Omega, y_0) = 0$. On the other hand, if this degree is different from zero, then the above equation has at least one root $x \in \Omega$, i.e., $y_0 \in f(\Omega)$. We therefore consider the degree as a measure for the intersection of y_0 with $f(\Omega)$ and define the intersection number $i(f(\Omega), y_0)$ by setting

$$i(f(\Omega), y_0) = d(f, \Omega, y_0). \tag{2.2}$$

To obtain an interpretation of the winding number $u(f(\partial\Omega), y_0)$ as intersection number we need the following theorem.

2.2. THEOREM. *Let f be a g.L.-S. map $\partial\Omega \to E$. In addition to (2.1) we assume that*

$$u(f(\partial\Omega), y_0) \neq 0. \tag{2.3}$$

Then every ray issuing from y_0 (see §1.23) intersects $f(\partial\Omega)$.

PROOF. Let $\overline{f}$ be a g.L.-S. extension of f to $\overline{\Omega}$. Then by definition and by (2.3)

$$d(\overline{f}, \Omega, y_0) = u(f(\partial\Omega), y_0) \neq 0. \tag{2.4}$$

On the other hand, $\overline{f}(\overline{\Omega})$ is a bounded set. Consequently, there exists an $R > 0$ such that the equation $\overline{f}(x) = y$ has no solution for $|y| > R$, and therefore

$$d(\overline{f}, \Omega, y) = 0 \quad \text{for } \|y\| > R. \tag{2.5}$$

Suppose now there were a ray issuing from y_0 and not intersecting $f(\partial\Omega)$. If we choose a point y_1 on the ray with $\|y_1\| > R$, we arrive at a contradiction. For on the one hand, the segment on this ray given by the points $y_t = (1-t)y_0 + ty_1$ for $0 \le t \le 1$ does not intersect $f(\partial\Omega)$. Therefore by subsection 6.8 in Chapter 3

$$d(f, \Omega, y_0) = d(f, \Omega, y_1). \tag{2.6}$$

On the other hand, (2.5) holds for $y = y_1$, and the two members of (2.6) are not equal as seen from (2.4) and from (2.5) (with $y = y_1$).

2.3. An obvious corollary to Theorem 2.2 is the assertion: if there exists a ray issuing from y_0 which does not intersect $f(\partial\Omega)$, then $u(f(\partial\Omega), y_0) = 0$.

2.4. Theorem 2.2 and its corollary motivate the following definition. If f is a g.L.-S. map: $\partial\Omega \to E$, if y_0 satisfies (2.1), and if h_0 is a ray issuing from y_0, then the intersection number $i(f(\partial\Omega), h_0)$ of $f(\partial\Omega)$ with h_0 is given by

$$i(f(\partial\Omega), h_0) = u(f(\partial\Omega), y_0). \tag{2.7}$$

A more detailed motivation for this definition can be given in the finite-dimensional case (see subsection 3.12 in Chapter 7).

2.5. If f is as in subsection 2.4 and if $\overline{f}$ is a g.L.-S. extension of f to $\overline{\Omega}$, then (2.2) holds with f replaced by $\overline{f}$. But with this substitution the right member of (2.2) is—by definition of the winding number—equal to the right member of (2.7). We therefore see that

$$i(f(\partial\Omega), h_0) = i(\overline{f}(\Omega), y_0). \tag{2.8}$$

2.6. We will use formula (2.8) to derive the "duality formula" (2.12) below. To do this we need some natural definitions. If y_1 is a point of h_0 different from y_0 satisfying $y_1 \notin f(\partial\Omega)$, we set $\partial(y_0, y_1) = y_1 - y_0$ and set

$$i(f(\Omega), \partial(y_0, y_1)) = i(f(\Omega), y_1) - i(f(\Omega), y_0). \tag{2.9}$$

We now subject y_1 to the additional condition $\|y_1\| > R$ where R is the number defined in the proof of Theorem 2.2. Then (2.5) holds with $y = y_1$. Therefore, by (2.2), $i(\overline{f}(\Omega), y_1) = 0$, and we see from (2.9) that for such y_1

$$i(\overline{f}(\Omega), \partial(y_0, y_1)) = -i(\overline{f}(\Omega), y_0). \tag{2.10}$$

We note next that by our choice of y_1 all intersection points of $f(\partial\Omega)$ with h_0 lie in the open segment (y_0, y_1) of h_0. Therefore we define

$$i(f(\partial\Omega), (y_0, y_1)) = i(f(\partial\Omega), h_0). \tag{2.11}$$

But (2.11), (2.8), and (2.10) together yield the formula

$$i(f(\partial\Omega), (y_0, y_1)) = -i(\overline{f}(\Omega), \partial(y_0, y_1)). \tag{2.12}$$

§3. Notes to Chapter 4

Notes to §1.

1. We use the notation of subsection 1.6 of this chapter. Let (r, ϕ) be polar coordinates of the point $x \in E^2$ with θ as pole. For $r \neq 0$, $(r, \phi) = (r, \phi_1)$ if and only if $\phi_1 - \phi$ is an integer multiple of 2π. If $s = \phi/2\pi$ we define a one-valued continuous map of the real s-axis onto S^1 by setting $x(s) = (1, s)$. Then

$$x(s_1) = x(s_2) \quad \text{if and only if } s_1 - s_2 \text{ is an integer.} \tag{1}$$

Given a continuous map $f: S^1 \to E^2 - \theta$, we set

$$g(s) = f(x(s)). \tag{2}$$

From (1) we see that if $s_1 - s_2$ is an integer, then

$$g(s_1) = g(s_2). \tag{3}$$

Now let $(\rho(s), \psi(s))$ be polar coordinates for $g(s)$:

$$g(s) = (\rho(s), \psi(s)), \tag{4}$$

where by (1), $\rho(s+1) = \rho(s) > 0$ and where again the polar angle $\psi(s)$ is determined up to integer multiples of 2π. Thus $\psi(s)$ is not necessarily continuous. However, it can be proved that there exists a real-valued function $A(s) = A_f(s)$, called an angle function for f, which has the following two properties: (a) $A(s)$ is one-valued and continuous for all real s, and (b) for each fixed s, $A(s)$ equals one of the values of the polar angle $\psi(s)$. For the verification of the existence of such a function, we refer the reader to [2, p. 463]. It is clear that such an A is not unique. For if $A(s)$ is an angle function, then for each integer k, $A_k(s) = A(s) + 2\pi k$ is also an angle function. From property (b) it is also clear that to each angle function A and each real s there corresponds an integer n such that

$$A(s+1) - A(s) = 2\pi n. \tag{5}$$

However, it is not hard to see (we refer again to [2, p. 463]) that the integer n depends neither on the choice of the angle function A nor on the value of s. It thus depends only on f and θ. Its intuitive interpretation is obviously that if x varies one time over S^1, then f winds n times about the point θ. The assertion (1.13) states that this number $n = n(f)$ equals the winding number $u(f(S^1), \theta)$ as defined in subsection 1.5 (with $E = E^2$). The proof of (1.13) will be given in three steps.

   Step I. We consider the special case

$$f_0(x(s)) = g(s) = (1, 2\pi n s), \tag{6}$$

where n is an integer. Then $A_0(s) = 2\pi n s$ is an angle function for f_0, and (5) is true with $f = f_0$. To prove (1.13) in this case, we recall that by definition

$$u(f_0(S^1), \theta) = d(\overline{f}_0, B(1, \theta), \theta) \tag{7}$$

for any continuous extension $\overline{f}_0$ of f_0 to $\overline{B(1, \theta)}$. We define $\overline{f}_0$ as follows: for $x = (r, \phi),\ 0 < r \le 1,\ 0 \le \phi < 2\pi$, we set

$$\overline{f}_0(x) = (r, n\phi) = (r, 2\pi n s), \tag{8}$$

and for $r = 0$ we set

$$\overline{f}_0(\theta) = \theta. \tag{9}$$

Our assertion (1.13) for $f = f_0$ is, by (7), equivalent to

$$d(\overline{f}_0, B(1, \theta), \theta) = n. \tag{10}$$

Now θ is not a regular value for $\overline{f}_0$. But we know that the left member of (10) equals $d(\overline{f}_0, B(1, \theta), y_0)$ if $\|y_0\|$ is small enough. It is therefore sufficient to prove

$$d(\overline{f}_0, B(1, \theta), y_0) = n \tag{11}$$

for $y_0 = (r_0, 0)$ and r_0 small enough (actually any $r_0 \in (0,1)$ would do). We consider first the case $n \neq 0$. The roots of the equation

$$\overline{f}_0(x) = y_0 \tag{12}$$

are the points $x_i = (r_0, (2\pi i/n))$, $i = 0, 1, \ldots, n-1$. Now by Definition 2.2 in Chapter 2

$$d(\overline{f}_0, B(1,\theta), y_0) = \sum_{i=0}^{n-1} j_i, \tag{13}$$

where j_i is the index of the linear map $D\overline{f}_0(x_i; h)$, and, by subsection 2.3 in Chapter 2, j_i is the sign of the Jacobian J_i of $\overline{f}_0$ at x_i. Now the Jacobian of the rectangular coordinates $x_1 = r\cos\phi$, $x_2 = r\sin\phi$ with respect to r, ϕ equals r and thus is positive for $r = r_0$. It therefore follows from classical properties of Jacobians (see, e.g., [16, p. 171]) that for the computation of $j_i = \operatorname{sign} J_i$ it does not matter whether we consider $\overline{f}_0$ as a function of the rectangular or polar coordinates. But in the latter case $\overline{f}_0$ is given by $r' = r$, $\phi' = n\phi$ and obviously

$$J_i = n \quad \text{for } i = 0, 1, \ldots, n-1.$$

Thus

$$j_i = \begin{cases} +1 & \text{if } n > 0, \\ -1 & \text{if } n < 0. \end{cases}$$

This together with (13) proves assertion (11) if $n \neq 0$.

Now suppose $n = 0$. Then, by (6), f_0 maps S^1 into the single point $(1,0)$ of S^1, and we obtain a continuous extension $\overline{f}_0$ of f_0 by mapping $\overline{B(\theta, 1)}$ into $(1, 0)$. The equation $\overline{f}_0(x) = \theta$ has no roots in Ω. Therefore $d(\overline{f}_0, B(1,\theta), \theta) = 0$, which proves (11) with $n = 0$.

Step II. Let $f_1(x(s)) = g_1(s) = (1, \Psi_1(s))$, where $\Psi_1(s)$ is a one-valued function satisfying

$$\Psi_1(s+1) - \Psi_1(s) = 2\pi n \tag{14}$$

for $-\infty < s < \infty$. For $0 \leq t \leq 1$ we set $\Psi_t(s) = (1 - t)2\pi n + t\Psi_1(s)$ and $f_t(x(s)) = g_t(s) = (1, \Psi_t(s))$. From (14) we see that $\Psi_t(s+1) - \Psi_t(s) = 2\pi n$. This shows that f_t is for each $t \in [0,1]$ a one-valued continuous map of S^1 into $S^1 \in E_2 - \theta$. It therefore follows from Theorem 1.7 that $u(f_1(S^1), \theta) = u(f_0(S^1), \theta)$. But here the right member equals n by Step I. We thus proved that $u(f_1(S^1), \theta) = n$.

Step III. *The general case.* Let $f(x(s)) = g(s) = (\rho(s), \Psi(s))$, where $\rho(s)$ is as in (4) and where $\Psi(s)$ is an angle function for f. Since ρ is continuous and periodic, there exist constants m and M such that $0 < m \leq \rho(s) \leq M$ for all s. Obviously we can choose m and M in such a way that also $1 \in [m, M]$. Consequently, if $\rho_t(s) = (1 - t)\rho(s) + t \cdot 1$, then $0 < m \leq r_t(s) \leq M$ for all $t \in [0, 1]$ and for $-\infty < s < \infty$. If moreover $f_t(x(s)) = (\rho_t(s), \Psi(s))$ it follows from Theorem 1.7 that $u(f(S^1), \theta) = u(f_1(S^1), \theta)$. But the right member equals n by Step II. This proves our assertion (1.13).

2. *Proof of the assertion in subsection* 1.11. Let y_0 be a point in the bounded open convex set Ω in the Banach space E. Let h be a ray issuing from y_0 and let y_1 be a point of h different from y_0. Then, by definition, h consists of the points $h(t) = (1-t)y_0 + ty_1$, $0 \le t < \infty$. Now $h(t) \in \Omega$ for t small enough since $h(0) = y_0 \in \Omega$. On the other hand, $\|h(t)\| \to \infty$ as $t \to \infty$. Thus $h(t)$ lies in the exterior of Ω for big enough t. Thus the least upper bound of those t for which $h(t) \in \Omega$ is a finite positive number t_0. Then obviously $h(t_0) \in \overline{\Omega}$ and $h(t_0) \notin \Omega$. Therefore $h(t_0) \in \partial\Omega$.

It remains to show that the intersection of $h(t)$ with $\partial\Omega$ is unique. For this it will be sufficient to prove (A): If $h(t_0) \in \partial\Omega$, then $h(t) \in \Omega$ for $0 \le t < t_0$.

Indeed if $h(t_0)$ and $h(t_1)$ are in $\partial\Omega$ with $t_0 \ne t_1$, we may suppose $0 < t_1 < t_0$. But then, by (A), $h(t_1) \in \Omega$ which contradicts the assumption $h(t_1) \in \partial\Omega$.

To prove (A) we note first that there exists a $\delta > 0$ such that

$$B(y_0, \delta) \subset \Omega, \tag{15}$$

since y_0 is a point in the open set Ω. Now let $0 < t_1 < t_0$. We have to show that $h(t_1) \subset \Omega$, i.e., that $h(t_1)$ is the center of a ball lying in $\overline{\Omega}$. We will prove indeed that

$$B(h(t_1), \delta(t_0 - t_1)/t_0) \subset \overline{\Omega}. \tag{16}$$

Now if y is an element of this ball, then

$$\|y - h(t_1)\| < \delta(t_0 - t_1)/t_0. \tag{17}$$

From the definition of $h(t)$, we see by elementary calculation that

$$h(t_1) = y_0(t_0 - t_1)/t_0 + h(t_0)t_1/t_0.$$

Substituting this in (17) we obtain

$$\|yt_0/(t_0 - t_1) - h(t_0)t_1/(t_0 - t_1) - y_0\| < \delta.$$

This shows that

$$z = yt_0/(t_0 - t_1) - h(t_0)t_1/(t_0 - t_1) \in B(y_0, \delta).$$

Therefore, by (15), $z \in \Omega \subset \overline{\Omega}$. But

$$y = z(t_0 - t_1)/t_0 + h(t_0)t_1/t_0.$$

We thus see that y is a convex linear combination of the points z and $h(t_0)$ which lie in $\overline{\Omega}$. Since the convexity of Ω implies that of $\overline{\Omega}$ (see, e.g., [18, p. 413]), it follows that $y \in \overline{\Omega}$. This proves (16).

CHAPTER 5

The Product Theorem
and Some of Its Consequences

§1. The product theorem

1.1. The product theorem establishes a connection between the degrees of two L.-S. maps f, g and the degree of composite map gf. In the case that the maps are also C'', the theorem is essentially a consequence of the chain rule for differentials (see §1.14) and the product theorem in subsection 6.1 of Appendix A for indices of linear nonsingular L.-S. maps. The proof for the general case requires an approximation procedure.

1.2. STATEMENT OF THE PRODUCT THEOREM. *Let Ω be a bounded open set in the Banach space E, let f be an L.-S. map $\overline{\Omega} \to E$, and let g be an L.-S. map $f(\overline{\Omega}) \to E$. Let $\{C_\beta\}$ be the set of the bounded components of $E - f(\partial\Omega)$ where $\beta \in B$, an index set (§§1.33, 1.34). Finally let z_0 be a point of E satisfying*

$$z_0 \notin gf(\partial\Omega). \tag{1.1}$$

Then (i) the degree $d(g, C_\beta, z_0)$ exists for all $\beta \in B$ for which $C_\beta \subset f(\overline{\Omega})$; (ii) either the degree $d(gf, \Omega, z_0)$ equals zero (which, in particular, is the case if the equation

$$gf(x) = z_0 \tag{1.2}$$

has no solutions in Ω), or there exists a finite subset B^0 of B such that

$$d(gf, \Omega, z_0) = \sum_{\beta \in B^0} d(g, C_\beta, z_0) \cdot d(f, \Omega, y_\beta),$$
$$C_\beta \subset f(\Omega) \quad \text{for } \beta \in B^0 \tag{1.3}$$

with $y_\beta \in C_\beta$, and where all terms in the sum are different from zero.

We will show later (see subsection 1.5) that the product theorem is a consequence of Theorem 1.4 below. In preparation of the latter theorem we state the following lemma.

1.3. LEMMA. *Let E, Ω, and f be as in subsection 1.2. Let Y be a bounded open connected subset of E for which*

$$f(\overline{\Omega}) \subset Y \tag{1.4}$$

and

$$\mathrm{dist}(f(\overline{\Omega}), \partial Y) > 0. \tag{1.5}$$

Let g be an L.-S. map $\overline{Y} \to E$, and let z_0 be a point of E which, in addition to (1.1), satisfies

$$z_0 \notin g(\partial Y). \tag{1.6}$$

Finally let $\{Y_\alpha\}$ be the set of components of $Y - f(\partial\Omega)$ where $\alpha \in A$, an index set (see §§1.33, 1.34). Then

(a)

$$\partial Y_\alpha \subset f(\partial\Omega) \cup \partial Y \quad \text{for all } \alpha \in A; \tag{1.7}$$

(b) *if $y \in \overline{Y}$ is a root of the equation*

$$g(y) = z_0, \tag{1.8}$$

then

$$y \notin f(\partial\Omega) \cup \partial Y; \tag{1.9}$$

(c) *if A' is the subset of those $\alpha \in A$ for which Y_α contains a root $y \in \overline{Y}$ of (1.8), then A' is finite;*

(d) *the degree $d(g, Y_\alpha, z_0)$ exists for all $\alpha \in A$;*

(e) *let A^0 be the subset of those $\alpha \in A'$ for which Y_α contains a root $y \in f(\overline{\Omega})$ of (1.8). It is asserted that if (1.2) has a root in Ω, then A^0 is not empty.*

PROOF. Assertion (a) follows directly from the definition of components.

Proof of (b). That a solution y of (1.8) does not lie on ∂Y is the assumption (1.6). Suppose now that $y \in f(\partial\Omega)$ with y satisfying (1.8). Then $y = f(\dot{x})$ for some $\dot{x} \in \partial\Omega$. But then $gf(\dot{x}) = g(y) = z_0$ which contradicts the assumption (1.1).

Proof of (c). By part (iii) of Lemma 10 in Chapter 1 the set S of solutions $y \in \overline{Y}$ of (1.8) is compact, and, by (b), $S \in Y - f(\partial\Omega) = \bigcup_\alpha Y_\alpha$. Thus the Y_α form an open covering of the compact set S. Consequently there exists a finite subcovering, i.e., there are $\alpha_\rho \in A$, $\rho = 1, 2, \ldots, r$, such that $S \subset \bigcup_{\rho=1}^{r} Y_{\alpha_\rho}$. If we denote by A' the subset of A whose elements are $\alpha_1, \alpha_2, \ldots, \alpha_\rho$, then Y_α has no point in common with S if $\alpha \notin A'$ since for such an α $Y_\alpha \cap Y_{\alpha_\rho}$ would not be empty for some α_ρ. This would contradict the fact that the components Y_α are disjoint.

Proof of (d). We have to show that $z_0 \notin g(\partial Y_\alpha)$. But this follows directly from assertions (a) and (b) together with the definition of A'.

Proof of (e). If $x_0 \in \Omega$ is a root of (1.2), then $y_0 = f(x_0)$ is a root in Y of (1.8). Therefore, by (b) it lies in some Y_{α_0}. Then $\alpha_0 \in A'$ by definition of A'. Again by definition $\alpha_0 \in A_0$ since $y_0 = f(x_0) \in f(\overline{\Omega})$.

1.4. THEOREM. *Making the same assumptions and using the same notations as in Lemma 1.3, we assert: if* (1.2) *has solutions in* Ω, *then*

$$d(gf, \Omega, z_0) = \sum_{\alpha \in A^0} d(g, Y_\alpha, z_0) \cdot d(f, \Omega, y_\alpha), \tag{1.10}$$

where $y_\alpha \in Y_\alpha$. *If* (1.2) *has no solutions in* Ω, *the right member of* (1.10) *is to be replaced by zero.*

1.5. Before proving Theorem 1.4, we show that it implies Theorem 1.2. Let the assumptions of 1.2 be satisfied. Since the assertion of that theorem is obvious if (1.2) has no solutions, we suppose that this equation does have solutions. It remains to prove assertions (i) and (ii) of Theorem 1.2.

Now the map $g(y) = y - G(y)$ is defined on the closed bounded set $f(\overline{\Omega})$. Since G is completely continuous, the set $G(f(\overline{\Omega}))$ is bounded. Thus there exists a bounded convex set $\tilde{C}$ such that

$$G(y) \subset \tilde{C} \tag{1.11}$$

for $y \in f(\overline{\Omega})$. Now, by §1.31, G can be extended to a completely continuous map $E \to E$ such that (1.11) holds for all $y \in E$ if the extended map is again denoted by G. Since $\tilde{C}$ is bounded we see that for some positive M

$$\|G(y)\| < M \tag{1.12}$$

for all $y \in E$. Now $g(y) = y - G(y)$ is an L.-S. extension to E of the originally given g. From (1.12) we see that

$$\|g(y)\| \geq \|y\| - \|G(y)\| > \|y\| - M. \tag{1.13}$$

Now let R_1 be an upper bound for $\|f(x)\|$, $x \in \overline{\Omega}$, and let R be a number satisfying

$$R > M + \|z_0\|, \qquad R > R_1. \tag{1.14}$$

Then with $Y = B(\theta, R)$ all assumptions of Lemma 1.3—and thus of Theorem 1.4—are satisfied: that (1.4) and (1.5) hold is obvious, and (1.6) follows from (1.14) and (1.13) since $\|y\| = R$ for $y \in \partial Y$.

Thus, by Theorem 1.4, the formula (1.10) holds. Now since Y is a ball, it follows from part (iv) of §1.34 that each Y_α, $\alpha \in A$, is either a bounded component of $E - f(\partial\Omega)$ or

$$Y_\alpha = Y \cap C_\infty, \tag{1.15}$$

where C_∞ denotes the unbounded component of $E - f(\partial\Omega)$. Now it is possible that (1.15) is true for some $\alpha \in A_0$.[1] But for such α

$$d(f, \Omega, y_\alpha) = 0, \qquad y_\alpha \in A_\alpha. \tag{1.16}$$

Indeed, since $f(\overline{\Omega})$ is bounded, there exists an M_1 such that the equation $f(x) = y$ has no solution in $\overline{\Omega}$ if $\|y\| > M_1$. Now C_∞ certainly contains a point y with $\|y_1\| > M$. Then $d(f, \Omega, y_1) = 0$. But $d(f, \Omega, y_\alpha) = d(f, \Omega, y_1)$ since y_α and y_1 lie in the same component (viz., C_∞) of $E - f(\partial\Omega)$. This proves (1.16), and we see that the only nonzero terms in the sum in (1.10) are among those for which Y_α is a bounded component C_β of $E - f(\partial\Omega)$. Thus, again, if $B = \{\beta \in A \mid C_\beta$ is bounded$\}$, then (i) follows from assertion (d) of Lemma 1.3, and (1.10) may be written as

$$d(gf, \Omega, z_0) = \sum_{\beta \in B'} d(g, C_\beta, z_0) \cdot d(f, \Omega, y_\beta),$$

where B' is a finite subset of B and $y_\beta \in C_\beta$. Now if all terms in the sum are zero, then the degree at the left equals zero. Otherwise let B^0 be the subset of these $\beta \in B'$ for which $d(g, C_\beta, z_0) \cdot d(f, \Omega, y_\beta) \neq 0$. Then the equality in (1.3) follows. To see that also the inclusion in (1.3) is true, we note that $d(f, \Omega, y_\beta) \neq 0$ for all $y_\beta \in C_\beta$ if $\beta \in B^0$. Consequently, the equation $f(x) = y_\beta$ has a root $x \in \Omega$. This proves the asserted inclusion. (Thus in (1.3) only the original g defined on $f(\overline{\Omega})$ appears, not in its extension which was used only for the proof.)

1.6. Starting the proof of Theorem 1.4, we note first that the assertion of that theorem in the case that (1.2) has no roots is an obvious one. We therefore assume that (1.2) has a solution. Then by part (e) of Lemma 1.3 the set A^0 is not empty, but, as a subset of A', it is finite by part (c) of that lemma. We note next that for each $\alpha \in A$ the degree $d(f, \Omega, y)$ is constant for $y \in Y_\alpha$. We therefore define $d(f, \Omega, Y_\alpha)$ by

$$d(f, \Omega, Y_\alpha) = d(f, \Omega, y), \qquad y \in Y_\alpha, \tag{1.17}$$

and, restricting α to A^0, we denote by K^0 the set of integers k such that

$$d(f, \Omega, Y_\alpha) = k \tag{1.18}$$

for at least one $\alpha \in A^0$. Moreover, for fixed $k \in K^0$ we denote by D_k the union of those Y_α with $\alpha \in A^0$ for which (1.18) holds. Since this is a finite union, it follows easily from part (d) of Lemma 1.3 that $z_0 \notin g(\partial D_k)$, i.e., that the degree $d(g, D_k, z_0)$ exists for $k \in K^0$ and that by the sum theorem

$$d(g, D_k, z_0) = \sum d(g, Y_{\alpha_k}, z_0), \tag{1.19}$$

where α_k varies over those $\alpha \in A^0$ for which (1.18) holds. Now if x_0 is a root of (1.2), then $y_0 = f(x_0) \in D_k$ for some $k \in K^0$, and for that k

$$x_0 \in \Omega_k = f^{-1}(D_k) = \{x \in \Omega \mid f(x) \in D_k\}.$$

Since D_k is open and f continuous it follows that Ω_k is open. We thus see that every root of (1.2) lies in $\bigcup_{k \in K^0} \Omega_k$. Therefore

$$D(gf, \Omega, z_0) = d\left(gf, \bigcup_{k \in K^0} \Omega_k, z_0\right). \tag{1.20}$$

Now the Ω_k are disjoint since the D_k are disjoint. In order to apply the sum theorem to the right member of (1.20), we have to show that

$$z_0 \notin gf(\partial\Omega_k). \tag{1.21}$$

Suppose (1.21) to be wrong. Then some

$$\dot{x} \in \partial \Omega_k \tag{1.22}$$

is a root of (1.2). Therefore $\dot{x} \in \Omega_l$ for some $l \in K^0$. Now $l \neq k$ by (1.22) since no point of $\partial \Omega_k$ belongs to the open set Ω_k. But Ω_l is open. Therefore there exists some neighborhood N of $\dot{x}$ which is contained in Ω_l. On the other hand, N contains points of Ω_k as follows from (1.22). Thus $\Omega_k \cap \Omega_l$ is not empty. This contradicts the fact that $k \neq l$ and therefore that Ω_k and Ω_l are disjoint. We are now able to apply the sum theorem and see from (1.20) that

$$d(gf, \Omega, z_0) = \sum_{k \in K^0} d(gf, \Omega_k, z_0). \tag{1.23}$$

1.7. For fixed $k \in K^0$ we now turn to the computation of $d(gf, \Omega_k, z_0)$. We do this first under the assumptions, added to the previous ones, that f and g, and therefore gf, are C'' maps and that z_0 is a regular value for gf. Then the number of roots in Ω_k of (1.2) is finite, and so is the number of images under f of these roots. Let $y_1, y_2, \ldots, y_s$ be these images and, for $\sigma = 1, 2, \ldots, s$, let $x_{\sigma 1}, x_{\sigma 2}, \ldots, x_{\sigma \beta(\sigma)}$ be the roots of

$$f(x) = y_\sigma. \tag{1.24}$$

Then the $x_{\sigma \tau}$, $\sigma = 1, 2, \ldots, s$, $\tau = 1, 2, \ldots, \beta(\sigma)$, are the roots in Ω_k of (1.2) and the y_σ are solutions of (1.8).

Now the differential

$$Dgf(x; h) \tag{1.25}$$

is nonsingular for every root x of (1.2) since z_0 is a regular value. We claim that also the differentials

$$\text{(a)} \quad Df(x; h), \qquad \text{(b)} \quad Dg(y; \eta) \tag{1.26}$$

are nonsingular for every root x of (1.2) and for $y = y_\sigma$, $\sigma = 1, 2, \ldots, s$. Indeed by the chain rule for differentials

$$Dgf(x; h) = Dg(f(x); Df(x; h)). \tag{1.27}$$

Suppose now that the differential (1.26a) is singular, i.e., that there exists an $h \neq \theta$ such that $Df(x; h) = \theta$. (Recall that the differentials are linear L.-S. maps.) Then, by (1.27), the differential $Dgf(x; h) = \theta$ for this h and thus is singular against assumption. Suppose next that the differential (1.26b) is singular for $y = y_\sigma$. Then $Dg(y; \eta_0) = \theta$ for some $\eta_0 \neq \theta$. Now from the nonsingularity of $Df(x; h)$ just proved, we conclude that the equation $Df(x; h) = \eta_0$ has a unique solution $h = h_0$. Obviously $h_0 \neq \theta$. But then $Dgf(x; h_0) = \theta$ by (1.27), and we arrive again at a contradiction to the nonsingularity of $Dfg(x; h)$.

Now since the differentials (1.26) with $x = x_{\sigma \tau}$ and $y = y_\sigma$ are nonsingular L.-S. maps, the indices j of the linear maps given by these differentials are defined. Therefore, by the definition of the degree at a regular value, by the chain rule (1.27), by (1.24), and by the product theorem 6.1 for indices in Appendix A, we see that

$$d(gf, \Omega_k, z_0) = \sum_{\sigma=1}^{s} \sum_{\tau=1}^{\beta(\sigma)} j(Dgf(x_{\sigma\tau}; h))$$

$$= \sum_{\sigma=1}^{s} \sum_{\tau=1}^{\beta(\sigma)} j(Dg(f(x_{\sigma\tau}); Df(x_{\sigma\tau}; h)))$$

$$= \sum_{\sigma=1}^{s} \sum_{\tau=1}^{\beta(\sigma)} j(Dg(y_\sigma; \eta)) \cdot j(Df(x_{\sigma\tau}; h))$$

$$= \sum_{\sigma=1}^{s} j(Dg(y_\sigma; \eta)) \sum_{\tau=1}^{\beta(\sigma)} j(Df(x_{\sigma\tau}; h)).$$

But here the inner sum equals $d(f, \Omega_k, y_\sigma)$ since for fixed σ the $x_{\sigma\tau}$ are the roots in Ω_k of (1.24). But $y = f(x_{\sigma\tau}) \in f(\Omega_k) = D_k$. Therefore $d(f, \Omega_k, y_\sigma) = k$, and we see from the above formula that

$$d(gf, \Omega_k, z_0) = k \cdot \sum_{\sigma=1}^{s} j(Dg(y_\sigma; \eta)). \tag{1.28}$$

If K^0 contains $k = 0$, we see that

$$d(gf, \Omega_0, z_0) = 0. \tag{1.29}$$

Now let $k \neq 0$. We know that $y_1, y_2, \ldots, y_s$ are roots in D_k of (1.8). We assert that they are the only ones. Indeed, let $\bar{y} \in D_k$ be a root of (1.8). Then, by definition of D_k, $d(f, \Omega, \bar{y}) = k \neq 0$. Therefore the equation $f(x) = \bar{y}$ has a root $\bar{x}$ in Ω. Since $f(\bar{x}) = \bar{y} \in D_k$, it follows that $\bar{x} \in f^{-1}(\bar{y}) = \Omega_k$. Moreover $gf(\bar{x}) = g(\bar{y}) = z_0$. Thus $\bar{x}$ is a root in Ω_k of (1.2), and therefore $\bar{x} = x_{\sigma\tau}$ for some $\sigma \in (1, \ldots, s)$ and $\tau \in (1, \ldots, \beta(\sigma))$. But then $\bar{y} = f(\bar{x}) = f(x_{\sigma\tau}) = y_\sigma$ by (1.24).

From the assertion just proved, the sum in (1.28) equals $d(g, D_k, z_0)$. Thus

$$d(gf, \Omega_k, z_0) = k \cdot d(g, D_k, z_0). \tag{1.30}$$

From this equality together with (1.23), we obtain

$$D(gf, \Omega, z_0) = \sum_{k \in K^0} k \cdot d(g, D_k, z_0). \tag{1.31}$$

Here the sum in the right member is a re-ordering of the sum of the right member of the assertion (1.10) as is seen from (1.18) and (1.19). We thus proved Theorem 1.4 under the additional assumption made at the beginning of this section. To complete the proof of that theorem, it remains to show that (1.31) holds without these additional assumptions.

1.8. Proof of (1.31) *without the assumption that z_0 is a regular value of gf but still with the assumption that g and f are C'' maps.* It will be sufficient to prove (1.30). As proved in subsection 1.6 the sets $gf(\partial\Omega_k)$ and $g(\partial D_k)$ do not contain z_0. Therefore there exists a positive r such that the ball $B(z_0, r)$ has no point in common with these sets. Taking into account Definition 4.3 in Chapter 2 and Theorem 4.2 in Chapter 2, this implies that for $z \in B(z_0, r)$

$$d(g, D_k, z) = d(g, D_k, z_0), \qquad d(gf, \Omega_k, z) = d(gf, \Omega_k, z_0). \tag{1.32}$$

Now by Theorem 4.1 in Chapter 2 we can choose a point $z_1 \in B(z_0, r)$ such that z_1 is a regular value for gf. Therefore by subsection 1.7 the relation (1.30) holds with z_0 replaced by z_1. But then, by (1.32), it also holds for z_0.

1.9. It remains to prove (1.31) without the additional assumption that f and g are C'' maps. We start by defining for each $k = 0, \pm 1, \pm 2, \ldots$, the set Δ_k by

$$\Delta_k = \{y \in Y - f(\partial\Omega) \mid d(f, \Omega, y) = k\}. \tag{1.33}$$

For some k the set Δ_k may be the empty set $\varnothing$. We define

$$d(g, \Delta_k, z_0) = 0 \quad \text{if } \Delta_k = \varnothing. \tag{1.34}$$

Moreover, as will be verified in the Notes,[2] the following relations hold:

$$k \cdot d(g, \Delta_k, z_0) = 0 \quad \text{if } \Delta_k \neq \varnothing \text{ and } k \notin K^0; \tag{1.35}$$

$$k \cdot d(g, \Delta_k, z_0) = k \cdot d(g, D_k, z_0) \quad \text{for } k \in K^0 \tag{1.36}$$

(see subsection 1.6 for the definition of K^0 and D_k). It follows from (1.34)–(1.36) that our assertion (1.31) is equivalent to

$$d(gf, \Omega, z_0) = \sum_k k \cdot d(g, \Delta_k, z_0), \tag{1.37}$$

where the summation is extended over all integers k. The proof of (1.37) will be based on the following approximation lemma whose proof will be given in the Notes.[3]

1.10. APPROXIMATION LEMMA. *With the notation of subsection 1.9 and Lemma 1.3 we assume that the maps f and g satisfy the assumptions of that lemma. Moreover, we assume that (1.37) is satisfied. Then there exist positive numbers γ and δ of the following property: if f^* and g^* are L.-S. maps $\bar{\Omega} \to E$ and $\bar{Y} \to E$ resp. and satisfy the inequalities*

$$\|g^*(y) - g(y)\| < \gamma \tag{1.38}$$

for all $y \in Y$ and

$$\|f^*(x) - f(x)\| < \delta \tag{1.39}$$

for $x \in \bar{\Omega}$, then

(i) f^* and g^* satisfy the assumptions of Lemma 1.3 with f and g replaced by f^* and g^*, and

(ii)

$$d(g^* f^*, \Omega, z_0) = \sum_k k \cdot d(g^*, D_k^*, z_0), \tag{1.40}$$

where K_*^0 and D_k^* are defined for f^*, g^* as K^0 and D_k were defined for f, g.

1.11. AN EXTENSION THEOREM. *We use the assumptions and notations of subsection 1.9 and of Lemma 1.3. Moreover, as in subsection 1.1 of Chapter 3, we denote by $C(\bar{\Omega})$ and $C(\bar{Y})$ the space of completely continuous maps $F: \bar{\Omega} \to E$ and $G: \bar{Y} \to E$ resp. Let $C_1(\bar{\Omega})$ and $C_1(\bar{Y})$ be linear subsets of $C(\bar{\Omega})$ and $C(\bar{Y})$ resp. Suppose the product formula (1.10) holds for all couples $f_1: \bar{\Omega} \to E$ and $g_1: \bar{Y} \to E$ for which $F_1(x) = x - f_1(x)$ and $G_1(y) = y - g_1(y)$ are elements of $C_1(\bar{\Omega})$ and $C_1(\bar{Y})$ resp. Then (1.10) holds for all couples f, g for which $F(x) = x - f(x)$ and $g(y) = y - G(y)$ lie in $\overline{C_1(\bar{\Omega})}$ and $\overline{C_1(\bar{Y})}$ resp.*

This theorem is an immediate consequence of the approximation lemma in subsection 1.10 and the fact that (1.37) implies (1.10).

1.12. If $C_2(\bar{\Omega}) = C(\bar{\Omega}) \cap C''(\bar{\Omega})$ for every bounded open set $\Omega \in E$, then, by subsections 1.11 and 1.8, Theorem 1.4 holds for any couple f, g for which $F(x) = x - f(x)$ and $G(y) = y - g(y)$ lie in $\overline{C_2(\bar{\Omega})}$ and in $\overline{C_2(\bar{Y})}$ resp. Thus Theorem 1.4 has been proved for all those Banach spaces E for which

$$\overline{C_2(\bar{\Omega})} = C(\bar{\Omega}) \tag{1.41}$$

for all bounded open subsets Ω of E. (1.41) is true in particular for finite-dimensional Banach spaces.

1.13. We know from subsection 2.2 in Chapter 3 that the property (1.41) does not hold in all Banach spaces. In subsection 1.14 we will prove our product formula for finite layer maps in arbitrary Banach spaces, and application of the extension theorem 1.11 will then, in subsection 1.15, yield the product formula (1.10) in the generality stated.

1.14. In addition to the assumptions made for our product formula (1.10) we suppose in the present section that f and g are finite layer maps. Then there exists a subspace E^n of E of finite dimension n such that not only are f and g layer maps with respect to E^n but also

$$z_0 \in E^n. \tag{1.42}$$

We set $\Omega^n = \Omega \cap E^n$, $Y^n = Y \cap E^n$ and denote by f^n and $(gf)^n$ the restriction of f and gf resp. to $\bar{\Omega}^n$. Likewise g^n denotes the restriction of g to $\bar{Y}^n$. Now $f(x) \in Y$ by (1.4) and $F(x) = x - f(x) \subset E^n$. Therefore

$$f^n(x) \in Y^n \subset E^n \quad \text{for } x \in \bar{\Omega}^n. \tag{1.43}$$

Moreover

$$g^n(y) \in E^n \quad \text{for } y \in \bar{Y}^n, \tag{1.44}$$

since $G(y) = y - g(y) \in E^n$. It follows that for $x \in \bar{\Omega}^n$

$$(gf)^n(x) = gf^n(x) = g^n f^n(x). \tag{1.45}$$

Moreover (see 1.4))

$$f^n(\bar{\Omega}^n) \subset Y^n. \tag{1.46}$$

Since $\partial\Omega^n \subset \partial\Omega$ and $\partial Y^n \subset \partial Y$, we see from (1.1), (1.6), and (1.45) that

$$z_0 \notin g^n f^n(\partial\Omega^n) \tag{1.47}$$

and

$$z_0 \notin g^n(\partial Y^n). \tag{1.48}$$

Since it is clear that (1.5) implies that $\mathrm{dist}(f^n(\overline{\Omega^n}), \partial Y^n) > 0$, we see that the assumptions of Lemma 1.3 (and therefore of our product formula) are satisfied for the continuous finite-dimensional maps $f^n: \overline{\Omega^n} \to E^n$, $g^n: \overline{Y^n} \to E^n$. Therefore (1.10) holds with f and g replaced by f^n and g^n resp. of subsection 1.14. Since (1.10) is equivalent to (1.37), we see that

$$d(g^n f^n, \Omega^n, z_0) = \sum_k k \cdot d(g^n, \tilde{\Delta}_k^n, z_0), \tag{1.49}$$

where (cf. (1.33))

$$\tilde{\Delta}_k^n = \{y \in Y^n \mid d(f^n, \Omega^n, y) = k\}. \tag{1.50}$$

It remains to prove (1.37) for the given layer maps f and g. Now from the definition in subsection 3.4 of Chapter 3 of the degree for finite layer maps and from (1.45) we see that

$$d(gf, \Omega, z_0) = d((gf)^n, \Omega^n, z_0) = d(g^n f^n, \Omega^n, z_0),$$

$$d(f, \Omega, y) = d(f^n, \Omega^n, y) \quad \text{for } y \in Y^n, \qquad d(g, \Delta_k, z_0) = d(g^n, \Delta_k^n, z_0), \tag{1.51}$$

where Δ_k is defined by (1.33) and where

$$\Delta_k^n = \Delta_k \cap E^n. \tag{1.52}$$

Now by (1.49) and (1.51)

$$d(gf, \Omega, z_0) = \sum_k k \cdot d(g^n, \tilde{\Delta}_k^n, z_0), \tag{1.53}$$

while by the last of the equalities (1.51) our assertion (1.37) reads

$$d(gf, \Omega, z_0) = \sum_k k \cdot d(g^n, \Delta_k^n, z_0).$$

This together with (1.53) shows that (1.37) will be proved once it is shown that

$$\tilde{\Delta}_k^n = \Delta_k^n. \tag{1.54}$$

Now, since f is a layer map with respect to E^n,

$$d(f, \Omega, y) = d(f^n, \Omega^n, y) \tag{1.55}$$

if $y \in E^n$ and $y \notin f(\partial\Omega)$. Now let $y \in \tilde{\Delta}_k^n$. Then by (1.50) the right member of (1.55) equals k. But then the left member equals k, which by (1.33) means that $y \in \Delta_k$. Since $y \in E^n$, it follows from (1.52) that $y \in \Delta_k^n$. We thus proved $\tilde{D}_k^n \subset \Delta_k^n$. Suppose now $y \in \Delta_k^n$. Then by (1.52), (1.55) and (1.51)

$$y \in \{y \in Y \mid d(f, \Omega, y) = k\} \cap E^n = \{y \in Y^n \mid d(f^n, \Omega^n, y) = k\} = \tilde{\Delta}_k^n.$$

This finishes the proof of (1.37), and therefore of (1.10) for finite layer maps.

1.15. The product formula (1.10) follows now from 1.13 and the second Leray-Schauder lemma in subsection 4.3 of Chapter 3 upon application of the extension theorem in subsection 1.11, with $C_1(\overline{\Omega})$ and $C_1(\overline{Y})$ being the set of completely continuous maps into E whose domains are $\overline{\Omega}$ and $\overline{Y}$ resp. and whose ranges are contained in finite-dimensional subspaces of E.

§2. The invariance of the domain

2.1. As is well known the property of a set to be open is not invariant under a continuous map. For instance $y = f(x) = x^2$ maps the open interval $(-1, +1)$ onto the interval $0 \le y < 1$ which is not open. However, Brouwer proved in 1912 for finite-dimensional spaces that openness is invariant under continuous maps which have a continuous inverse. His theorem is know as the theorem of the invariance of the domain (Gebietsinvarianz). Leray [34] generalized this theorem to L.-S. maps in Banach spaces. He based his proof on the product theorem 1.4. The proof of Leray's generalization is the main object of this chapter. We first need the following theorem.

2.2. THEOREM. *Let Ω be a bounded open set in the Banach space E, and let f be an L.-S. map which maps $\overline{\Omega}$ onto $f(\overline{\Omega})$ in a one-to-one fashion. Then*

$$d(f, \Omega, y) = \pm 1 \quad \text{for } y \in f(\Omega). \tag{2.1}$$

PROOF. We note first of all that since f is one-to-one on $\overline{\Omega}$, a $y \in f(\Omega)$ does not lie in $f(\partial\Omega)$. Therefore the left member of (2.1) is defined for the y indicated. We note next that by assertion (iv) of §1.10 the assumptions of our theorem imply that f has an inverse g which is an L.-S. map: $f(\overline{\Omega})$ onto $\overline{\Omega}$. Thus

$$g(y) = y - G(y), \qquad y \in f(\overline{\Omega}),$$

where G is completely continuous. Now

$$gf(z_0) = z_0 \in \Omega \tag{2.2}$$

for every $z_0 \in \Omega$. This shows that assumption (1.1) of the product theorem 1.2 is satisfied for every $z_0 \in \Omega$. Consequently by this theorem the equality (1.3) holds where by (2.2) the left member equals one. We know that only a finite number of the terms on the right member of (1.3) are different from zero, say for $\beta = 1, 2, \ldots, r_0$. Then

$$1 = \sum_{\beta=1}^{r_0} d(g, C_\beta, z_0) \cdot d(f, \Omega, y_\beta), \tag{2.3}$$

where $y_\beta \in C_\beta$, and

$$d(g, C_\beta, z_0) \cdot d(f, \Omega, y_\beta) \ne 0 \quad \text{for } \beta = 1, 2, \ldots, r_0. \tag{2.4}$$

From the first factor being $\ne 0$, we conclude that there exists a $y_\beta \in C_\beta$ such that

$$g(y_\beta) = z_0, \tag{2.5}$$

and from the second factor in (2.4) being $\neq 0$, we conclude the existence of an $x_\beta \in \Omega$ for which

$$f(x_\beta) = y_\beta. \tag{2.6}$$

From the last two equalities we see that $z_0 = gf(x_\beta) = x_\beta$ and therefore by (2.6)

$$f(z_0) = f(x_\beta) = y_\beta \in C_\beta. \tag{2.7}$$

But since the C_β are disjoint, the inclusion $f(z_0) \in C_\beta$ determines β uniquely, say $\beta = 1$, and since (2.7) is a consequence of (2.4), we see that $r_0 = 1$. Thus, by (2.3) and (2.7),

$$1 = d(g, C_1, z_0) \cdot d(f, \Omega, y_1) = d(g, C_1, z_0) \cdot d(f, \Omega, f(z_0)).$$

This proves assertion (2.1) since $y = f(z_0)$ varies over $f(\Omega)$ as z_0 varies over Ω.

2.3. THEOREM OF THE INVARIANCE OF THE DOMAIN. *Under the assumption of Theorem 2.2 it is asserted that* (i) $f(\Omega)$ *is open, and* (ii)

$$f(\partial\Omega) = \partial f(\Omega). \tag{2.8}$$

PROOF. (i). If $y_0 \in f(\Omega)$, we have to prove the existence of a neighborhood $N(y_0)$ of y_0 such that

$$N(y_0) \in f(\Omega). \tag{2.9}$$

Now, as shown at the beginning of the proof for Theorem 2.2, $y_0 \notin f(\partial\Omega)$. Therefore there exists a neighborhood $N(y_0)$ of y_0 which does not intersect $f(\partial\Omega)$. We know that then

$$d(f, \Omega, y) = d(f, \Omega, y_0) \tag{2.10}$$

for $y \in N(y_0)$. But here, by Theorem 2.2, the right member is different from zero. Consequently so is its left member. But this implies that for $y \in N(y_0)$ the equation $f(x) = y$ has a solution $x \in \Omega$, i.e., that $y \in f(\Omega)$.

(ii) We note first that

$$f(\overline{\Omega}) = \overline{f(\Omega)}. \tag{2.11}$$

Indeed the inclusion $f(\overline{\Omega}) \subset \overline{f(\Omega)}$ follows from the continuity of f. On the other hand, $\overline{f(\Omega)} \subset \overline{f(\overline{\Omega})} = f(\overline{\Omega})$ since $f(\overline{\Omega})$ is closed, f being an L.-S. map. Now since Ω is open, $\overline{\Omega} = \Omega \cup \partial\Omega$, and therefore $f(\overline{\Omega}) = f(\Omega) \cup f(\partial\Omega)$. But by assertion (i) $f(\Omega)$ also is open. Therefore $\overline{f(\Omega)} = f(\Omega) \cup \partial f(\Omega)$. Using (2.11) we see that

$$f(\Omega) \cup f(\partial\Omega) = f(\Omega) \cup \partial f(\Omega).$$

Since f is one-to-one on $\overline{\Omega}$ this implies the assertion (2.8).

2.4. REMARK. An obvious consequence of Theorem 2.3 is the following assertion A: if the assumptions of that theorem are satisfied, if $x_0 \in \Omega$ and if $y_0 = f(x_0)$, then there exist neighborhoods $U(x_0)$ and $V(y_0)$ of x_0 and y_0 resp. such that for $y \in V(y_0)$ the equation $f(x) = y$ has one and only one solution $x \in U(x_0)$. Comparing this assertion with the inversion theorem in §1.21 we note: (i) the latter requires a differentiability assumption while assertion A does not; (ii) the assumption for A that f is one-to-one is in the inversion theorem replaced by the assumption that the differential of the L.-S. map f at x_0 is a one-to-one map. We also note that the "Fredholm alternative" for linear L.-S. maps (cf. part (iv) of Lemma 12 in Chapter 1) asserts: "one-to-one implies existence." Assertion A asserts this implication for not necessarily linear L.-S. maps (in the small).

2.5. COROLLARY TO THEOREMS 2.2 AND 2.3. *Under the assumptions and with the notation of Theorem 1.2, it is asserted*
(i) *if Ω is connected, if f is one-to-one on $\overline{\Omega}$ and $z_0 \in gf(\Omega)$, then*

$$d(gf, \Omega, z_0) = d(g, f(\Omega), z_0) \cdot d(f, \Omega, y) = \pm d(g, f(\Omega), z_0), \qquad (2.12)$$

where $y \in f(\overline{\Omega})$.
(ii) *Let $g\colon f(\overline{\Omega}) \to g(f(\overline{\Omega}))$ be one-to-one, let $z_0 \in g(f(\Omega))$, and let y_0 be the unique root of (1.8). Then*

$$d(gf, \Omega, z_0) = \pm d(f, \Omega, y_0). \qquad (2.13)$$

If, in particular,

$$y_0 \notin C_\infty, \qquad (2.14)$$

then

$$d(gf, \Omega, z_0) = d(g, C_{\beta_0}, z_0) \cdot d(f, \Omega, y_0) \qquad (2.15)$$

for some unique $\beta_0 \in B$. Finally

$$d(f, \Omega, y_0) = 0 \qquad (2.16)$$

if

$$y_0 \in \left(\bigcup_{\beta \in B - B_0} C_\beta \right) \cup C_\infty. \qquad (2.17)$$

PROOF. (i) Since Ω is connected, so is $f(\Omega)$. Moreover, by Theorem 2.3, Ω is open and (2.8) holds. It follows that $f(\Omega)$ is one of the bounded components of $E - f(\partial\Omega)$, say $f(\Omega) = C_{\beta_0}$. Suppose first that $d(g, f(\Omega), z_0) \neq 0$. Then for $y \in f(\Omega)$

$$d(g, f(\Omega), z_0) \cdot d(f, \Omega, y) \neq 0, \qquad (2.18)$$

since by Theorem 2.2

$$d(f, \Omega, y) = \pm 1. \qquad (2.19)$$

This shows that the product (2.18) is, by definition of B^0, an element of the sum of the right member of (1.3). On the other hand, it is the only term in that sum since for $y \in C_\beta$ with $\beta \neq \beta_0$ the equation $f(x) = y$ has no solution in Ω. Therefore for $y \in C_\beta$ with $\beta \neq \beta_0$

$$d(f, \Omega, y) = 0 \qquad (2.20)$$

(if a $C_\beta \neq C_{\beta_0} = f(\Omega)$ exists). This proves the first part of (2.12) under our assumption $d(g, f(\Omega), z_0) \neq 0$.

Suppose now

$$d(g, f(\Omega), z_0) = 0. \tag{2.21}$$

This equality together with (2.20) shows that for all $\beta \in B$

$$d(g, C_\beta, z_0) \cdot d(f, \Omega, y_\beta) = 0 \tag{2.22}$$

with $y_\beta \in C_\beta$. Therefore the set B^0 is empty, and

$$d(gf, \Omega, z_0) = 0 \tag{2.23}$$

by Theorem 1.2. But (2.23) and (2.21) show that the first part of (2.12) is again true. The second part of (2.12) follows now from Theorem 2.2.

Proof of (ii). Since $y_0 \notin f(\partial\Omega)$ and since the components of $E - f(\partial\Omega)$ are disjoint, there are two cases:

(a) $y_0 \in C_{\beta_0}$ for some unique $\beta_0 \in B$;
(b) $y_0 \in C_\infty$.

In case (a) the equation (1.8) has no solution $y \in C_\beta$ for $\beta \neq \beta_0$, and therefore

$$d(g, C_\beta, z_0) = 0 \tag{2.24}$$

for $\beta \neq \beta_0$. Now if $\beta_0 \in B^0$ the sum in (1.3) consists of exactly one term, viz. $\beta = \beta_0$, and (1.3) is identical with assertion (2.15). But if $\beta_0 \notin B^0$ then, because of (2.24), B^0 is empty. Therefore by Theorem 1.2

$$d(gf, \Omega, z_0) = 0, \tag{2.25}$$

and, by definition of B^0,

$$d(g, C_{\beta_0}, z_0) \cdot d(f, \Omega, y_0) = 0 \tag{2.26}$$

since $y_0 \in C_{\beta_0}$. But by (2.26) and (2.25) the equality (2.15) again holds. This finishes the proof of (2.15). But this equality implies (2.13) since $d(g, C_{\beta_0}, y_0) = \pm 1$ by Theorem 2.2.

In case (b) the equality (1.8) has no solution in any C_β. Thus we have $d(g, fC_\beta, z_0) = 0$ for all $\beta \in B$. Therefore B^0 is empty, and we see from Theorem 1.2 that (2.25) holds. But the right member of (2.13) also equals zero as is seen by the argument given for the proof of (1.16).

Finally, the proof of (2.17) is contained in the preceding arguments.

2.6. Let O be an open (not necessarily bounded) subset of E. Let $\phi: \overline{\Omega} \to E$ be an L.-S. map which is one-to-one. Then, for any bounded open set Ω for which $\overline{\Omega} \subset O$,

$$d(\phi, \Omega, y_0) = \pm 1 \tag{2.27}$$

by Theorem 2.2 provided $y_0 \in \phi(\Omega)$. We call ϕ a coordinate transformation (in O) if the right member of (2.27) is independent of the choice of Ω and of the point $y_0 \in \phi(\Omega)$. An example of a coordinate transformation in $E = O$ is $\phi = l$, where l is a linear nonsingular L.-S. map $E \to E$. In this case for $y \in \phi(\Omega)$

$$d(\phi, \Omega, y) = j(\phi), \tag{2.28}$$

where $j(\phi)$ is the index of ϕ (see subsection 1.4 of Chapter 2). In a generalization of the linear case, we define for any coordinate transformation ϕ (in O) the index $j(\phi) = j(\phi, O)$ by (2.28). Note that $|j(\phi)| = 1$ by (2.27).

2.7. LEMMA. *If ϕ is a coordinate transformation in O, then ϕ^{-1} is a coordinate transformation in $\phi(O)$ and*

$$j(\phi) = j(\phi^{-1}). \tag{2.29}$$

PROOF. Let Ω be a bounded open set whose closure $\overline{\Omega}$ is a subset of O. Then since $\phi^{-1}\phi$ is the identity map on $\overline{\Omega}$,

$$d(\phi^{-1}\phi, \Omega, x_0) = +1 \quad \text{for } x_0 \in \Omega. \tag{2.30}$$

On the other hand, by part (ii) of Corollary 2.5 (with $g = \phi^{-1}$, $f = \phi$), we see that for $x_0 \in \Omega$ and $y_0 \in \phi(\Omega)$

$$d(\phi^{-1}\phi, \Omega, x_0) = d(\phi^{-1}, \phi(\Omega), x_0) \cdot d(\phi, \Omega, y_0).$$

Thus by (2.30) and (2.28)

$$1 = d(\phi^{-1}, \phi(\Omega), x_0) \cdot j(\phi). \tag{2.31}$$

Since $\phi(\Omega)$ varies over all bounded open subsets of $\phi(O)$ as Ω varies over all open bounded subsets of O, (2.31) shows that ϕ^{-1} is a coordinate transformation in $\phi(O)$ and that the first factor on the right member of (2.31) equals $j(\phi^{-1})$. Consequently (2.31) implies (2.29).

2.8. THEOREM. *The Leray-Schauder degree is invariant under a coordinate transformation. More precisely, let Ω be a bounded open connected set in E, let f be an L.-S. map $\overline{\Omega} \to E$, let O be an open set which contains all the bounded components of $E - f(\partial\Omega)$ and for which*

$$\overline{\Omega} \subset O, \qquad \overline{f(\Omega)} \subset O, \tag{2.32}$$

let y_0 be a point in E satisfying

$$y_0 \notin f(\partial\Omega), \tag{2.33}$$

and let ϕ_0 be a coordinate transformation in O for which $\phi(O) \subset O$. Then

$$d(f, \Omega, y_0) = d(\phi f \phi^{-1}, \phi(\Omega), \phi(y_0)). \tag{2.34}$$

PROOF. Obviously

$$f(x) = y_0 \tag{2.35}$$

has a solution x in Ω if and only if

$$\phi f \phi^{-1}(y) = \phi(y_0) \tag{2.36}$$

has a solution $y \in \phi(\Omega)$. This shows that (2.34) is true if (2.35) has no solution in Ω, for then both members of that equality are zero. Let us assume then that (2.35) has a solution in Ω, i.e., that

$$y_0 \in f(\Omega). \tag{2.37}$$

Now with Ω the set $\phi(\Omega)$ is connected, and by Theorem 2.3 the latter set is also open. Therefore we can apply part (i) of Corollary 2.5 to obtain

$$d(f\phi^{-1}, \phi(\Omega), y_0) = d(f, \phi^{-1}\phi(\Omega), y_0) \cdot d(\phi^{-1}, \phi(\Omega), x_0),$$

where

$$x_0 \in \phi^{-1}\phi(\Omega) = \Omega \tag{2.37}$$

and where $d(\phi^{-1}, \phi(\Omega), x_0) = j(\phi^{-1})$ by Lemma 2.7. Thus

$$d(f\phi^{-1}, \phi(\Omega), y_0) = d(f, \Omega, y_0) \cdot j(\phi^{-1}). \tag{2.38}$$

To continue the proof of (2.34) we note that with $f_1 = f\phi^{-1}$ the right member of that equality equals $d(\phi f_1, \phi(\Omega), \phi(y_0))$. To compute this degree we apply part (ii) of Corollary 2.5 (with $g = \phi$ and with f replaced by f_1). Now by Theorem 2.3 and by definition of f_1, $f_1(\partial\phi(\Omega)) = f_1(\phi(\partial\Omega)) = f(\partial\Omega)$. Thus

$$E - f_1(\partial\phi(\Omega)) = E - f(\partial\Omega). \tag{2.39}$$

We first consider the case that y_0 is not contained in the unbounded component C'_∞ of the set (2.39). Then y_0 is contained in some bounded component C_{β_0} and we see from (2.15) (with the proper substitutions) that

$$d(\phi f_1, \phi(\Omega), \phi(y_0)) = d(\phi, C_{\beta_0}, \phi(y_0)) \cdot d(f_1, \phi(\Omega), y_0).$$

Here the first factor of the right member equals $j(\phi)$ and the second factor is given by (2.38). Thus

$$d(\phi f\phi^{-1}, \phi(\Omega), \phi(y_0)) = j(\phi) \cdot d(f, \Omega, y_0) \cdot j(\phi^{-1}),$$

and the assertion (2.34) follows from Lemma 2.7.

We now consider the case that

$$\phi(y_0) \in C'_\infty. \tag{2.40}$$

We use (2.13) (with the proper substitutions) and (2.38) to obtain

$$d(\phi f_1, \phi(\Omega), \phi(y_0)) = \pm d(f_1, \phi(\Omega), y_0) = \pm d(f, \Omega, y_0) \cdot j(\phi^{-1}). \tag{2.41}$$

But from (2.17) we see that (2.40) implies that $d(f, \Omega, y_0) = 0$. Therefore by (2.41)

$$d(\phi f_1, \phi(\Omega), \phi(y_0)) = d(f, \Omega, y_0) = 0,$$

which by definition of f_1 proves the assertion (2.34).

2.9. THEOREM. *Let Ω be an open bounded connected set in E, let $\Omega^- = \{x \in E \mid -x \in \Omega\}$, and let f be an L.-S. map $\overline{\Omega \cup \Omega^-} \to E$. Let $f_1(x) = f(-x)$. Then*

$$d(-f_1, \Omega^-, -y_0) = d(f, \Omega, y_0). \tag{2.42}$$

PROOF. If we set $\phi(x) = -x$ for every $x \in E$, then the assertion (2.42) becomes identical with the equality (2.34) of Theorem 2.8, one of whose assumptions was that ϕ is an L.-S. map. Now $\phi(x) = -x$ is L.-S. if and only if E is finite dimensional. Thus for finite-dimensional spaces the assertion (2.42) is a direct consequence of that theorem. To prove the validity of (2.42) for arbitrary Banach spaces we note first that $-f_1$ is L.-S., for if $f(x) = x - F(x)$ then $-f_1(x) = -f(-x) = x + F(-x)$. Now let ε be a positive number which is smaller than the distance from y_0 to $f(\partial\Omega)$ and from $-y_0$ to $f(\partial\Omega^-)$. Then by subsection 5.5 in Chapter 3 there exists a subspace E^n of E of finite dimension n containing y_0 (and therefore $-y_0$) of the following property: if $\Omega_n = \Omega \cap E^n$ and $\Omega_n^- = \Omega^- \cap E^n$, and if f^n and f_1^n are the restrictions of f and f_1 to Ω_n and Ω_n^- resp., then

$$d(f, \Omega, y_0) = d(f^n, \Omega_n, y_0), \qquad d(-f_1, \Omega^-, -y_0) = d(-f_1^n, \Omega_n^-, -y_0).$$

But by the above argument the right members of these equalities are equal to each other. This proves (2.42).

2.10. AN EXISTENCE THEOREM. *Let f be an L.-S. map $E \to E$. Suppose that f is one-to-one and that*

$$\lim_{x \to \infty} \|f(x)\| = \infty. \tag{2.43}$$

Then $f(E) = E$.

PROOF. The only subsets of E which are closed and open are E and the empty set. Since $f(E)$ is not empty, it will be sufficient to prove: (a) $f(E)$ is closed; (b) $f(E)$ is open.

Proof of (a). Let y_0 be a point of the closure of $f(E)$. We have to prove the existence of a point x_0 such that

$$y_0 = f(x_0). \tag{2.44}$$

Now by assumption there exists a sequence of points $y_1, y_2, \ldots$ in $f(E)$ such that

$$\lim_{i \to \infty} y_i = y_0 \tag{2.45}$$

and points $x_1, x_2, \ldots$ such that

$$y_i = f(x_i), \qquad i = 1, 2, \ldots. \tag{2.46}$$

The y_i are bounded by (2.45), and by (2.43) this implies that the x_i are bounded. Since $F(x) = x - f(x)$ is completely continuous, it follows that for some subsequence $n_1, n_2, \ldots$ of the integers the points $F(x_{n_i})$ converge to some point F_0. Since $x_{n_i} = y_{n_i} + F(x_{n_i})$ by (2.46), we see from (2.45) that the x_{n_i} converge to the point $x_0 = y_0 + F_0$. It is now clear from (2.46) and (2.45) (applied to the subsequence n_i) that x_0 satisfies (2.44).

Proof of (b). Let $y_0 \in f(E)$. We have to show the existence of a neighborhood $N(y_0)$ of y_0 such that

$$N(y_0) \subset f(E). \tag{2.47}$$

Now by assumption there exists a unique point $x_0 \in E$ such that $y_0 = f(x_0)$. Then let Ω_0 be a bounded open subset of E containing x_0 and let f_0 be the restriction of f to $\overline{\Omega}_0$. Then, by Theorem 2.3, $f_0(\Omega_0)$ is open. But $y_0 \in f_0(\Omega_0) = f(\Omega_0) \subset f(E)$. This proves (2.47) with $N(y_0) = f(\Omega_0)$.

§3. The Jordan-Leray theorem

3.1. Let K be a circle of positive radius r in the plane E^2, and let $J \subset E^2$ be a homeomorphic image of K, i.e., a one-to-one continuous image of K. Then

J is called a Jordan curve. The original Jordan theorem states: *if J is a Jordan curve, then the set $E^2 - J$ consists of exactly two components* (see §1.33 for the definition of component). It is obvious that the set $E^2 - K$ has exactly two components, namely the bounded component $\{x \in E^2 \mid \|x\| < r\}$ and the unbounded component $\{x \in E^2 \mid \|x\| > r\}$ (if we suppose the center of K to be the zero point θ of E^2). Therefore the Jordan theorem may be stated as follows: *the number of components of $E^2 - K$ is invariant under a homeomorphic change of K*. Theorem 3.2 below, due to Leray, is a generalization in two directions: E^2 is replaced by an arbitrary Banach space E, and the circle K by an arbitrary closed bounded set in E. The admitted homeomorphisms are L.-S. maps.

3.2. THE JORDAN-LERAY THEOREM. *Let K and J be closed bounded sets in the Banach space E. Suppose there exists a one-to-one L.-S. map of K onto J. Then the cardinal number of components of $E - J$ equals the cardinal number of components of $E - K$.*

PROOF. Let A and B be index sets which are in one-to-one correspondence with sets of bounded components of $E - K$ and $E - J$ resp. The bounded component of $E - K$ corresponding to the element α of A will be denoted by K_α, and the bounded component of $E - J$ corresponding to the element β of B will be denoted by J_β. The unbounded components of $E - K$ and $E - J$ will be denoted by K_∞ and J_∞ resp. We have to prove

$$c(B) = c(A) \tag{3.1}$$

where, as in §1.39, $c(S)$ denotes the cardinal number of the set S.

For the proof of (3.1) we introduce the vector spaces V and W generated by A and B, resp. (see §1.36). Now by §1.37, A and B are bases for V and W resp. Therefore by §1.39 the assertion (3.1) will be proved once it is shown that

$$V \text{ and } W \text{ are linearly isomorphic.} \tag{3.2}$$

We will construct linear maps $g^* : W \to V$ and $f^* : V \to W$ such that

$$g^* f^* = I_V, \qquad f^* g^* = I_W, \tag{3.3}$$

where I_V and I_W denote the identities on V and W resp. By §1.35 the existence of such f^*, g^* implies (3.2).

3.3. *The definition of f^* and g^*.* We recall our assumption that there exists a one-to-one L.-S. map $f : K$ onto J. By part (iv) of §1.10 this implies that $g = f^{-1}$ is an L.-S. map J onto K. By Lemma 31 in Chapter 1 there exist L.-S. extensions of f and g to all of E. We choose one pair of such extensions and call them again f and g resp. Before defining f^* we need to verify the following facts (i)–(iii):

(i) the degree $d(f, K_\alpha, y)$ exists for every $y \in E - J$;

(ii) $d(f, K_\alpha, y_\beta) = $ const. for $y_\beta \in J_\beta$, $d(f, K_\alpha, y_\infty) = $ const. for $y_\infty \in J_\infty$;

(i) and (ii) allow us to define

$$d(f, K_\alpha, J_\beta) = d(f, K_\alpha, y_\beta) \quad \text{for } y_\beta \in J_\beta,$$
$$d(f, K_\alpha, J_\infty) = d(f, K_\alpha, y_\infty) \quad \text{for } y_\infty \in J_\infty;$$

(3.4)

(iii)

$$d(f, K_\alpha, J_\infty) = 0,$$

(3.5)

and at most a finite number of K_α contain a root of $f(x) = y_\beta$ for $y_\beta \in J_\beta$, and therefore $d(f, K_\alpha, J_\beta) = 0$ except for at most a finite number of $\alpha \in A$.

For the proof of assertion (i) we have to show that $y \in f(\partial K_\alpha)$ only if $y \in J$. Now $\partial K_\alpha \in K$; therefore $f(\partial K_\alpha) \in f(K) = J$.

Assertion (ii) follows from subsection 4.4 of Chapter 3.

Proof of (iii). (3.5) follows from an argument used earlier: we can choose $y \in J_\infty$ in such a way that $f(x) = y$ has no solution in $\overline{K}_\alpha$, and therefore $d(f, K_\alpha, y) = 0$. The remaining part of (iii) follows from an argument similar to the ones used in the proof of part (c) of Lemma 1.3. Suppose for some $y_\beta \in J_\beta$ the equation $f(x) = y_\beta$ has at least one solution $x_i \in K_{\alpha_i}$ where $\alpha_1, \alpha_2, \ldots$ is an infinite subsequence of elements of A. Since the K_{α_i} are disjoint, $x_i \neq x_j$ for $i \neq j$. Now the infinite sequence $x_1, x_2, \ldots$ is bounded since it is contained in the union of the bounded components of $E - K$ (cf. assertion (iii) of §1.34). Since f is L.-S., it follows that the sequence contains a convergent subsequence. Let x_0 be the limit of such a subsequence. Then $y_\beta = f(x_0)$. Now x_0 lies either in $E - K$ or K. If $x_0 \in E - K$, then $x_0 \in K_{\alpha_0}$ for some $\alpha_0 \in A$. But then infinitely many $x_i \in K_{\alpha_0}$ and therefore $K_{\alpha_i} \cap K_{\alpha_0}$ is not empty for infinitely many $\alpha_c \in A$, which leads to a contradiction since the K_α are disjoint. If $x_0 \in K$ then $y_\beta = f(x_0) \in f(K) = J$. This also is impossible since $y_\beta \in J_\beta \subset E - J$.

This finishes the proof of (iii). It follows that for each $\beta \in B$ the sum

$$\sum_{\alpha \in A} \alpha \cdot d(f, K_\alpha, J_\beta)$$

is finite and therefore an element of the vector space V generated by A. Since B is a base for the vector space W generated by B, we can define a linear map $g^*(W): W \to V$ as follows: for $\beta \in B$ we set

$$g^*(\beta) = \sum_{\alpha \in A} \alpha \cdot d(f, K_\alpha, J_\beta)$$

(3.6)

and extend this map linearly to all of W, i.e., if $w = \sum_1^n r_i \beta_i$ with $\beta_i \in B$ and with r_i real, we set

$$g^*(w) = \sum_{i=1}^n r_i g^*(\beta_i).$$

(3.7)

If we denote by $(\tilde{\text{i}})$, $(\tilde{\text{ii}})$, $(\tilde{\text{iii}})$ the assertions obtained from (i), (ii), (iii) by replacing A, K, K_α, V, v, f by B, J, J_α, W, w, g resp. and use the definition obtained from (3.4) by the same replacement, we see from our definitions and assumptions that (ĩ), (ĩĩ), (ĩĩĩ) are true. Consequently we obtain a linear map $f^* : V \to W$ by setting for $\alpha \in A$

$$f^*(\alpha) = \sum_{\beta \in B} \beta \cdot d(g, J_\beta, K_\alpha) \tag{3.8}$$

and extending this map linearly from A to V.

3.4. Proof of the relations (3.3).   We will prove the first of these relations. The second one then follows by the replacements defined above. Now for $\alpha' \in A$ we see from (3.8), (3.6) and (3.7) that

$$g^* f^*(\alpha') = \sum_{\beta \in B} g^*(\beta) \cdot d(g, J_\beta, K_{\alpha'}) = \sum_{\alpha \in A} \alpha \sum_{\beta \in B} d(g, J_\beta, K_{\alpha'}) \cdot d(f, K_\alpha, J_\beta).$$

Since A is a base for V and since $g^* f^*$ is linear, it follows that the first of the assertions (3.3) is equivalent to the equality

$$\sum_{\beta \in B} d(g, J_\beta, K_{\alpha'}) \cdot d(f, K_\alpha, J_\beta) = \delta_{\alpha \alpha'}, \tag{3.9}$$

where $\delta_{\alpha \alpha'}$ denotes the Kronecker delta. To prove this equality we note first that

$$d(gf, K_\alpha, K_{\alpha'}) = \delta_{\alpha \alpha'}. \tag{3.10}$$

Indeed, gf is the identity on K and, therefore, on ∂K_α since $\partial K_\alpha \subset K$. Thus the identity I_α on $\overline{K}_\alpha$ is an L.-S. extension of the restriction of gf to ∂K_α. Therefore by Theorem 1.3 in Chapter 4

$$d(gf, K_\alpha, x) = d(I_\alpha, K_\alpha, x) = \begin{cases} 1 & \text{if } x \in K_\alpha, \\ 0 & \text{if } x \notin \overline{K}_\alpha. \end{cases}$$

This proves (3.10) since K_α and $K_{\alpha'}$ are disjoint for $\alpha' \neq \alpha$.

We now apply the product theorem 1.2 to the left member of (3.10) to obtain

$$\sum_{\gamma \in \Gamma_\alpha} d(g, K_{\alpha \gamma}, K_{\alpha'}) \cdot d(f, K_\alpha, K_{\alpha \gamma}) = \delta_{\alpha \alpha'}. \tag{3.11}$$

Here Γ_α is an index set which is in one-to-one correspondence with the bounded components $K_{\alpha \gamma}$ of $E - f(\partial K_\alpha)$. On account of (3.11) the proof of our assertion (3.9) is reduced to verifying that

$$\sum_{\beta \in B} d(g, J_\beta, K_{\alpha'}) \cdot d(f, K_\alpha, J_\beta) = \sum_{\gamma \in \Gamma_\alpha} d(g, K_{\alpha \gamma}, K_{\alpha'}) \cdot d(f, K_\alpha, K_{\alpha \gamma}). \tag{3.12}$$

To establish this relation between the components J_β of $E - J$ and the components $K_{\alpha \gamma}$ of $E - f(\partial K_\alpha)$ we prove first that

$$\begin{array}{ll} \text{either} & J_\beta \in K_{\alpha \gamma} \quad \text{for some unique } \gamma \in \Gamma_\alpha, \\ \text{or} & J_\beta \in K_{\alpha \infty}, \end{array} \tag{3.13}$$

where $K_{\alpha \infty}$ is the unbounded component of $E - f(\partial K_\alpha)$: as noticed before (see the paragraph following (3.10)) $\partial K_\alpha \in K$, and therefore $f(\partial K_\alpha) \subset f(K) = J$. Consequently,

$$J_\beta \subset E - J = J_\infty \cup \left(\bigcup_{\beta \in B} J_\beta \right) \subset E - f(\partial K_\alpha) = K_{\alpha \infty} \cup \left(\bigcup_{\gamma \in \Gamma_\alpha} K_{\alpha \gamma} \right).$$

Since J_β is an open connected set in $E - f(\partial K_\alpha)$, the assertion (3.13) now follows from the definition of the $K_{\alpha\infty}$ and $K_{\alpha\gamma}$ as maximal open connected sets in $E - f(\partial K_\alpha)$

(3.13) allows us to re-order the J_β (for fixed α) by putting together those which are contained in a fixed component of $E - f(\partial K_\alpha)$: let $B_\infty = \{\beta \in B \mid J_\beta \in K_{\alpha\infty}\}$, and for $\gamma \in \Gamma_\alpha$ let $B_\gamma = \{\beta \in B \mid J_\beta \in K_{\alpha\gamma}\}$. Then

$$B = B_\infty \cup \left(\bigcup_{\gamma \in \Gamma_\alpha} B_\gamma \right).$$

Our next step in proving (3.12) consists in establishing the following relations:

$$d(f, K_\alpha, J_\beta) = \begin{cases} d(f, K_\alpha, K_{\alpha\gamma}) & \text{for } \beta \in B_\gamma, \\ d(f, K_\alpha, K_{\alpha\infty}) = 0 & \text{for } \beta \in B_\infty, \end{cases} \tag{3.14}$$

and, for $x \in E - K$

$$d(g, K_{\alpha\gamma}, x) = \sum_{\beta \in B_\gamma} d(g, J_\beta, x), \tag{3.15}$$

where

$$d(g, J_\beta, x) = 0 \tag{3.16}$$

except for at most a finite number of $\beta \in B_\gamma$.

PROOF. Let

$$x \in J_\beta. \tag{3.17}$$

Then, by definition, $d(f, K_\alpha, J_\beta) = d(f, K_\alpha, x)$. Now if $\beta \in B_\gamma$ then, by (3.13), $J_\beta \in K_{\alpha\gamma}$ and, by (3.17), $x \in K_{\alpha\gamma}$. Therefore $d(f, K_\alpha, x) = d(f, K_\alpha, K_{\alpha\gamma})$. This proves the first equality in (3.14), and the left part of the second equality in (3.14) follows the same way. The right part of the latter equality follows from the fact that we can choose a $y \in K_{\alpha\infty}$ of norm so large that the equation $f(x) = y$ has no root in $\overline{K}_\alpha$.

For the proof of (3.15) we note first: if $x \in E - K$ and if ξ is a root of

$$g(\xi) = x, \tag{3.18}$$

then

$$\xi \notin J \tag{3.19}$$

since for $\xi \in J$, $x = g(\xi) \subset g(J) = K$. By (3.19) $\xi \in J_\infty \cup (\bigcup_{\beta \in B} J_\beta)$. If ξ is a root of (3.18) which lies in $K_{\alpha\gamma}$, then since the $J_{\alpha\infty}$ and J_β are disjoint, by (3.13) ξ lies in some J_β which is a subset of $K_{\alpha\gamma}$. Thus $\xi \in \bigcup_{\beta \in B_\gamma} J_\beta$. Let us now denote by B_γ^* the set of those $\beta \in B_\gamma$ for which J_β contains a root of (3.18). The set B_γ^* is finite. The proof for this is quite similar to the one for the second part of assertion (iii) and will therefore be omitted. It follows that

$$d(g, K_{\alpha\gamma}, x) = d \left(g, \bigcup_{\beta \in B_\gamma^*} J_\beta, x \right).$$

Application of the sum theorem yields the assertion (3.15) since (3.16) holds for all $\beta \notin B^*_\gamma$.

To finally prove (3.12) we note that by (3.15) with $x \in K_{\alpha'}$, and by (3.14) the right member of this equality equals

$$\sum_{\gamma \in \Gamma_\alpha} \sum_{\beta \in B_\gamma} d(g, J_\beta, K_{\alpha'}) \cdot d(f, K_\alpha, J_\beta). \tag{3.20}$$

Here the double summation is extended over all $\beta \in B - B_\infty$. But if $\beta \in B_\infty$ (i.e., $J_\beta \in K_{\alpha\infty}$), then by (3.14), $d(f, K_\alpha, J_\beta) = 0$. Thus the sum (3.20) equals the left member of (3.12).

§4. Notes to Chapter 5

Notes to §1.

1. This note refers to the paragraph containing (1.15). It may happen that for an $x_0 \in \Omega$ which is a root of

$$gf(x) = z_0 \tag{1}$$

the point $y_0 = f(x_0)$ lies in the unbounded component C_∞ of $E - f(\partial\Omega)$.

EXAMPLE. Let E be the two-dimensional Euclidean plane, and let $\Omega \in E$ be the open circular disc with center θ and radius 3. Let $f: \overline{\Omega} \to E$ be defined as follows: f is the identity map for $\|x\| \le 2$. For $2 < \|x\| \le 3$ we set

$$f(x) = x\|x\|^{-1}(4 - \|x\|). \tag{2}$$

Then $f(\partial\Omega)$ is the circle $\|x\| = 1$. The bounded component C_1 of $E - f(\partial\Omega)$ is the open disc $0 \le \|x\| < 1$, and the unbounded component C_∞ of that set consists of those $x \in E$ for which $\|x\| > 1$. Let g be the identity map on E, and let z_0 be a point satisfying $1 < \|z_0\| < 2$. The roots of (1) then are $x_0 = z_0$ and $x_1 = z_0\|z_0\|^{-1}(4 - \|z_0\|)$, and for $i = 0, 1$, $y_i = f(x_i) = gf(x_i) = z_0 \in C_\infty$. Note that by the argument given in the text $d(f, \Omega, y) = 0$ for $y \in C_\infty$. In particular $0 = d(f, \Omega, z_0) = j_0 + j_1$ where j_0 and j_1 are the indices at x_0 and x_1. Thus $j_1 = -j_0 = -1$.

The reader will also note the geometric interpretation of (2): the "ring" $2 \le \|x\| \le 3$ is folded at the rim $\|x\| = 2$ into the disc $\|x\| \le 2$, and that $f(\partial\Omega) \ne \partial f(\Omega)$.

2. *Proof of* (1.35) *and* (1.36). These relations are obviously true if $k = 0$. Suppose then

$$k \ne 0. \tag{3}$$

Now if $k \notin K^0$ and

$$d(f, \Omega, Y_\alpha) = k, \tag{4}$$

then by definition $\alpha \notin A^0$. We claim

$$\alpha \notin A'. \tag{5}$$

Indeed, (3) and (4) imply that $y \in f(\Omega)$ for every $y \in Y_\alpha$. But this would imply that $\alpha \in A^0$ if (5) were not true. Now (5) thus proved implies (1.35) since Δ_k is the union of Y_α satisfying (4).

Since D_k is an open subset of Δ_k, it will for the proof of (1.36) be sufficient to show that if $y \in \Delta_k$ is a root of

$$g(y) = z_0, \tag{6}$$

then $y \in D_k$. Now $y \in Y_\alpha \subset \Delta_k$ for some $\alpha \in A$. But by (6), $\alpha \in A'$. Moreover $d(f, \Omega, y) = k$. By (3) this implies that $y \in f(\Omega)$. It follows that $\alpha \in A^0$. Thus $y \in D_k$ as we wanted to show.

3. *Proof of the approximation lemma* 1.10.

Proof of assertion (i). Since g, f, and gf are L.-S. maps, we see from (1.1) and (1.6) that there exists a positive γ such that

$$\mathrm{dist}(z_0, gf(\partial\Omega)) \cup g(\partial Y) > 2\gamma. \tag{7}$$

We assert first that if γ satisfies (7) and if the L.-S. map g^* satisfies (1.38), then no roots of the equation

$$g^*(y) = z_0 \tag{8}$$

lie on $f(\partial\Omega) \cup g(\partial Y)$. Indeed, by (1.38)

$$\|g(f(x)) - g^*(f(x))\| < \gamma \tag{9}$$

for all $x \in \bar\Omega$. In particular for $x \in \partial\Omega$ we see from (7) and (9) that

$$\|g^*(f(x)) - z_0\| \geq \|g(f(x)) - z_0\| - \|g^*(f(x)) - g(f(x))\| > \gamma. \tag{10}$$

On the other hand, we see from (7) and (1.38) that for $y \in \partial Y$

$$\|g^*(y) - z_0\| \geq \|g(y) - z_0\| - \|g^*(y) - g(y)\| > \gamma. \tag{11}$$

(10) and (11) together show that, as asserted, the set $(g^*)^{-1}(z_0)$ is disjoint from the set $f(\partial\Omega) \cup \partial Y$. Since the first of these sets is compact (see Lemma 10 in Chapter 1) and the second one is closed, there exists a $\delta > 0$ such that

$$\mathrm{dist}((g^*)^{-1}(z_0), f(\partial\Omega) \cup g(\partial Y)) > 2\delta. \tag{12}$$

Because of (1.5) we may choose δ in such a way that, in addition to (12),

$$\mathrm{dist}(f(\bar\Omega), \partial Y) > 2\delta. \tag{13}$$

To prove assertion (i) of our lemma we have to show that (1.1) and (1.6)–(1.8) remain valid if f and g are replaced by f^* and g^* resp., provided that (1.38) and (1.39) are satisfied. To prove that

$$z_0 \notin g^* f^*(\partial\Omega), \tag{14}$$

we suppose this relation to be wrong. Then $z_0 = g^* f^*(\dot x)$ for some $\dot x \in \partial\Omega$. Let $y^* = f^*(\dot x)$. Then

$$\|f^*(\dot x) - f(\dot x)\| = \|y^* - f(\dot x)\| > 2\delta \tag{15}$$

by (12) since $y^* \in (g^*)^{-1}(z_0)$. But (15) contradicts (1.39).

That $f^*(\overline{\Omega}) \subset Y$ and that $\text{dist}(f^*(\overline{\Omega}), \partial Y) > 0$ (cf. (1.4) and (1.5)) follow from (13) and (1.39). Finally, to prove the relation corresponding to (1.6), we note that for any $\dot{y} \in \partial Y$

$$\|z_0 - g^*(\dot{y})\| \geq \|z_0 - g(\dot{y})\| - \|g(\dot{y}) - g^*(\dot{y})\| > 0 \quad \text{by (7) and (1.38)}.$$

We now turn to the proof of assertion (ii) of Lemma 1.10. Since (1.37) is true by assumption, the assertion (1.40) will be proved once it is shown that

$$d(g^* f^*, \Omega, z_0) = d(gf, \Omega, z_0) \tag{16}$$

and that for each integer k

$$k \cdot d(g^*, \Delta_k^*, z_0) = k \cdot d(g, \Delta_k, z_0). \tag{17}$$

For the proof of (16) we note that by (9) and (7) for $\dot{x} \in \partial\Omega$

$$\|gf(x) - g^* f(x)\| < \text{dist}(z_0, gf(\partial\Omega)) \leq \|z_0 - gf(\dot{x})\|.$$

By the Rouché theorem 1.9 in Chapter 4 this implies $d(gf, \Omega, z_0) = d(g^* f, \Omega, z_0)$. It therefore remains to prove that

$$d(g^* f^*, \Omega, z_0) = d(g^* f, \Omega, z_0). \tag{18}$$

Now if for $t \in [0, 1]$ we set $f_t(x) = (1 - t)f(x) + tf^*(x)$, then (18) will be proved once it is shown that

(a) gf_t is L.-S. for every $t \in [0, 1]$, and
(b)
$$z_0 \notin g^* f_t(\partial\Omega) \quad \text{for } t \in [0, 1]. \tag{19}$$

Now with $f(x) = x - F(x)$, $g^*(y) = y - G^*(y)$ and $f^*(x) = x - F^*(x)$, elementary calculation shows that

$$g^*(f_t(x)) = x - [(1 - t)F(x) + tF^*(x) + G^*(f_t(x))],$$

which obviously proves (a). For the proof of (b) let us suppose that $z_0 = g^* f_{t_0}(\dot{x})$ for some $t_0 \in [0, 1]$ and some $\dot{x} \in \partial\Omega$. Then $y_0 = f_{t_0}(\dot{x}) \in (g^*)^{-1}(z_0)$. Therefore

$$\text{dist}((g^*)^{-1}(z_0), f(\partial\Omega)) \leq \|f_{t_0}(\dot{x}) - f(\dot{x})\| \leq t_0\|f^*(\dot{x}) - f(\dot{x})\| < \delta.$$

But this contradicts (12). Thus (b) is proved.

Turning to the proof of (17) we set $R^* = (g^*)^{-1}(z_0) = \{y \in \overline{Y} \mid g^*(y) = z_0\}$ and claim first that

$$d(f^*, \Omega, y) = d(f, \Omega, y) \tag{20}$$

for $y \in R^*$. To prove this we note that for such y and for $\dot{x} \in \partial\Omega$ by (12) and (1.39)

$$\|y - f(\dot{x})\| \geq \text{dist}(R^*, f(\partial\Omega)) > \|f(\dot{x}) - f^*(\dot{x})\|. \tag{21}$$

This inequality proves (20) by the Rouché theorem.

Furthermore we will need the equality

$$R^* \cap \Delta_k = R^* \cap \Delta_k^*. \tag{22}$$

We prove first that

$$R^* \cap \Delta_k \subset R^* \cap \Delta_k^*. \tag{23}$$

This inclusion is trivial if the left member is empty. Now let $y \in R^* \cap \Delta_k$. Then $d(f, \Omega, y) = k$ since $y \in \Delta_k$. But also (20) holds since $y \in R^*$. It follows that $d(f^*, \Omega, y) = k$. Thus $y \in \Delta_k^*$. This proves (23). To prove that

$$R^* \cap \Delta_k^* \subset R^* \cap \Delta_k, \tag{24}$$

we may again assume the left member contains an element y. Now $d(f^*, \Omega, y) = k$. But again (20) holds, y being an element of R^*. Thus $d(f, \Omega, y) = k$, i.e., $y \in \Delta_k$. This proves (24), and (22) is established.

To prove (17) we may assume $k \neq 0$. We will show that

$$d(g^*, \Delta_k^*, z_0) = d(g^*, \Delta_k, z_0) \tag{25}$$

and that

$$d(g^*, \Delta_k, z_0) = d(g, \Delta_k, z_0). \tag{26}$$

We first consider the case that Δ_k is empty. Then the right member of (25) and both members of (26) are zero due to our convention (1.34). But by (22), $R^* \cap \Delta_k^*$ is empty if Δ_k is empty. By definition of R^* this implies that the equation $g^*(y) = z_0$ has no roots in Δ_k^*. It follows that the left member of (25) is also zero. This proves (25) and (26) for empty Δ_k.

Suppose Δ_k not to be empty. Now R^* has no point in common with $\partial \Delta_k$ since $\partial \Delta_k \subset f(\partial \Omega) \cap \partial Y$. Since Δ_k is open and R^* is compact, it follows that there exists an open set V_k such that

$$\Delta_k \cap R^* \subset V_k \subset \Delta_k. \tag{27}$$

Similarly one sees that there exists an open set V_k^* such that

$$\Delta_k^* \cap R^* \subset V_k^* \subset \Delta_k^*. \tag{28}$$

Now for every couple of open sets V_k and V_k^* satisfying (27) and (28)

$$d(g^*, \Delta_k, z_0) = d(g^*, V_k, z_0), \qquad d(g^*, \Delta_k^*, z_0) = d(g^*, V_k^*, z_0). \tag{29}$$

But by (22) the left members of (27) and (28) are equal. Consequently these two inclusions remain valid if V_k and V_k^* are both replaced by their intersection $V_k \cap V_k^*$, and the equations (29) also remain valid under this replacement. Since with this replacement the right members of (29) become identical, the asserted equality (25) follows.

To finally prove (26) we note that, for any $y \in f(\partial \Omega) \cup \partial Y$,

$$\|y - z_0\| \geq \operatorname{dist}(z_0, f(\partial \Omega) \cup \partial Y).$$

Therefore by (7) and (1.38)

$$\|y - z_0\| > \|g(y) - g^*(y)\|. \tag{30}$$

This inequality holds in particular if y lies in the subset $\partial \Delta_k$ of $f(\partial \Omega) \cup \partial Y$. Therefore (30) implies (26) by the theorem of Rouché.

CHAPTER 6

The Finite-Dimensional Case

§1. Some elementary prerequisites

1.1. DEFINITION. Let E^n be a Banach space of finite dimension n. Then $r + 1$ points $a_0, a_1, \ldots, a_r$ in E^n are called dependent or independent according to whether the r points $a_1 - a_0$, $a_2 - a_0, \ldots, a_r - a_0$ are linearly dependent or linearly independent. (We assume the reader to be familiar with the notions of linear dependence and linear independence of a finite number of points in a vector space, with the notion of the linear space spanned by a finite number of elements in such a space, and with the notion of a base for a finite-dimensional vector space.)

1.2. REMARK. In Definition 1.1 the index zero was distinguished. But we would have obtained an equivalent definition by considering the r points

$$a_0 - a_i, \; a_1 - a_i, \ldots, \; a_{i-1} - a_i, \; a_{i+1} - a_i, \ldots, \; a_r - a_i$$

for any integer $i \leq r$.[1] These points span an r-dimensional subspace of E^n if the points $a_0, a_1, \ldots, a_r$ are independent. It is easy to see that this space is independent of i. Thus a set of $r + 1$ independent points in E^n determines uniquely an E^r.

1.3. REMARK. If $r > n$, any $r + 1$ points $a_0, a_1, \ldots, a_r$ in E^n are dependent (cf. 1.5).

1.4. LEMMA. *The points $a_0, a_1, \ldots, a_r$ of E^n are independent if and only if the relations*

$$\sum_{i=0}^{r} a_i \lambda_i = \theta, \qquad \sum_{i=0}^{r} \lambda_i = 0, \qquad \lambda_i \; real, \tag{1.1}$$

imply that

$$\lambda_0 = \lambda_1 = \cdots = \lambda_r = 0. \tag{1.2}$$

The proof will be given in the Notes.[2]

1.5. DEFINITION. An r-space in E^n is an r-dimensional linear subspace of E^n. An r-dimensional plane, shortly an r-plane, Π^r in E^n is the translate of an r-subspace E^r of E^n. If $a_0 \in \Pi^r$ and if $b_1, b_2, \ldots, b_r$ are a base for E^r, then the points x of Π^r are given by

$$x = a_0 + \sum_{i=1}^{r} b_i \alpha_i, \qquad \alpha_i \text{ real.} \tag{1.3}$$

We sometimes call the b_i also a base for Π^r.

1.6. LEMMA. *The points $a_0, a_1, \ldots, a_r$ with $r \leq n$ of E^n are independent if and only if they do not lie in a plane of dimension less than r.*

The proof will be given in the Notes.[3]

1.7. DEFINITION. Let Π^r and Π^s be r- and s-planes in E^n resp. The hull H of Π^r and Π^s is defined as the plane of smallest dimension containing Π^r and Π^s. The intersection $\Pi^r \cap \Pi^s$ (which is obviously a plane unless empty) will be denoted by K. The dimensions of H and K will be denoted by h and k resp.

1.8. LEMMA. *Using the notations of subsection 1.7 we assert: if K is not empty, then*

$$k + h = r + s. \tag{1.4}$$

The proof will be given in Notes.[4]

1.9. DEFINITION. A finite or countably infinite set of points in E^n is said to be in general position if for $r \leq n$ any $r + 1$ points of S are independent (cf. Definition 1.1).

1.10. Note that by Lemma 1.6 the points of S are in general position if for $r \leq n$ no $r + 1$ points of S lie in a plane of dimension $< r$. Note also that in E^1 any set of disjoint points not containing θ is in general position.

1.11. THEOREM. *Let $b_0, b_1, \ldots, b_r$ be independent points in E^n ($n \geq 1$), and let $a_0, a_1, \ldots$ be a finite or infinite sequence of not necessarily different points in E^n. Let ε be a given positive number. Then there exist points $a'_0, a'_1, \ldots$ in E^n such that*

$$\|a'_i - a_i\| < \varepsilon, \qquad i = 0, 1, 2, \ldots, \tag{1.5}$$

and such that the points $b_0, b_1, \ldots, b_r, a'_0, a'_1, a'_2, \ldots$ are in general position.

PROOF. We give the proof for $r = 1$. If $n = 1$ we choose for a'_0 a point different from b_0 and b_1 which satisfies (1.5) for $i = 0$. For a'_1 we choose a point different from b_0, b_1, a'_0 which satisfies (1.5) for $i = 1$. Going on this way we obtain points $a'_0, a'_1, a'_2, \ldots$ of the desired properties.

Now let $n > 1$. We choose for a'_0 a point in $B(a_0, \varepsilon)$ which does not lie in the one-plane (straight line) containing b_0 and b_1. Then b_0, b_1, a'_0 are in general position. Suppose we have chosen $a'_0, a'_1, \ldots, a'_{N-1}$ in such a way that the points

$$b_0, b_1, a'_0, a'_1, \ldots, a'_{N-1} \tag{1.6}$$

are in general position and such that (1.5) is satisfied for $i = 0, 1, \ldots, N - 1$. We now consider the points

$$b_0, b_1, a'_0, a'_1, \ldots, a'_{N-1}, x, \tag{1.7}$$

where x is different from the points (1.6) and satisfies

$$x \in B(a_N, \varepsilon). \tag{1.8}$$

We have to show that x can be chosen to satisfy (in addition to (1.8)) the following condition C: for any integer $r \leq \min(n, N+2)$ and for any choice of r different points of the set (1.6), say $c_0, c_1, \ldots, c_{r-1}$, the $r+1$ points $c_0, c_1, \ldots, c_{r-1}, x$ are independent, i.e., they do not lie in a plane of dimension $\leq r-1$ (cf. Lemma 1.6). In other words, condition C says that x does not lie in the $(r-1)$-plane $\Pi(c_0, c_1, \ldots, c_{r-1})$ in which the independent points $c_0, c_1, \ldots, c_{r-1}$ lie. Now, as r and the c_ρ vary subject to the above conditions, there are only a finite number of these $(r-1)$-planes. But since $r-1 < n$ a finite number of $(r-1)$-planes does not cover a ball in E^n (see the Notes).[5] Therefore there exists a point x satisfying condition C and (1.8). Taking such a point for a'_N finishes the induction proof of our theorem.

1.12. LEMMA. *Let* $a_0, a_1, \ldots, a_r$ *be* $r+1$ *independent points in* E^n. *We assert:* (i) *the points*

$$x = \sum_{i=0}^{r} p_i a_i \tag{1.9}$$

vary over an r-plane Π^r as the real numbers i vary subject to the condition

$$\sum_{i=0}^{r} p_i = 1; \tag{1.10}$$

(ii) *the p_i in the representation (1.9) are uniquely determined.*

PROOF. (i)

$$p_0 = 1 - \sum_{i=1}^{r} p_i \quad \text{by (1.10).}$$

Hence by (1.9)

$$x = a_0 + \sum_{i=1}^{r} p_i(a_i - a_0). \tag{1.11}$$

Here the sum varies over the vector space whose base is the $a_i - a_0$, $a_2 - a_0, \ldots, a_r - a_0$ as the real numbers $p_1, p_2, \ldots, p_r$ vary arbitrarily. This proves assertion (i) (cf. 1.5).

(ii) The $a_i - a_0$ $(i = 1, 2, \ldots, r)$ are linearly independent. Therefore $p_1, p_2, \ldots, p_r$ are uniquely determined by (1.11). But then p_0 is uniquely determined by (1.10).

1.13. DEFINITION. Let $a_0, a_1, \ldots, a_r$ be as in subsection 1.12, and let Π^r be the r-plane determined by (1.9) and (1.10). The subset of Π^r consisting of those points for which

$$p_i \geq 0, \qquad i = 0, 1, \ldots, r, \tag{1.12}$$

is called an r-simplex and will be denoted by $|\sigma^r| = |a_0, a_1, \ldots, a_r|$. The points $a_0, a_1, \ldots, a_r$ are called the vertices of $|\sigma^r|$. The simplex $|\sigma^r|$ is a closed set. The open set $\sigma^r - \partial\sigma^\theta$ is called an open simplex. For $r \geq 1$ and $\rho = 0, 1, \ldots, r$, the r points $a_0, \ldots, a_{\rho-1}, a_{\rho+1}, \ldots, a_r$ are vertices of an $(r-1)$-simplex $|\sigma|_\rho^{r-1}$ called the ρth $(r-1)$-face of $|\sigma|^r$. Correspondingly if we omit two of the vertices of $|\sigma|^r$ we obtain an $(r-2)$-simplex called an $(r-2)$-face of $|\sigma|^r$, and so on. Note that $\partial|\sigma|^{r+1}$ is the union of the $(r-1)$-simplices $|\sigma|_\rho^{r-1}$, $\rho = 0, 1, \ldots, r$.

1.14. LEMMA. *The simplex*

$$|\sigma^r| = |a_0, a_1, \ldots, a_r| \tag{1.13}$$

is the closed convex hull (see §§1.28, 1.23) of the set V of its vertices.

For a proof see the Notes.[6]

1.15. By Definition 1.13 the simplex (1.13) is an r-dimensional set in E^n and

$$|\sigma^r| = |a_{i_0}, a_{i_1}, \ldots, a_{i_r}|, \tag{1.14}$$

where $i_0, i_1, \ldots, i_r$ is a permutation of the numbers $0, 1, \ldots, r$. We now define as follows the notion of an oriented simplex in which the order of the vertices plays a role.

1.16. *Orientation of a simplex.* Let $a_0, a_1, \ldots, a_r$ be $r+1$ independent points in E^n where $r \leq n$. We consider first the case $r \geq 1$. Then let

$$i_0, i_1, \ldots, i_r \tag{1.15}$$

be a fixed permutation of $0, 1, \ldots, r$. Then the oriented simplex

$$\sigma^r = (a_{i_0}, a_{i_1}, \ldots, a_{i_r}) \tag{1.16}$$

is the point set (1.14) (or what is the same (1.13)) together with the ordering of the vertices given by the permutation (1.15) or an even permutation thereof. Thus $(a_{j_0}, a_{j_1}, \ldots, a_{j_r}) = (a_{i_0}, a_{i_1}, \ldots, a_{i_r})$ if $j_0, j_1, \ldots, j_r$ is an even permutation of $i_0, i_1, \ldots, i_r$. In contrast the simplex $|\sigma^r|$ defined in subsection 1.13 is sometimes referred to as an absolute simplex. Every absolute simplex $|\sigma^r|$ gives rise to two oriented simplices. If $|\tau^r| = |\sigma^r|$ but σ^r and τ^r have different orientation, we write $\tau^r = -\sigma^r$. Now let $r = 0$. Then $|\sigma^0|$ consists of a single point a_0. To introduce an orientation also in this case, we associate the symbols $+$ and $-$ to a_0 and call $+a_0$ and $-a_0$ the positive and negative orientation of $|\sigma^0|$. We make the convention $-(+a_0) = -a_0, -(-a_0) = +a_0$.

For $r \geq 1$ the oriented $(r-1)$-faces σ_ρ^{r-1} $(\rho = 0, 1, \ldots, r)$ of the oriented simplex $\sigma^r = (a_0, a_1, \ldots, a_r)$ are defined as follows: if $r \geq 2$ then $\sigma_\rho^{r-1} = (-1)^\rho(a_0, a_1, \ldots, a_{\rho-1}, a_{\rho+1}, \ldots, a_r)$. If $r = 1$, i.e., $\sigma^r = (a_0, a_1)$, we set $\sigma_0^0 = +a_1$ and $\sigma_1^0 = -a_0$.

1.17. DEFINITION. Let E^r and E^p be linear spaces of finite dimension r and p resp. An affine map $A: E^r \to E^p$ is a map of the form $A(x) = c + l(x)$, where c is a point of E^p and l a linear map $E^r \to E^p$ called the linear part of A. A is called nonsingular if l is nonsingular. By determinant A we mean determinant l. (Note that c and l are uniquely determined by A.)

1.18. LEMMA. *Let $a_0, a_1, \ldots, a_r$ be independent points in E^n. Let the plane $\Pi^r \in E$ and the numbers $p_0, p_1, \ldots, p_r$ be as in Lemma 1.12. Let A be an affine map with domain E^n. Then for $x \in \Pi^r$*

$$A(x) = \sum_{i=0}^{r} A(a_i)p_i. \tag{1.16}$$

PROOF. By Definition 1.17, by (1.9) and (1.10)

$$A(x) = c \sum_{i=0}^{r} p_i + l\left(\sum_{i=0}^{r} a_i p_i\right) = \sum_{i=0}^{r} (c + l(a_i))p_i = \sum_{i=0}^{r} A(a_i)p_i.$$

1.19. Let E^n and E_1^n be linear spaces of dimension n. Let $a_0, a_1, \ldots, a_n$ be independent points in E^n and let $b_0, b_1, \ldots, b_n$ be arbitrary, not necessarily different points in E_1^n. Then

(a) there exists an affine map $A\colon E^n \to E_1^n$ such that

$$A(a_i) = b_i, \qquad i = 0, 1, \ldots, n; \tag{1.17}$$

(b) the affine map A is uniquely determined by the conditions (1.17).

The proof will be given in the Notes.[7]

1.20. In addition to the assumptions of subsection 1.19 we suppose that the points $b_0, b_1, \ldots, b_n$ are independent. Then the affine map A determined by (1.17) is nonsingular and maps the simplex $|\sigma^n| = |a_0, a_1, \ldots, a_n|$ onto the simplex $|\sigma^n| = |b_0, b_1, \ldots, b_n|$.

For proof see the Notes.[8]

1.21. DEFINITION. Let $\sigma^n = (a_0, a_1, \ldots, a_n)$ and $\tau^n = (b_0, b_1, \ldots, b_n)$ be two n-simplices in E^n. Let A be the uniquely determined affine map which satisfies (1.17). We then say that σ^n and τ^n have the same or opposite orientation according to whether $\det A$ is positive or negative. (For the definition of $\det A$ see subsection 1.17; note that $\det A \neq 0$ since, by subsection 1.20, A is nonsingular.)

1.22. It follows from Definition 1.21 that a single oriented n-simplex σ^n in E^n determines what is the same orientation for every n-simplex $|\tau^n|$ of E^n. We therefore say: σ^n induces an orientation in E^n. If, in particular, $e_1, e_2, \ldots, e_n$ is a base (in this order) for E^n, then $\sigma^n = (\theta, e_1, e_2, \ldots, e_n)$ is an oriented simplex. Thus an ordered base in E^n determines an orientation of E^n, and we see from Definition 1.21 that if $e_1', e_2', \ldots, e_n'$ is another base for E^n, then the two bases determine the same orientation if and only if the unique linear map l which maps e_i into e_i', $i = 1, 2, \ldots, n$, has a positive determinant. If an orientation for E^n is given, then an oriented simplex τ^n is called positive or negative according to whether the orientation of E^n induced by τ^n agrees with the given one or not, and the terms positive and negative basis are defined correspondingly.

1.23. Let $A(x) = c + l(x)$ be a nonsingular affine map $E^n \to E^n$, let y_0 be an arbitrary point in E^n, let $x_0 \in E^n$ be the unique solution of $A(x) = y_0$, and let Ω be an open bounded domain in E^n containing y_0. Then, by Definition 2.2 in Chapter 2, $d(A, \Omega, y_0) = j(DA(x_0; h))$. But $DA(x_0; h) = l(h)$. Therefore

$d(A, \Omega, y_0)$ equals $j(l)$. Thus the degree $d(A, \Omega, y_0)$ is seen to be independent of y_0 and Ω (provided that $x_0 = A^{-1}(y_0) \in \Omega$) and will be denoted by $j(A)$. Then $j(A) = j(l)$.

1.24. DEFINITION. Let Π^{n-1} be an $(n-1)$-plane in E^n and let $x_1, x_2, \ldots,$ x_n be coordinates in E^n for which Π^{n-1} is given by $x_n = c$, a constant. Then the two open sets $\{x = (x_1, \ldots, x_n) \in E \mid x_n > c\}$ and $\{x = (x_1, \ldots, x_n) \in E \mid x_n < c\}$ are called the halfspaces into which Π^{n-1} decomposes E^n (shortly, halfspaces with respect to Π^{n-1}).

1.25. Let $|\sigma^{n-1}| = |(a_0, a_1, \ldots, a_{n-1})|$ be an $(n-1)$-simplex in E^n. Then (i) the plane of smallest dimension containing $|\sigma|^{n-1}$ is an $(n-1)$-plane Π^{n-1}; (ii) let H_1 and H_2 be the two halfspaces of E^n with respect to Π^{n-1}, and let a_n and b_n be points in H_1 and H_2 resp. Then the simplices $\sigma_1^n = (a_0, a_1, \ldots, a_{n-1}, a_n)$ and $\sigma_2^n = (a_0, a_1, \ldots, a_{n-1}, b_n)$ have different orientation. (iii) If E^n is oriented, then one of these simplices is positive and the other is negative.

Since $a_0, a_1, \ldots, a_{n-1}$ are n independent points, assertion (i) is obvious from (1.3) with $b_i = a_i - a_0$ and $r = n - 1$. Assertion (iii) is an obvious consequence of (ii) and subsection 1.22. The proof of (ii) will be given in the Notes.[9]

1.26. If $|\sigma^n| = |(a_0, a_1, \ldots, a_n)|$ is an n-simplex in E^n, then the (point set-) boundary $\partial|\sigma^n|$ is obviously the union of the $(n-1)$-simplices

$$|\sigma_i^{n-1}| = |(a_0, a_{i-1}, a_{i+1}, \ldots, a_n)|, \qquad i = 0, 1, \ldots, n.$$

The $(n-1)$-face $\sigma_i^{n-1} = (a_0, a_1, \ldots, a_{i-1}, a_{i+1}, \ldots, a_n)$ is called a positive or negative face of the oriented simplex $\sigma^n = (a_0, a_1, \ldots, a_n)$ according to whether i is even or odd.

§2. The degree for mappings between finite-dimensional spaces of the same dimension

2.1. A Leray-Schauder map f in a Banach space E is a completely continuous perturbation of the identity map on this space. It therefore was defined only for maps whose domain and range lie in the same space. However, as noted in subsection 2.4 of Chapter 3, the Leray-Schauder maps are identical with continuous maps if E is finite dimensional. This fact makes it easy to extend the degree theory to maps between different spaces if these spaces have the same finite dimensions.

2.2. Let E_1^n and E_2^n be two oriented spaces of finite dimension n (see subsection 1.22). Let Ω be a bounded open set in E_1^n, and let f be a continuous map $\overline{\Omega} \to E_2^n$. We want to define a degree $d(f, \Omega, y_0)$ for every $y_0 \in E_2^n$ which satisfies the condition

$$y_0 \notin f(\partial\Omega) \tag{2.1}$$

To this end we consider positive bases (see subsection 1.22) $b_1, b_2, \ldots, b_n$ and $\beta_1, \beta_2, \ldots, \beta_n$ for E_1^n and E_2^n resp. Let h be the linear (and therefore continuous) map which maps the point $y = \sum_1^n \eta_i \beta_i$ of E_2^n into the point $x = \sum_1^n \eta_i b_i$ of E_1^n.

Then the map $f_1 = hf$ is continuous and maps $\bar{\Omega}$ into E_1^n. Moreover, since h is one-to-one, the relation (2.1) implies that

$$h(y_0) \notin hf(\partial\Omega). \tag{2.2}$$

Therefore the degree $d(hf, \Omega, h(y_0))$ is defined. Moreover it can be verified that this degree does not depend on the special choices of the positive bases $b_1, b_2, \ldots, b_n$ and $\beta_1, \beta_2, \ldots, \beta_n$ (see the Notes).[10] Thus the following definition is legitimate.

2.3. DEFINITION. With the notations and assumptions of subsection 2.2 we define

$$d(f, \Omega, y_0) = d(hf, \Omega, h(y_0)). \tag{2.3}$$

2.4. DEFINITION. Let E_1^n, E_2^n, Ω be as in subsection 2.2. Let $f \colon \bar{\Omega} \to E_2^n$ be a C' map, i.e., $f \in C'(\bar{\Omega})$. In analogy to §1.17 we define: the point $y_0 \in E_2^n$ is called a regular value for f if the equation

$$f(x) = y_0 \tag{2.4}$$

either has no solution or if for every solution x of (2.4) the linear map $l_x(h) = Df(x; h)$ is nonsingular (equivalently, if $\det l_x \neq 0$).

2.5. If the point y_0 of E_2^n is a regular value for f and satisfies (2.1), then the equation (1.4) has at most a finite number of solutions.

The proof is quite analogous to the one given for part (a) of Lemma 22 in Chapter 1 and will therefore be omitted.

2.6. LEMMA. *Let E_1^n, E_2^n, f, and h be as in Definition 2.3. Let $f_1 = hf$. Suppose $f \in C'(\bar{\Omega})$. Then* (i) $Df_1(x; \eta) = h(Df(x; \eta))$;
(ii) $\det Df_1(x; \eta) = \det h \cdot \det Df(x; h)$;
(iii) *if $\det Df_1(x; \eta) \neq 0$, then* sign $\det Df_1(x; \eta) =$ sign $\det Df(x; h)$;
(iv) y_0 *is a regular value for f if and only if $h(y_0)$ is a regular value for f_1.*

PROOF. (i) follows from the chain rule for differentials together with the fact that h is linear. (ii) is a consequence of (i), and (iii) follows from (ii) together with the fact that

$$\det h > 0. \tag{2.5}$$

Finally (iv) follows from (ii) and (2.5) together with the fact that x is a root of (2.4) if and only if x is a root of the equation

$$f_1(x) = h(y_0). \tag{2.6}$$

2.7. LEMMA. *Let the assumptions of Definition 2.3 be satisfied. Then* (i) *if the equation (2.4) has no roots, then $d(f, \Omega, y_0) = 0$;*
(ii) *if, in addition, $f \in C'(\bar{\Omega})$, if y_0 is a regular value for f, and if $x_1, x_2, \ldots, x_r$ are the roots of (2.4), then*

$$d(f, \Omega, y_0) = \sum_{\rho=1}^{r} \text{sign } \det Df(x_\rho; \cdot). \tag{2.7}$$

PROOF. (i) If (2.4) has no roots, then (2.6) has no roots. Consequently the right member of (2.3) equals zero.

(ii) As above let $f_1 = hf$. Then by Definition 2.3 and by (2.3) in Chapter 2

$$d(f, \Omega, y_0) = d(hf, \Omega, h(y_0)) = \sum_{\rho=1}^{r} j(Df_1(x_\rho; \cdot)) = \sum_{\rho=1}^{r} \text{sign det } Df_1(x_\rho; \cdot).$$

But here the right member equals by assertion (iii) of Lemma 2.6 the right member of our assertion (2.7).

2.8. LEMMA. *The assertions* (ii)–(vii) *of Lemma 2.4 of Chapter 2 are true for the degree defined in subsection 2.3. Assertion* (i) *of that lemma is replaced by the following one: if h is as in subsection 2.2 and $l = h^{-1}$, then*

$$d(l, \Omega, y_0) = \begin{cases} 1 & \text{if } y_0 \in l(\Omega), \\ 0 & \text{if } y_0 \notin l(\overline{\Omega}). \end{cases}$$

The proof of this lemma is omitted since it follows trivially from Definition 2.3, from the corresponding assertions of Lemma 2.4 of Chapter 2, and from subsection 2.4 of Chapter 3.

The proof of the following theorem is also omitted since it is an obvious consequence of Definition 2.3 and subsections 2.4 of Chapter 3 and 1.9 of Chapter 4.

2.9. THEOREM. *Let E_1^n, E_2^n, and Ω be as in subsection 2.2. It is asserted:*

(a) *Let $f_t(x) = f(x, t)$ be a continuous map $\overline{\Omega} \times [0, 1] \to E_2^n$. Let y_0 be a point in E_2^n satisfying $y_0 \notin f_t(\partial\Omega)$ for all $t \in [0, 1]$. Then $d(f_t, \Omega, y_0)$ is independent of t.*

(b) *Let f be a continuous map $\overline{\Omega} \to E_2^n$. Let $y = y(t), 0 \le t \le 1$, be a continuous curve in E_2^n which does not intersect $f(\partial\Omega)$. Then $d(f, \Omega, y(t))$ is independent of t.*

(c) *Theorem of Rouché. Let f and g be continuous maps $\overline{\Omega} \to E_2^n$. Suppose that the point $y_0 \in E_2^n$ does not lie on $f(\partial\Omega)$. Suppose moreover that, for all $x \in \partial\Omega$,*

$$\|f(x) - g(x)\| < \|f(x) - y_0\|.$$

Then

$$d(g, \Omega, y_0) = d(f, \Omega, y_0).$$

We remark that the latter equality holds in particular if $f(x) = g(x)$ for $x \in \partial\Omega$. This allows us to define the following.

2.10. DEFINITION. Let f_0 be a continuous map $\partial\Omega \to E_2^n$. Let y_0 be a point in E_2^n not lying on $f_0(\partial\Omega)$. Then the winding number $u(f_0(\partial\Omega), y_0)$ is defined by

$$u(f_0(\partial\Omega), y_0) = d(f, \Omega, y_0),$$

where f is an arbitrary continuous extension of f_0 to $\overline{\Omega}$ (cf. subsection 1.5 in Chapter 4).

The next lemma will be used in the proof of the product theorem in subsection 2.12.

2.11.   LEMMA. *Let E_1^n, E_2^n, E_3^n be three oriented Banach spaces of finite dimension n. Let b_i, β_i, $i = 1, 2, \ldots, n$, and h be as in subsection 2.2. Let Ω_2 be a bounded open domain in E_2^n, and let ϕ be a continuous map $\overline{\Omega}_2$ into E_3^n. Then*

$$d(\phi h^{-1}, h\Omega_2, z_0) = d(\phi, \Omega_2, z_0) \quad \text{for } z_0 \in E_3 - \phi(\partial\Omega_2). \tag{2.8}$$

PROOF. Since the continuous map ϕ can be uniformly approximated by a map in $C'(\overline{\Omega}_2)$ we may, by part (a) of Lemma 2.9, assume that $\phi \in C'(\overline{\Omega})$, and by part (b) of that lemma together with the Sard-Smale theorem in subsection 4.1 of Chapter 2 and part (iv) of Theorem 2.6 we may assume that z_0 is a regular value for ϕ. Now $y \in \Omega_2$ is a root of

$$\phi(y) = z_0 \tag{2.9}$$

if and only if $x = h(y)$ satisfies

$$\phi h^{-1}(x) = z_0. \tag{2.10}$$

We consider first the case that (2.9) has no roots in Ω_2. Then (2.10) has no roots $x \in h(\Omega_2)$. This proves (2.8), both members of this equality being zero. Now suppose $y_1, y_2, \ldots, y_r$ are the roots of (2.9). Then $x_\rho = h(y_\rho)$, $\rho = 1, 2, \ldots, r$, are the roots of (2.10), and, by Lemma 2.7

$$d(\phi, \Omega_2, z_0) = \sum_{\rho=1}^{r} \text{sign det } D\phi(y_\rho; \cdot) \tag{2.11}$$

and

$$d(\phi h^{-1}, h\Omega_2, z_0) = \sum_{\rho=1}^{r} \text{sign det } D\phi h^{-1}(x_\rho; \cdot). \tag{2.12}$$

But by the chain rule and by the linearity of h^{-1}

$$\text{det } D\phi h^{-1}(x_\rho; \cdot) = \text{det } D\phi(y_\rho; \cdot) \cdot \text{det } h^{-1}.$$

From this and (2.5) we see that the right members of (2.11) and (2.12) are termwise equal.

2.12. THE PRODUCT THEOREM.   (A) *Let E_1^n, E_2^n, E_3^n be as in Lemma 2.11. Let Ω be a bounded open set in E_1^n, let f be a continuous map $\overline{\Omega} \to E_2^n$, and let g be a continuous map $f(\overline{\Omega}) \to E_3^n$. Let $\{C_\beta\}$ be the set of bounded components of $E_2^n - f(\partial\Omega)$, where $\beta \in B$, an index set. Finally, let z_0 be a point of E_3^n not lying on $gf(\partial\Omega)$. Then*

$$d(gf, \Omega, z_0) = \sum_{\beta} d(g, C_\beta, z_0) \cdot d(f, \Omega, y_\beta), \tag{2.13}$$

where y_β is an arbitrary point of C_β and where the summation is extended over a finite number of $\beta \in B$. (Note that all C_β occurring in (2.13) are contained in $f(\overline{\Omega})$ since $d(f, \Omega, y_\beta) = 0$ for $y_\beta \notin f(\overline{\Omega})$.)

(B) *Let $E_1^n, E_2^n, E_3^n, \Omega$ and f be as in (A). Let Y be an open bounded set in E_2^n for which (1.4) in Chapter 5 (and therefore (1.5) in Chapter 5) holds. Let g*

be a continuous map $\overline{Y} \to E_3^n$. Let z_0 be a point of E_3^n for which (1.1) and (1.6), both in Chapter 5, hold. Let $\{Y_\alpha\}$ be the set of components of $Y - f(\partial\Omega)$, where $\alpha \in A$, an index set. Then provided that the equation $gf(x) = z_0$ has solutions in Ω

$$d(gf, \Omega, z_0) = \sum_{\alpha \in A} d(g, Y_\alpha, z_0) \cdot d(f, \Omega, y_\alpha),$$

where $y_\alpha \in Y_\alpha$ and where only a finite number of terms in the sum are different from zero. (The degree at the left obviously equals zero if $gf(x) = z_0$ has no solutions.)

PROOF. (A) For $i = 1, 2, \ldots, n$ let b_i and β_i be as in subsection 2.11, and let $\gamma_i, \gamma_2, \ldots, \gamma_n$ be a positive base for E_3^n. Let h_1 be the linear map $E_2^n \to E_1^n$ which maps β_i into b_i, and h_2 the linear map $E_3^n \to E_2^n$ which maps γ_i into β_i. Then $h_1 h_2$ is the linear map $E_3^n \to E_1^n$ which maps $\gamma_i \to b_i$. Therefore by Definition 2.3

$$d(gf, \Omega, z_0) = d(h_1 h_2 gf, \Omega, h_1 h_2 z_0). \tag{2.14}$$

We define for $x \in \overline{\Omega}$ and $y \in h_1 f(\overline{\Omega})$

$$f_1(x) = h_1 f(x), \qquad g_1(y) = h_1 h_2 g h_1^{-1}(y), \qquad x_0 = h_1 h_2(z_0). \tag{2.15}$$

Then by (2.14)

$$d(gf, \Omega, z_0) = d(g_1 f_1, \Omega, x_0). \tag{2.16}$$

Since the maps f_1 and g_1 both take place in E_1^n, we may apply the product theorem in subsection 1.2 of Chapter 5 to the right member of (2.16) to obtain

$$d(g_1 f_1, \Omega, x_0) = \sum_\beta d(g_1, C_\beta', x_0) \cdot d(f_1, \Omega, y_\beta'), \tag{2.17}$$

where the C_β' are bounded components of $E_1^n - f_1(\partial\Omega)$, where $y_\beta' \in C_\beta'$, and where the sum is a finite one. Now by (2.15) and Lemma 2.11

$$d(g_1, C_\beta', x_0) = d(h_1 h_2 g h^{-1}, C_\beta', x_0) = d(h_1 h_2 g, h_1^{-1} C_\beta', x_0). \tag{2.18}$$

But $h_1^{-1} C_\beta' = C_\beta$ is a bounded component of $E_2^n - f(\partial\Omega)$. Therefore by (2.18), (2.15), and Definition 2.3,

$$d(g_1, C_\beta', x_0) = d(h_1 h_2 g, C_\beta, h_1 h_2 z_0) = d(g, C_\beta, z_0). \tag{2.19}$$

Now

$$d(f_1, \Omega, y_\beta') = d(h_1 f, \Omega, h_1 h_1^{-1} y_\beta'),$$

and if $y_\beta' \in C_\beta'$ then $y_\beta = h_1^{-1} y_\beta' \in C_\beta$. Thus

$$d(f_1, \Omega, y_\beta') = d(h_1 f, \Omega, h_1 y_\beta) = d(f, \Omega, y_\beta). \tag{2.20}$$

The assertion (2.13) now follows from (2.16), (2.17), (1.19) and (2.20).

Assertion (B) follows from Theorem 1.4 of Chapter 5 in the same way assertion (A) followed from Theorem 1.2 of Chapter 5. Its proof is therefore omitted.

2.13. THEOREM. *Let Ω be a bounded open set in E_1^n, and let f be a continuous map $\overline{\Omega} \to E_2^n$ which is one-to-one. Then $d(f, \Omega, y) = \pm 1$ for $y \in f(\Omega)$.*

PROOF. The right member of Definition 2.3 equals ± 1 by Theorem 2.2 of Chapter 5.

2.14. THEOREM (INVARIANCE OF THE DOMAIN). *Under the assumptions of Theorem 2.13, $f(\Omega)$ is open and $f(\partial\Omega) = \partial f(\Omega)$.*

PROOF. The map $f_1 = h_1 f$ of Definition 2.3 has by Theorem 2.3 of Chapter 5 the properties asserted for f. This implies that $f = h_1^{-1} f_1$ has these properties.

2.15. THEOREM. *Let $E_1^n, E_2^n, E_3^n, \Omega, f, g, z_0$ be as in Theorem 2.12. It is then asserted:*

(i) *if Ω is connected, if f is one-to-one on $\overline{\Omega}$, and if $z_0 \in gf(\Omega)$, then*

$$d(gf, \Omega, z_0) = d(g, f(\Omega), z_0) \cdot d(f, \Omega, y) = \pm d(g, f(\Omega), z_0), \qquad (2.21)$$

where $y \in f(\Omega)$;

(ii) *let $g: f(\overline{\Omega}) \to gf(\overline{\Omega})$ be one-to-one, let $z_0 \in g(f(\Omega))$, and let y_0 be the unique root of $g(y) = z_0$. Then*

$$d(gf, \Omega, z_0) = \pm d(f, \Omega, y_0). \qquad (2.22)$$

If, in particular,

$$y_0 \notin C_\infty, \qquad (2.23)$$

the unbounded component of $E_2^n - f(\partial\Omega)$, then

$$d(gf, \Omega, z_0) = d(g, C_{\beta_0}, z_0) \cdot d(f, \Omega, y_0) \qquad (2.24)$$

for some unique $\beta_0 \in B$. Finally

$$d(f, \Omega, y_0) = 0 \qquad (2.25)$$

if

$$y_0 \in \left(\bigcup_{\beta \in B - \beta_0} C_\beta \right) \cup C_\infty \qquad (2.26)$$

PROOF. We use the definitions (2.15) and formula (2.16). To the right member of the latter we apply part (i) of subsection 2.5 in Chapter 5. Using Definition 2.3, Lemma 2.11, and (2.20) we obtain assertion (i).

Assertion (ii) is proved similarly by reduction to part (ii) of Theorem 2.5 in Chapter 5.

§3. Simplicial mappings

3.1. Again let Ω be a bounded open set in E_1^n, f a continuous map $\overline{\Omega} \to E_2^n$, and y_0 a point of E_2^n for which the degree $d(f, \Omega, y_0)$ is defined. Formerly we approximated the continuous f by a differentiable map f_1 which has the following two properties: (i) $d(f_1, \Omega, y_0) = d(f, \Omega, y_0)$, and (ii) if y_0 is a regular value for f_1, then the equation $f_1(x) = y_0$ has at most a finite number of solutions (see §1.22). There is another class of mappings which has these two properties, the simplicial ones defined below. That these enjoy property (ii) (with the proper definition of "regular value") will be shown in the present section. For the proof of property (i) given in §6.5 we will need "subdivisions" treated in §6.4. Historically the approach to finite-dimensional degree theory via simplicial maps was the first one used (see [8, 2]).

3.2. DEFINITION. Let E^n be a Banach space of finite dimension n, and let ν be a nonnegative integer $\leq n$. A subset $|K|$ of E^n is called a simplicial set of dimension ν if it is the union of a finite number of ν-simplices (see Definition 1.13) subject to the condition that for $\nu \geq 1$ the intersection of two of these simplices is either empty or a common $(\nu - 1)$-face of the two simplices. If E^n is oriented and $\nu = n$, we orient the simplicial set $|K^n|$ by giving each of its n-simplices the positive orientation (see subsection 1.22). The thus oriented simplicial set will be denoted by K^n.

3.3. LEMMA. *Let $|K^n|$ be a simplicial set of dimension n in E^n. Then the union of the ν-faces of $|K^n|$, $\nu \leq n$, form a simplical set of dimension ν, called $|K^\nu|$. For a proof see the Notes.*[11]

3.4. Let E_1^n be oriented, and let $|K^n|$ be a simplicial set in E_1^n of dimension n. Since by definition each simplex is closed, it follows that $|K^n|$ is closed and that $|\overset{\circ}{K}{}^n| = |K^n| - \partial|K^n|$ is open and bounded. Consequently if E_2^n is an n-dimensional oriented Banach space and if the map $f : |K^n| \to E_2^n$ is continuous, then by Definition 2.3 the degree $d(f, |\overset{\circ}{K}{}^n|, y_0)$ is defined for any $y_0 \in E_2^n$ provided that

$$y_0 \notin f(\partial|K^n|). \tag{3.1}$$

3.5. DEFINITION. (a) Let E_1^n, $|K_1^n|$, and E_2^n be as in subsection 3.4. Then a continuous map $f : |K_1^n| \to E_2^n$ is called simplicial (with respect to $|K_1^n|$) if for each simplex $|\sigma^n|$ of $|K_1^n|$ the restriction of f to $|\sigma^n|$ equals the restriction $A\sigma^n$ of an affine map $A : E_1^n \to E_2^n$ to $|\sigma^n|$. The simplicial map f is called nondegenerate if all these affine maps are nonsingular (cf. subsection 1.17). (b) The corresponding definition holds if we deal with spaces $E_1^{n_1}$, $E_2^{n_2}$, whose respective dimensions n_1, n_2 are not necessarily equal. (c) Let $|K_1^{n_1}|$ and $|K_2^{n_2}|$ be simplicial sets in $E_1^{n_1}$ and $E_2^{n_2}$ resp. Then a map $f : |K_1^{n_1}| \to |K_2^{n_2}|$ is called simplicial with respect to $(K_1^{n_1}, K_2^{n_2})$ if it has the following two properties: (i) if $v_0, v_1, \ldots, v_{n_1}$ are the vertices of a simplex $\sigma_1^{n_1}$ of K^{n_1}, then the points $w_0 = f(v_0)$, $w_1 = f(v_1), \ldots, w_{n_1} = f(v_{n_1})$ are vertices of a simplex $\sigma_2^{n_2}$ of $K_2^{n_2}$, in other words, the points $w_0, w_1, \ldots, w_{n_1}$ are vertices of a ν-face of $\sigma_2^{n_2}$ ($\nu \leq \min n_1, n_2$); (ii) the restriction of f to a simplex $\sigma_1^{n_1}$ of K^{n_1} is the restriction of the unique affine map mapping $v_i \to w_i$, $i = 0, 1, \ldots, n_1$, (cf. subsection 1.19). Obviously f is continuous.

3.6. DEFINITION. Let $\sigma_1^n, \sigma_2^n, \ldots, \sigma_r^n$ be the simplices whose union is $|K^n| \subset E_1^n$. Let f be a continuous map $E_1^n \to E_2^n$. Then a point $y_0 \in E_2^n$ is called a regular value for f (with respect to $|K^n|$) if the equation

$$f(x) = y_0 \tag{3.2}$$

has no roots lying on $\partial \sigma_\rho^n$ for $\rho = 1, 2, \ldots, r$.

3.7. With the above notations we assume that f is a simplicial nondegenerate map $|K^n| \to E_2^n$, that the simplices σ_ρ^n, $\rho = 1, 2, \ldots, r$, are positive, and that y_0 is a regular value for f (with respect to $|K^n|$). Then

(i) each solution x of (3.2) lies in the interior of one of the simplices $|\sigma_1^n|, |\sigma_2^n|, \ldots, |\sigma_r^n|$;

(ii) each $|\sigma_\rho^n|$, $\rho = 1, 2, \ldots, r$, contains at most one solution of (3.2);

(iii) the number s of solutions of (3.2) is at most r.

(iv) Suppose $s > 0$. Denoting the roots of (3.2) by $x_1, x_2, \ldots, x_s$, we may assume that $x_\rho \in \sigma_\rho^n$, $\rho = 1, 2, \ldots, s$. Let p be the number of those $\rho \leq s$ for which $f(\sigma_\rho^n)$ is a positive simplex in E_2^n, and let q be the number of those ρ for which $f(\sigma_\rho^n)$ is a negative simplex of E_2^n.

Then

$$d(f, |\overset{\circ}{K}{}^n|, y_0) = p - q. \tag{3.3}$$

PROOF. (i) follows directly from the definition of regularity of y_0; (ii) follows from the assumption that f is not degenerate and that therefore the affine map A_ρ, which is the restriction of f to σ_ρ^n, is nonsingular. (iii) follows from (ii). To prove (iv) we note first that (2.7) holds with $\Omega = |K^n| - \partial |K^n|$ and with r replaced by s. Now for $x \in \sigma_\rho^n$, $\det Df(x; \cdot) = \det A_\rho$. But for $\rho = 1, 2, \ldots, s$, $\det A_\rho$ equals $+1$ if $f(\sigma_\rho^n) = A_\rho(\sigma_\rho^n)$ is a positive simplex, and -1 if that simplex is a negative one.

We finish this section by some remarks and definitions which will be useful later on.

3.8. Let E^m and E^{n+1} be Banach spaces of finite dimension m and $n + 1$ resp. with $n \geq m$. Let $|K^m|$ be a simplicial set in E^m and σ^{n+1} be an $(n + 1)$-simplex in E^{n+1}. Suppose that to every vertex v of K^m is assigned a vertex $w = \phi(v)$ of σ^{n+1}. Then there exists a unique map $\bar{\phi} \colon K^m \to \partial \sigma^{n+1}$ which is simplicial with respect to (K^m, σ^{n+1}) (see part (c) of subsection 3.5), and which is an extension of ϕ. Indeed, let $v_0, v_1, \ldots, v_m$ be the vertices of an m-simplex σ_1^m of K^m, and let $w_0, w_1, \ldots, w_\nu$ be the different ones among the vertices $\phi(v_0), \phi(v_1), \ldots, \phi(v_m)$ of σ^{n+1}. Now $\nu \leq m \leq n$. Therefore the points $w_0, w_1, \ldots, w_\nu$ are the vertices of a ν-face σ^ν of $\partial \sigma^{n+1}$. Thus the assignment $v_i \to \phi(v_i)$ determines a unique affine map $\sigma_1^m \to \sigma^\nu \in \partial \sigma^{n+1}$ (cf. subsection 1.19). In this way we obtain the unique map having the required properties.

3.9. Let K^n be a simplicial set in E^n. We want to define a certain covering of K^n. If σ^n is an n-simplex of K^n and $u_0, u_1, \ldots, u_n$ are its vertices, then the $(n - 1)$-face of σ^n opposite to the vertex u_i is defined as the $(n - 1)$-simplex $(u_0, u_1, \ldots, u_{i-1}, u_{i+1}, \ldots, u_n)$. Now let $v_1, v_2, \ldots, v_N$ be the vertices of K^n. Then for $i = 1, 2, \ldots, N$ we denote by $\tilde{U}_i$ the union of those simplices of K^n having v_i as a vertex, and by U_i the set $\tilde{U}_i$ minus the $(n-1)$-faces opposite to v_i

of the n-simplices of $\tilde{U}_i$. The set U_i is called the (open) star of v_i and denoted by $\mathrm{st}(v_i, K^n)$. U_i is open (with respect to $|K^n|$). Thus the U_i, $i = 1, 2, \ldots, N$, form an open finite covering of the compact set $|K^n|$. It is well known that this implies the existence of a positive δ such that each subset S of $|K^n|$ of diameter less than δ is contained in one of the sets U_i (see, e.g., [2, p. 100, Satz IV]).

3.10. If σ^{n+1} is an $(n+1)$-simplex in E^{n+1}, we define a covering of $\partial\sigma^{n+1}$ in a way similar to the one used in the preceding section: let $v_0, v_1, \ldots, v_{n+1}$ be the vertices of σ^{n+1}. Then for $i = 0, 1, \ldots, n+1$ let $\tilde{U}_i$ be the union of all n-simplices of $\partial\sigma^{n+1}$ which have v_i as vertex. The set $U_i = \mathrm{st}(v_i, \sigma^{n+1})$ is then defined as the set $\tilde{U}_i$ minus the $(n-1)$-faces opposite to v_i of the n-simplices belonging to $\tilde{U}_i$. Again the sets U_i, $i = 0, 1, \ldots, n+1$, form a finite covering of $\partial\sigma^{n+1}$ which is open with respect to $\partial\sigma^{n+1}$.

§4. On subdivisions

4.1. DEFINITION. Let E^n be a Banach space of finite dimension n. If $\sigma^\nu = (a_0, a_1, \ldots, a_\nu)$, $0 \leq \nu \leq n$, is a ν-simplex in E^n, then the point

$$\beta = \frac{1}{\nu + 1} \sum_{i=0}^{\nu} a_i \qquad (4.1)$$

is called the barycenter of σ^ν. If K^n is a simplicial set of dimension n in E^n and if K^ν is the simplicial set consisting of the ν-faces of the n-simplices of K^n $(1 \leq \nu \leq n)$ (see subsection 3.3), we will, by induction on ν, define the barycentric subdivision BK^ν of K^ν: $B|K^0| = |K^0|$ ($|K^0|$ is the set of vertices of $|K^n|$). Now suppose $BK^{\nu-1}$ has been defined for some integer ν in the interval $[1, n]$. Then let σ^ν be a ν-simplex of K^ν, and let $\sigma^{\nu-1}$ be a $(\nu-1)$-face of σ^ν. Then by the induction assumption $B\sigma^{\nu-1}$ is defined. Now let $b_0, b_1, \ldots, b_{\nu-1}$ be one of the $(\nu-1)$-simplices of $B\sigma^{\nu-1}$. Now if β is the barycenter of σ^ν, the ν-simplex $\beta, b_0, \ldots, b_{\nu-1}$ is defined as a ν-simplex of BK^ν, and every ν-simplex of BK^ν is defined in this way. Thus BK^n is a finite union of n-simplices of E^n. If E^n is oriented, we give each n-simplex of BK^n the positive orientation.

4.2. LEMMA. *BK^n is a simplicial set as will be verified in the Notes.*[12]

4.3. DEFINITION. Using Definition 4.1 and Lemma 4.2 we define for every nonnegative integer r the rth barycentric subdivision $B^r K^n$ of the simplicial set K^n by setting $B^0 K^n = K^n$, $B^1 K^n = BK^n$, and for $r \geq 2$, $B^r K^n = B^1(B^{r-1} K^n)$. If E^n is oriented, we again orient $B^r K^n$ by giving each of its n-simplices the positive orientation.

4.4. LEMMA. *Let $|\sigma^n|$ be an n-simplex in E^n, and let $|\sigma_r^n|$ be a simplex of the rth barycentric subdivision of $|\sigma^n|$.*
Then

$$\delta(|\sigma_r^n|) \leq \delta(|\sigma^n|)(n/(n+1))^r, \qquad (4.2)$$

where for any bounded set $S \subset E^n$

$$\delta(S) = \sup_{x', x'' \in S} |x' - x''| = \text{diameter of } S.$$

For a proof see the Notes.[13]

4.5. LEMMA. *Let K^n be a simplicial set in E^n, and let ε be a positive number. Then there exists an integer $r_1 \geq 1$ such that*

$$\delta(|\sigma_1^n|) < \varepsilon \tag{4.3}$$

for every n-simplex $|\sigma_1^n|$ of the simplicial set $|K_1^n| = B^{r_1}|K^n|$.

PROOF. Since $|K^n|$ contains only a finite number of simplices and since $[n(n+1)^{-1}]^r$ tends to zero as $r \to \infty$, the lemma is an immediate consequence of Lemma 4.4.

4.6. Simplicial decomposition of a cube. Let $x_1, x_2, \ldots, x_n$ be the coordinates of the point $x \in E^n$ (with respect to a given base). Let $a_1, a_2, \ldots, a_n$ be the coordinates of a given point $a \in E^n$, and let l be a positive number. Then the cube $Q^n = Q^n(a, 2l)$ of center a and side length $2l$ is defined as the set of those $x \in E^n$ whose coordinates satisfy the inequalities

$$|x_i - a_i| \leq l, \qquad i = 1, 2, \ldots, n. \tag{4.4}$$

It is not hard to see that Q^n can be subdivided into n-simplices in such a way that the union of these simplices is a simplicial set K^n which as a point set equals Q^n (see the Notes).[14]

§5. Simplicial approximations

5.1. THEOREM. *Let $|K^n|$ be a simplicial set of dimension n in E_1^n. Let f be a continuous map $|K^n| \to E_2^n$, and let ε be a positive number. Then there exist a positive integer r_1 and a map $\phi \colon |K^n| \to E_2^n$ of the following properties:*

(a) *ϕ is simplical with respect to $|K_1^n| = B^{r_1}|K^n|$ (see Definition 3.5);*

(b) *if $u_1, u_2, \ldots, u_N$ are the vertices of $|K_1^n|$, then the points $w_1 = \phi(u_1)$, $w_2 = \phi(u_2), \ldots, w_N = \phi(u_N)$ of E_2^n are in general position;*

(c) *$\|f(x) - \phi(x)\| < \varepsilon$ for all $x \in |K^n|$.*

PROOF. The domain $|K^n|$ of the continuous map f is a closed and bounded subset of E_1^n. Consequently f is uniformly continuous on $|K^n|$, and there exists a $\gamma > 0$ such that

$$\|f(x') - f(x'')\| < \varepsilon/5 \tag{5.1}$$

for any couple of points x', x'' of $|K^n|$ for which

$$\|x' - x''\| < \gamma. \tag{5.2}$$

Now by Lemma 4.5 there exists a positive integer r_1 such that

$$\text{diameter } |\sigma_1| < \gamma \tag{5.3}$$

for every simplex $|\sigma_1|$ of $|K_1^n| = B^{r_1} K^n$. Now let

$$v_i = f(u_i), \qquad i = 1, 2, \ldots, N, \tag{5.4}$$

where as above the u_i are the vertices of $|K_1^n|$. By Theorem 1.11 there exist points $w_1, w_2, \ldots, w_N$ in E_2^n which are in general position and satisfy the inequalities

$$\|w_i - v_i\| < \varepsilon/5, \qquad i = 1, 2, \ldots, N. \tag{5.5}$$

We define

$$\phi(u_i) = w_i, \qquad i = 1, 2, \ldots, N. \tag{5.6}$$

Now if $|\sigma^n| = |a_0, a_1, \ldots, a_n|$ is an n-simplex of $|K_1^n|$, then the a_i are a subset of the u_i and the $\phi(a_i)$ are a subset of the w_i. Therefore the points $b_0 = \phi(a_0)$, $b_1 = \phi(a_1), \ldots, b_n = \phi(a_n)$ are in general position and are thus vertices of an n-simplex $|\tau_1^n|$ in E_2^n. For $x \in |\sigma_1^n|$ we define $\phi(x)$ as the unique affine map determined by

$$\phi(a_i) = b_i \tag{5.7}$$

(see subsection 1.20). Defining ϕ in this way for every n-simplex of $|K^n|$, we obtain a map having the properties (a) and (b) of our assertion. To prove assertion (c) let $|\phi_1^n|$ be as above and let x be an arbitrary point of $|\sigma_1^n|$. Then

$$\|f(x) - \phi(x)\| \le \|f(x) - f(a_0)\| + \|f(a_0) - \phi(a_0)\| + \|\phi(a_0) - \phi(x)\|. \tag{5.8}$$

Since x and a_0 lie in the simplex $|\sigma_1^n|$ whose diameter is, by (5.3), less than γ, it follows from (5.1) that $\|f(s) - f(a_0)\| < \varepsilon/5$. But also $\|f(a_0) - \phi(a_0)\| < \varepsilon/5$ as follows from (5.5) since $f(a_0) = v_i$ for some i and $\phi(a_0) = w_i$ for the same i. We now see from (5.8) that assertion (c) will be proved once it is shown that for $x \in |\sigma_1^n|$

$$\|\phi(x) - \phi(a_0)\| < 3\varepsilon/5. \tag{5.9}$$

Now by Definition 1.13

$$x = \sum_{i=0}^{n} p_i a_i, \qquad \sum_{i=0}^{n} p_i = 1, \ p_i \ge 0, \tag{5.10}$$

and by Lemma 1.18

$$\phi(x) = \sum_{i=0}^{n} p_i \phi(a_i).$$

Therefore

$$\phi(x) - \phi(a_0) = \sum_{i=0}^{n} p_i(\phi(a_i) - \phi(a_0)). \tag{5.11}$$

But

$$\|\phi(a_i) - \phi(a_0)\| \le \|\phi(a_i) - f(a_i)\| + \|f(a_i) - f(a_0)\| + \|f(a_0) - \phi(a_0)\|. \tag{5.12}$$

Here the first and the last term of the right member are each less than $\varepsilon/5$ as is seen from (5.4)–(5.6), and the middle term has the same bound as seen from (5.1)–(5.3). The assertion (5.9) follows now from (5.11) and (5.12) in conjunction with the properties of the p_i stated in (5.10).

5.2. COROLLARY TO THEOREM 5.1. (i) *The map ϕ is nondegenerate in the sense of Definition 3.5.*

(ii) *If y_0, y_1 are two disjoint points of E_2^n, then assertion* (b) *of our theorem may be replaced by the assertion that the points $y_0, y_1, w_1, w_2, \ldots, w_N$ are in general position.*

The corollary is an obvious consequence of Theorem 5.1 and Theorem 1.11.

5.3. THEOREM. *Let $|K^n|$ and f be as in Theorem 5.1. Let y_0 be a point of E_2^n satisfying*

$$y_0 \notin f(\partial|K^n|). \tag{5.13}$$

Then there exist a positive integer r_1 and a map $\phi\colon |K^n| \to E_2^n$ which is a nondegenerate simplicial map with respect to $B^{r_1}|K^n|$ and for which

$$d(f, \overset{\circ}{K}{}^n, y_0) = d(\phi, \overset{\circ}{K}{}^n, y_0), \tag{5.14}$$

where $\overset{\circ}{K}{}^n = |K^n| - \partial|K^n|$.

PROOF. The assumption (5.13) implies the existence of a positive ε for which

$$\|y_0 - f(x)\| > \varepsilon \quad \text{for } x \in \partial|K^n|. \tag{5.15}$$

Now let ϕ be a map having the properties (a), (b), (c) of Theorem 5.1 with an ε satisfying (5.15). Then by property (c)

$$\|f(x) - \phi(x)\| < \|y_0 - f(x)\| \tag{5.16}$$

for $x \in \partial|K^n|$. But by the theorem of Rouché (see subsection 2.9) the inequality (5.16) implies the assertion (5.14).

§6. Notes to Chapter 6

Notes to §1.

1. *Proof of Remark 1.2.* Suppose, e.g., $i = 1$. Then for any numbers $x_1, x_2, \ldots, x_r$

$$\sum_{j=1}^{r} x_j(a_j - a_0) = \sum_{j=1}^{r} [x_j(a_j - a_1) - x_j(a_0 - a_1)]$$

$$= (a_0 - a_1) \sum_{j=1}^{r} (-x_j) + \sum_{j=2}^{r} x_j(a_j - a_1).$$

This identity shows that the r elements $a_j - a_0$, $j = 1, 2, \ldots, r$, are linearly dependent if and only if the r elements $a_j - a_1$, $j = 0, 2, \ldots, r$, are linearly dependent.

2. *Proof of Lemma 1.4.* Suppose the points a_i are independent, and suppose that (1.1) holds. Subtracting the second equality (1.1) multiplied by a_0 from the first one, we see that

$$\sum_{i=1}^{r} (a_i - a_0)\lambda_i = 0.$$

Since the $a_i - a_0$ are linearly independent, this implies that $\lambda_1 = \lambda_2 = \cdots = \lambda_r = 0$. Therefore by the second equality (1.1) also $\lambda_0 = 0$. Thus (1.2) is satisfied.

Suppose now that (1.1) implies (1.2). By Definition 1.1 we have to show that the $a_i - a_0$ are linearly independent or that the relation

$$\sum_{i=1}^{r} \mu_i(a_i - a_0) = 0, \qquad \mu_i \text{ real},\tag{1}$$

implies that

$$\mu_1 = \mu_2 = \cdots = \mu_r = 0.\tag{2}$$

Now (1) may be written as

$$a_0 \sum_{i=1}^{r}(-\mu_i) + \sum_{i=1}^{r} a_i\mu_i = 0.$$

Then both relations (1.2) are satisfied with

$$\lambda_0 = -\sum_{j=1}^{r} \mu_j, \qquad \lambda_i = \mu_i \text{ for } i = 1, 2, \ldots, r.\tag{3}$$

Then, by our assumption, (1.2) is satisfied with $\lambda_0, \lambda_1, \ldots, \lambda_r$ given by (3). But (1.2) together with (3) implies the assertion (2).

3. *Proof of Lemma* 1.6. We will show that the points $a_0, a_1, \ldots, a_r$ are dependent if and only if they lie in a plane of dimension less than r.

Suppose that $a_0, a_1, \ldots, a_r$ are dependent. Then by definition the points $a_1 - a_0, a_2 - a_0, \ldots, a_r - a_0$ are linearly dependent, i.e.,

$$\sum_{i=1}^{r} \mu_i(a_i - a_0) = 0\tag{4}$$

for some μ_i not all zero. Suppose, e.g., that $\mu_r \neq 0$. Then

$$a_r = a_0 + \sum_{i=1}^{r-1} \lambda_i(a_i - a_0), \qquad \lambda_i = -\frac{\mu_i}{\mu_0}.$$

This shows that the points $a_0, a_1, \ldots, a_r$ lie in the plane Π which is the translate by a_0 of the subspace of E^n spanned by the $r-1$ elements $a_1 - a_0, \ldots, a_{r-1} - a_0$. This subspace is of dimension $\leq r - 1 < r$. Thus Π is of dimension $< r$.

Now suppose $a_0, \ldots, a_r$ lie in a plane Π^k of dimension $k \leq r - 1$. Then there are only k among the elements $a_1 - a_0, \ldots, a_r - a_0$ which are linearly independent. Suppose they are $a_1 - a_0, \ldots, a_k - a_0$. Then

$$a_{k+1} - a_0 = \sum_{i=1}^{k} \mu_i(a_i - a_0)$$

for some μ_i. This implies that the elements $a_1 - a_0, a_2 - a_0, \ldots, a_r - a_0$ are linearly dependent. Consequently points $a_0, a_1, \ldots, a_r$ are dependent.

4. For the proof of (1.4) we will make use of the following well-known fact of linear algebra: if V^n is a vector space of dimension n and if $b_1, b_2, \ldots, b_r$ are arbitrary $r < n$ linearly independent elements of V^n, then there exist elements $b_{r+1}, \ldots, b_n$ of V^n such that the $b_1, b_2, \ldots, b_n$ form a base for E^n. Now let k_0 be an element of K, and let $x_1, \ldots, x_r, y_1, \ldots, y_s, z_1, \ldots, z_k$ be bases for Π^r, Π^s, and K resp. (see 1.5). Then the elements x of Π^r, y of Π^s, z of K may be written as

$$x = k_0 + \sum_{i=1}^{r} \alpha_i x_i, \qquad y = k_0 + \sum_{i=1}^{s} \beta_i y_i, \qquad z = k_0 + \sum_{i=1}^{k} \gamma_i z_i,$$

where the $\alpha_i, \beta_i, \gamma_i$ are real numbers. We consider first the special case that

$$s = k. \tag{5}$$

In this case $\Pi^s = K \subset \Pi^r = H$ and $r = h$. This together with (5) proves the assertion (1.4). Suppose now $s \neq k$. Then $s > k$. Since $K \subset \Pi^s$ we can by the remark made at the beginning of this proof choose $y_1, y_2, \ldots, y_s$ in such a way that $y_i = z_i$ for $i = 1, 2, \ldots, k$. Since $K \subset \Pi^r$ it is clear that with this choice the $r + s - k$ elements

$$x_1, x_2, \ldots, x_r, y_{k+1}, \ldots, y_s \tag{6}$$

span $H - K$. Thus we obtain the inequality

$$h \leq r + s - k \tag{7}$$

for the dimension h of H, and our assertion (1.4) will be proved once it is shown that the equality sign holds in (7), i.e., that the elements (6) are linearly independent. Now a linear relation

$$\sum_{i=1}^{r} \lambda_i x_i = \sum_{i=k+1}^{s} \mu_i y_i \tag{8}$$

implies that all $\mu_i = 0$. For otherwise the right member of (8), let's call it y, would be different from θ since the y_i are linearly independent. Now $y + k_0 \in \Pi^s \cap \Pi^r = K$ by (8). But this contradicts the definition of $y_{k+1}, y_{k+2}, \ldots, y_s$. Thus indeed $\mu_1 = \mu_2 = \cdots = \mu_r = 0$, and therefore the left member of (8) equals zero. But this implies that all $\lambda_i = 0$ since the x_i are linearly independent. We thus proved that (8) implies that all λ_i and $\mu_i = 0$, i.e., we proved the linear independence of the elements (6).

5. That the union of a finite number of $(n-1)$-planes in E^n does not contain an open ball of E^n is a consequence of Baire's theorem [18, p. 20], since obviously a single $(n-1)$-plane does not contain an open ball of E^n.

6. *Proof of Lemma* 1.14. We have to prove:
(a) $|\sigma^r|$ is a convex set;
(b) if Q is a convex set containing the points $a_0, a_1, \ldots, a_r$, then $|\sigma^r| \subset Q$.

(a) Let $x = \sum_0^r p_i a_i$ and $y = \sum_0^r q_i a_i$ be two points of $|\sigma^r|$. Then

$$\sum_{i=0}^r p_i = \sum_{i=0}^r q_i = 1, \tag{9}$$

and

$$p_i \geq 0, \quad q_i \geq 0, \qquad i = 0, 1, \ldots, r. \tag{10}$$

We have to show that

$$(1-t)x + ty \in |\sigma^r| \tag{11}$$

for all $t \in [0,1]$. Now

$$(1-t)x + ty = \sum_{i=0}^r r_i a_i, \tag{12}$$

where $r_i = (1-t)p_i + tq_i$, and therefore by (9)

$$\sum_{i=0}^r r_i = (1-t)\sum_{i=0}^r p_i + t\sum_{i=0}^r q_i = 1,$$

and, by (10), $r_i \geq 0$. This proves (11).

(b) We will prove by induction for $0 \leq k \leq r$ that if Q_k is a convex set containing the points $a_0, a_1, \ldots, a_k$, then

$$|\sigma^k| = |a_0, a_1, \ldots, a_k| \subset Q_k. \tag{13}$$

The assertion is trivial for $k = 0$ since $|\sigma^0| = a_0$ is contained in any set containing a_0. Suppose then the assertion (13) is true if k is replaced by $k-1$. We will prove it for k. Now the convex set Q_k certainly contains the points $a_0, a_1, \ldots, a_{k-1}$. Therefore by the induction assumption

$$|\sigma^{k-1}| = |a_0, a_1, \ldots, a_{k-1}| \subset Q_k. \tag{14}$$

We have to show that if $x \in |\sigma^k|$, then

$$x \in Q_k. \tag{15}$$

By (14) this is clear if $x \in |\sigma^{k-1}|$. Suppose now that

$$x = \sum_{i=0}^k p_i a_i \in |\sigma^k| - |\sigma^{k-1}|.$$

Then

$$p_k \neq 0, \qquad \sum_{i=0}^k p_i = 1, \qquad p_i \geq 0. \tag{16}$$

Consider first the case that $p_k = 1$. Then $p_0 = p_1 = \cdots = p_{k-1} = 0$ and $x = a_k \in Q_k$. Thus (15) is true in this case. Suppose $p_k \neq 1$. Then

$$x = \sum_{i=0}^{k-1} p_i a_i + p_k a_k = (1-p_k)\sum_{i=0}^{k-1} s_i a_i + p_k a_k, \tag{17}$$

where

$$s_i = p_i/(1 - p_k) \geq 0.$$

Therefore by (16) $\sum_{i=0}^{k-1} s_i = 1$. This shows that the point $x_0 = \sum_{i=0}^{k-1} s_i a_i$ lies in $|\sigma^{k-1}|$, and therefore by (14) in Q_k. Since $a_k \in Q_k$ the assertion (15) now follows from (17) since Q_k is convex.

7. *Proof of* 1.19. By our assumptions the plane Π^n spanned by $a_0, a_1, \ldots, a_n$ is the whole space E^n. Therefore every $x \in E^n$ has the unique representation (1.9) (with $r = n$) subject to the condition (1.10). For x given by (1.9) we set

$$A(x) = \sum_{i=0}^{n} b_i p_i. \tag{18}$$

Since the p_i are uniquely determined by x, (18) implies (1.17). We now verify that the A given by (18) is affine: from (1.9) and (1.10) (with $r = n$) we see that

$$x = a_0 + \sum_{i=1}^{n} (a_i - a_0) p_i, \tag{19}$$

and, taking into account (18), that

$$A(x) = b_0 + \sum_{i=1}^{n} (b_i - b_0) p_i. \tag{20}$$

Since the $a_i - a_0$ form a base for E^n, there exists a unique linear map $l \colon E^n \to E_1^n$ such that $l(a_i - a_0) = b_i - b_0$. Then

$$\sum_{i=1}^{n} (b_i - b_0) p_i = l \left(\sum_{i=1}^{n} (a_i - a_0) p_i \right), \tag{21}$$

and we see from (19)–(21) that $A(x) = b_0 - l(a_0) + l(x)$. Thus A is affine.

To prove the uniqueness assertion (b) let

$$A_1(x) = c_1 + l_1(x), \qquad c_1 = \text{const. and } l_1 \text{ linear}, \tag{22}$$

be an affine map $E^n \to E_1^n$ satisfying (1.17) with A replaced by A_1. If x is again given by (1.9) (with $r = n$), then

$$A_1(x) = c_1 + \sum_{i=1}^{n} p_i l_1(a_i) \tag{23}$$

and, in particular, $b_j = A_1(a_j) = c_1 + l_1(a_j)$ for $j = 0, 1, \ldots, n$. Multiplying by p_j, adding over j, and taking into account that $\sum p_j = 1$, we see that

$$\sum_{j=1}^{n} b_j p_j = c_1 + l_1 \left(\sum_{j=0}^{n} a_j p_j \right) = c_1 + l_1(x) = A_1(x).$$

Comparison with (18) shows that $A_1(x) = A(x)$ as asserted.

8. *Proof of* 1.20. By Definition 1.12 the point x given by (1.9) (subject to (1.10) and with $r = n$) lies in $|\sigma^n|$ if and only if $p_i \geq 0$. But the same condition is necessary and sufficient so that the point $A(x)$ given by (1.16) with $A(a_i) = b_i$ lies in the simplex $|\tau^n|$. That A is not singular follows from the fact (shown in the Notes[7]) that the linear part l of A maps $a_i - a_0$ into $b_i - b_0$ for $i = 1, 2, \ldots, n$.

9. *Proof of part* (ii) *of* 1.25. Every $x \in E^n$ has the unique representation

$$x = \sum_{i=0}^{n} p_i a_i, \qquad \sum_{i=0}^{n} p_i = 1. \tag{24}$$

Since H_1 and H_2 are convex, it is clear that two points of $E^n - \Pi^{n-1}$ given in the form (24) lie in the same or different halfspaces H_1 and H_2 according to whether the coefficients of a_n in (24) for these points have the same sign or not. If we assume $p_n > 0$ for $x = a_n$, then since a_n and b_n lie in different halfspaces we see that if

$$b_n = \sum_{i=0}^{n} a_i q_i, \qquad \sum_{i=0}^{n} q_i = 1, \tag{25}$$

then

$$q_n < 0. \tag{26}$$

Now to prove our assertion (ii) we have to show that the affine map A which maps the vertices of σ_1^n onto the vertices of σ_2^n (in the given order) has a negative determinant (cf. Definition 1.21). Now by (1.19) for x given by (24)

$$y = A(x) = \sum_{i=0}^{n-1} p_i a_i + p_n b_n,$$

or by (25)

$$y = A(x) = \sum_{i=0}^{n-1} (p_i + q_i p_n) a_i + p_n q_n a_n.$$

The linear part of A is then given by

$$y - a_0 = \sum_{i=1}^{n-1} (p_i + q_i p_n)(a_i - a_0) + p_n q_n (a_n - a_0). \tag{27}$$

Now $p_1, p_2, \ldots, p_n$ are the coordinates of $x - a_0$ with respect to the base $a_1 - a_0$, $a_2 - a_0, \ldots, a_n - a_0$. If we denote by $y_1, y_2, \ldots, y_n$ the coordinates of $y - a_0$ with respect to the same basis, we see from (27) that the matrix of the corresponding coordinate transformation has only zeros below the main diagonal, has ones in the first $n - 1$ elements of that diagonal, and q_n as its nth element. Its value is therefore q_n. By (26) this finishes the proof.

Notes to §2.

10. *Proof of the independence of* $d(hf, \Omega, h(y_0))$ *of the chosen base* (see subsection 2.2). In addition to the notations used in subsection 2.2, let $b_1', b_2', \ldots, b_n'$

and $\beta_1', \beta_2', \ldots, \beta_n'$ denote positive bases for E_1^n and E_2^n resp., and let h' be the linear map $E_2^n \to E_1^n$ which maps the point $y = \sum \eta_i' \beta_i' \in E_2^n$ into the point $x = \sum \eta_i' b_i' \in E_1^n$. We have to show that $d(h'f, \Omega, h'(y_0)) = d(hf, \Omega, h(y_0))$, or that

$$d(gf_1, \Omega, z_0) = d(f_1, \Omega, x_0), \qquad (28)$$

where

$$f_1 = hf, \quad g = h'h^{-1}, \quad x_0 = h(y_0), \quad z_0 = g(x_0). \qquad (29)$$

To prove (28) we use part (ii) of Corollary 2.5 in Chapter 5 with f replaced by f_1 and y_0 by x_0. With this replacement we see from (2.13) and (2.16), both in Chapter 5, that if x_0 lies in the unbounded component of $E_1^n - f_1(\partial\Omega)$, then (28) holds, both members being zero. But if x_0 lies in a bounded component C_{β_0} of $E_1^n - f_1(\partial\Omega)$, then by (2.15) in Chapter 5

$$d(gf_1, \Omega, z_0) = d(g, C_{\beta_0}, z_0) \cdot d(f_1, \Omega, x_0),$$

and for the proof of (28) it remains to show that $d(g, C_{\beta_0}, z_0) = +1$, i.e., by (29) that the determinant of the linear map $g = h'h^{-1}$ is positive. Now since $b_1, b_2, \ldots, b_n$ is a base for E_1^n, we may write

$$b_j' = \sum_{k=1}^{n} c_{jk} b_k, \qquad j = 1, 2, \ldots, n,$$

where the c_{jk} are real numbers. But since the b_i and the b_i' are both positive bases for E_1^n, it follows from subsection 1.22 that the determinant $|C|$ of the matrix C of the c_{ij} is positive. Similarly

$$\beta_j = \sum_{k=1}^{n} \gamma_{jk} \beta_k', \qquad j = 1, 2, \ldots, n,$$

where the determinant $|\Gamma|$ of the matrix Γ of the γ_{jk} is positive. Now if $\eta_1, \eta_2, \ldots, \eta_n$ are the coordinates of $x \in E_1^n$ with respect to the base $b_1, b_2, \ldots, b_n$ and if $\eta_1', \eta_2', \ldots, \eta_n'$ are the coordinates with respect to the same base of the point $x' = g(x) = h'h^{-1}(x)$, then direct computation shows that

$$\eta_j' = \sum_i \eta_i \sum_k \gamma_{ik} c_{kj} = \sum_i \eta_i (\Gamma C)_{ij}, \qquad j = 1, 2, \ldots, n, \qquad (30)$$

where $(\Gamma C)_{ij}$ denotes the term in the ith row and the jth column of the matrix ΓC. Therefore the determinant of the linear map $h'h^{-1}$ is positive, the determinants of Γ and C being positive.

Notes to §3.

11. *Proof of Lemma 3.3.* If $|\sigma^\nu| = |a_{i_0}, a_{i_1}, \ldots, a_{i_\nu}|$ is a ν-face of the simplex $|\sigma^n| = |a_0, a_1, \ldots, a_n|$, then the point

$$x = \sum_{i=0}^{n} p_i a_i, \qquad \sum_{i=0}^{n} p_i = 1, \; p_i \geq 0, \qquad (21)$$

of $|\sigma^n|$ belongs to $|\sigma^\nu|$ if and only if $p_i = 0$ for all indices i not contained in the set $i_0, i_1, \ldots, i_\nu$. Noting that the representation (21) of a point $x \in \sigma^n$ is unique (Lemma 1.12), we conclude that if $|\sigma^\mu| = |a_{i_0'}, a_{i_1'}, \ldots, a_{i_\mu'}|$ is a μ-face of $|\sigma^n|$, then the intersection $|\sigma^\nu \cap \sigma^\mu|$ is empty if the index sets $i_0, i_1, \ldots, i_\nu$ and $i_0', i_1', \ldots, i_\mu'$ have no point in common. If they have common elements, we may assume these are $i_0 = i_0', \ i_1 = i_1', \ldots, i_t = i_t'$ with $0 \leq t \leq \min(\mu, \nu)$. Then $|\sigma^\nu| \cap |\sigma^\mu|$ is the common face $|a_{i_0}, a_{i_1}, \ldots, a_{i_t}|$ of $|\sigma^\nu|$ and $|\sigma^\mu|$.

Notes to §4.

12. *Proof of Lemma 4.2.* By Definition 3.2 we have to show that the intersection of two n-simplices of BK^n is either empty or a common face of the two simplices. For this purpose we will prove the following assertion A_ν for $\nu = 0, 1, 2, \ldots, n$: if s_1^ν and s_2^ν are two ν-simplices of BK^n (i.e., ν-faces of n-simplices of BK^n), then either

A_ν^1: $s_1^\nu \cap s_2^\nu$ is empty, or

A_ν^2: $s_1^\nu \cap s_2^\nu$ is a common face of s_1^ν and s_2^ν.

A_0 is trivially true since s_1^0 and s_2^0 are vertices of K^n. Suppose then that $A_0, A_1, \ldots, A_{\nu-1}$ are true for some $\nu \leq n$. We want to prove A_ν. Now for $i = 1, 2$, the simplex s_i^ν is of the form

$$s_i^\nu = (\beta^i, b_0^i, b_1^i, \ldots, b_{\nu-1}^i), \tag{22}$$

where β^i is the barycenter of a ν-simplex σ_1^ν of K^n and where $(b_0^i, b_1^i, \ldots, b_{\nu-1}^i) = \sigma_i^{\nu-1}$ is a $(\nu-1)$-simplex of the barycentric subdivision of a $(\nu-1)$-face of σ^ν (i.e., of $\partial \sigma_i^\nu$). But by induction assumption either

$$\sigma_1^{\nu-1} \cap \sigma_2^{\nu-1} \quad \text{is empty,} \tag{23}$$

or

$$\sigma_1^{\nu-1} \cap \sigma_2^{\nu-1} = u^\mu \tag{24}$$

is a common μ-face of $\sigma_1^{\nu-1}$ and $\sigma_2^{\nu-1}$ with $0 \leq \mu \leq \nu - 1$. Now in case (23) we see that

$$\sigma_1^\nu \cap \sigma_2^\nu = \begin{cases} \text{the empty set} & \text{if } \beta' \neq \beta^2, \\ \beta' & \text{if } \beta' = \beta^2. \end{cases}$$

This proves the assertion A_ν in case (23). Now suppose (24) takes place. In this case we may assume the indexing to be such that

$$u^\mu = (b_0' = b_0^2, b_1' = b_1^2, \ldots, b_\mu' = b_\mu^2).$$

Then by (24)

$$s_1^\nu \cap s_2^\nu = \begin{cases} u^\mu & \text{if } \beta_1 \neq \beta_2, \\ (\beta_1, b_0', b_1', \ldots, b_\mu') & \text{if } \beta_1 = \beta_2. \end{cases}$$

13. The proof of Lemma 4.4 is based on the following.

LEMMA A. *Let* $|\sigma^n| = |a_0, a_1, \ldots, a_n|$. *Then*

$$\delta(|\sigma^n|) = \sup_{i,j=0,1,\ldots,n} \|a_i - a_j\|. \tag{25}$$

PROOF. Since $a_i \in |\sigma^n|$ for $i = 0, 1, \ldots, n$, it is clear that the right member of (25) is not greater than the left member. Thus it remains to prove

$$\delta(|\sigma^n|) \leq \sup_{i,j=0,1,\ldots,n} \|a_i - a_j\|. \tag{26}$$

Now if (26) were not true, there would exist a δ_0 such that for $i, j = 0, 1, \ldots, n$

$$\delta(|\sigma^n|) > \delta_0 > \|a_i - a_j\|. \tag{27}$$

To show that this inequality leads to a contradiction, we will show that it implies

$$\delta(|\sigma^n|) > \delta_0 > \|x - y\| \quad \text{for all } x, y \in |\sigma^n|, \tag{28}$$

which indeed leads to the contradiction

$$\delta(|\sigma^n|) > \delta_0 \geq \sup_{x,y \in |\sigma^n|} \|x - y\| = \delta(|\sigma^n|).$$

To prove that (27) implies (28), we note that the first parts of these inequalities are identical. It thus remains to show that (27) implies the second part of (28). To show this we choose an arbitrary but fixed i among the integers $0, 1, \ldots, n$. Then, by (27), $a_j \in B(a_i, \delta_0)$ for every $j = 0, 1, \ldots, n$. Thus the ball $B(a_i, \delta_0)$ is a convex set containing all vertices of $|\sigma^n|$. By Lemma 1.14 this ball contains $|\sigma^n|$. Thus $\|x - a_i\| < \delta_0$ for every $x \in |\sigma^n|$, or $a_i \in B(x, \delta_0)$ for each such x. But a_i was an arbitrary vertex of $|\sigma^n|$. Therefore for fixed $x \in |\sigma^n|$ the convex set $B(x, \delta_0)$ contains all vertices of $|\sigma^n|$. So, again by Lemma 1.14, $|\sigma^n| \in B(x, \delta_0)$. This obviously implies the second part of (28), and Lemma A is proved.

Turning to the proof of (4.2) we note that it will be sufficient to prove the assertion for $r = 1$ since the general case then follows by iteration. Then let $\sigma_1^n = (b_0, b_1, \ldots, b_n)$ be a simplex of BK^n. To prove (4.2) for $r = 1$ it will, by Lemma A, be sufficient to show that

$$\|b_i - b_j\| \leq \frac{n}{n+1} \delta(|\sigma^n|) \tag{29}$$

for $i, j = 0, 1, \ldots, n$, $i \neq j$. If as above $|\sigma^n| = |a_0, a_1, \ldots, a_n|$, it is no loss of generality to assume that

$$b_0 = a_0, \quad b_1 = (a_0 + a_1)/2, \ldots, \quad b_i = (a_0 + a_1 + \cdots + a_i)/(i+1), \ldots$$

Then for $i \neq j$

$$b_i - b_j = \frac{1}{i+1} \left[\sum_{\alpha=0}^{i} a_\alpha - (i+1) \frac{1}{j+1} \sum_{\beta=0}^{j} a_\beta \right]$$

$$= \frac{1}{i+1} \sum_{\alpha=0}^{i} \left[a_\alpha - \frac{1}{j+1} \sum_{\beta=0}^{j} a_\beta \right] = \frac{1}{i+1} \sum_{\alpha=0}^{i} \frac{1}{j+1} \sum_{\beta=0}^{j} (a_\alpha - a_\beta). \tag{30}$$

Now $i \neq j$ and we may suppose $j > i$. The double sum at the right member of (30) contains formally $(i+1)(j+1)$ terms and exactly $i+1$ of these equal zero. Thus the sum contains $(i+1)(j+1) - (i+1) = (i+1)j$ nonzero terms. Therefore we see from (30) and Lemma A that

$$\|b_i - b_j\| \leq \frac{(i+1)j}{(i+1)(j+1)} \delta(|\sigma^n|) = \frac{j}{j+1} \delta(|\sigma^n|). \tag{31}$$

Now $j/(j+1) \leq n/(n+1)$ since $j \leq n$. Therefore (31) implies the assertion (29).

14. To obtain a simplicial decomposition of Q^n (cf. subsection 4.6), we first define r-faces of Q^n for $r = n, n-1, \ldots, 1, 0$. The n-face of Q^n is defined as Q^n. To define $(n-1)$-faces we write (4.4) in the form

$$\left.\begin{array}{c} x_i - a_i \\ a_i - x_i \end{array}\right\} \leq l, \qquad i = 1, 2, \ldots, n. \tag{32}$$

Then a point x belongs to an $(n-1)$-face if and only if for its coordinates at least one of the inequalities (32) is replaced by equality: e.g., the set of x satisfying

$$x_n = a_n + l, \qquad \left.\begin{array}{c} x_i - a_i \leq l, \\ a_i - x_i \leq l, \end{array}\right\} \quad \text{for } i = 1, 2, \ldots, n-1,$$

is an $(n-1)$-face. It is an $(n-1)$-cube of side length $2l$ and center

$$(a_1, a_2, \ldots, a_{n-1}, a_n + l)$$

and lies in the plane $x_n = a_n + l$ of E^n. The number of $(n-1)$-faces is $2n$, and their union is obviously the boundary of Q^n. One obtains the $(n-2)$-faces by replacing exactly two of the inequalities (32) by equalities and so on. Each r-face is a cube in some r-plane of E^n and its boundaries consist of $(r-1)$-faces.

The desired subdivision of Q^n into simplices is constructed as follows by induction. The subdivision of a 0-face Q^0 (i.e., a vertex of Q^n) is Q^0. Suppose now that for some $r \leq n-1$ the r-faces of Q^n are subdivided into r-simplices, and let Q^{r+1} be an $(r+1)$-face of Q^n. The boundary of Q^{r+1} consists of a finite number of r-faces of Q^n, say $Q_1^r, Q_2^r, \ldots, Q_\nu^r$. Let v_0 be the center of Q^{r+1}. Then the simplices of the subdivision of Q^{r+1} are of the form $(v_0, v_1, \ldots, v_{r+1})$, where $(v_1, v_2, \ldots, v_{r+1})$ is a simplex of the simplicial subdivision of one of the cubes $Q_1^r, \ldots, Q_\nu^r$. In this way we obtain a decomposition of Q^{r+1} into simplices. The proof that the union of the simplices of the subdivision is a simplicial set will be omitted (cf. the corresponding proof in Note 12; cf. also [2, part 8 of Chapter III, §2]).

CHAPTER 7

On Spheres

Elementary properties and orientation of spheres.
Degree of mappings between spheres

1.1. DEFINITION. Let E^{n+1} be the $(n+1)$-dimensional Euclidean space whose points are the $(n+1)$-tuples $x = (x_1, x_2, \ldots, x_{n+1})$ of real numbers and whose norm $\|x\|$ is given by

$$\|x\|^2 = \sum_{i=1}^{n+1} x_i^2 = (x, x), \tag{1.1}$$

where (x, y) denotes the usual scalar product of the elements x, y of E^{n+1}. Let $c = (c_1, c_2, \ldots, c_{n+1})$ be a point of E^{n+1} and let r be a positive number. Then the n-sphere $S^n(c, r)$ of center c and radius r is defined by

$$S^n(c, r) = \partial B^{n+1}(c, r), \tag{1.2}$$

where as usual $B^{n+1}(c, r)$ denotes the (open) ball with center c and radius r. Equivalently,

$$S^n(c, r) = \left\{ x \in E^{n+1} \,\middle|\, \sum_{i=1}^{n+1} (x_i - c_i)^2 = r^2 \right\}. \tag{1.3}$$

The sphere $S^n(\theta, 1)$ is called the unit sphere in E^{n+1} and is often denoted by S^n. (Note that for $n \geq 1$ the sphere $S^n(c, r)$ is a connected set, while $S^0(c, r)$ consists of the points $c - r$, $c + r$ in E^1 with $r > 0$ and is thus disconnected.)

1.2. LEMMA. *If s is an integer $< n$, then the intersection of $S^n(c, r)$ with an $(s+1)$-plane Π^{s+1} is either empty or a single point or an s-sphere.*

For a proof see the Notes.[1]

1.3. If the $(s+1)$-plane Π^{s+1} of Lemma 1.2 contains the center c of $S^n(c, r)$, then the intersection $S^n(c, r) \cap \Pi^{s+1}$ (which by that lemma is an s-sphere) is called a "great sphere on $S^n(c, r)$." If $s = 1$, the great sphere is a great circle. If $s = 0$, then the 0-sphere $S^n(c, r) \cap \Pi^1$ consists of the two points in which the line Π^1 through c intersects $S^n(c, r)$. These two points are called antipodal to each other.

1.4. LEMMA. *Let N be a point of $S^n(c, r)$. Then the intersection of all great circles through N consists of N and its antipode S.[2]*

1.5. DEFINITION. Let x be a point of $S^n(c, r)$. Then the tangent plane T_x to $S^n(c, r)$ at x is defined as the set of all $t \in E^{n+1}$ for which $t - x$ is orthogonal to $x - c$. Obviously T_x is an n-plane containing x. The tangent space E_x^n at x is the subspace of E^{n+1} of which T_x is a translate, i.e., $y \in E_x^n$ if and only if $t = y + x \in T_x$.

1.6. *Stereographic projection.* Let n_0 and n_0^* be two antipodal points on $S^n(c, r)$. Let x be a point of $S^n(c, r)$ different from n_0, and let Π^1 be the 1-plane (straight line) determined by n_0 and x. Then the pair of these two points is the intersection $S^n(c, r) \cap \Pi^1$ (cf. Lemma 1.2), and Π^1 intersects the tangent plane $T_{n_0^*}$ in a single point $t = t(x)$ (as follows, e.g., from Lemma 1.8 of Chapter 6). Conversely, if t is a given point on $T_{n_0^*}$, then the straight line determined by n_0 and t intersects $S^n(c, r)$ in n_0 and in exactly one point $x \neq n_0$. We thus obtain a one-to-one map of $S^n(c, r) - \{n_0\}$ onto $T_{n_0^*}$. This map is called the stereographic projection of $S^n(c, r)$ with pole n_0. It is easily seen to be continuous (see the Notes).[3] The stereographic projection will be the basis for extending the notion of the degree for mappings between linear spaces to mappings between spheres.

1.7. *Orientation of spheres.* Let E^{n+1} be an oriented Euclidean $(n + 1)$-space. Orienting a sphere $S^n(c, r)$ in E^{n+1} means to give an orientation to the tangent space E_x^n to $S^{n+1}(c, r)$ for each x in that sphere. We define the orientation of E_x^n induced by the one of E^{n+1} as follows: let τ^n be an n-simplex in T_x^n with vertices $t_0 = x, t_1, t_2, \ldots, t_n$. Then the points $b_1 = t_1 - t_0$, $b_2 = t_2 - t_0, \ldots, b_n = t_n - t_0$ form a base for E_x^n, and the points $b_0 = c - t_0$, $b_1, \ldots, b_n$ form a base for E^{n+1}. We then call the base $b_1, b_2, \ldots, b_n$ a positive base for E_x^n if and only if $b_0, b_1, \ldots, b_n$ form a positive base for E^{n+1} (cf. subsection 1.22 in Chapter 6). This determines an orientation of E_x^n provided it can be shown that this definition does not depend on the particular choice of the points $t_1, t_2, \ldots, t_n$. This verification will be given in the Notes.[4] The orientation of T_x is defined as that of E_x^n. When it is necessary to distinguish between the oriented and unoriented T_x, we denote the former by $\vec{T}_x$. The orientation of $S^n(x, r)$ defined above will be referred to as its positive orientation. We conclude this section by the following remarks. (a) The simplex $(t_0, t_1, \ldots, t_n)$ is a positive simplex of T_x if the simplex $(t_0, c, t_1, \ldots, t_n)$ is a positive simplex of E^{n+1}. This follows from subsections 1.22 and 1.26 of Chapter 6. (b) In our definition for the orientation of T_x, the center c of $S^n(c, r) = \partial B^{n+1}(c, r)$ may be replaced by an arbitrary point of $B(c, r) - \{x\}$. This follows easily from (a) and subsection 1.25 of Chapter 6. (c) Let x_1 and x_2 be points of $S^n(c, r)$; let Ω_1 be a bounded open set in $\vec{T}_{x_1}$, and let f be a continuous map of $\overline{\Omega}_1 \to \vec{T}_{x_2}$. Let t_2 be a point in T_{x_2} satisfying $t_2 \notin f(\partial \Omega_1)$. Then the degree $d(f, \Omega_1, t_2)$ is defined as the degree of the corresponding map between the tangent *spaces* to $S^n(c, r)$ at x_1 and x_2 resp.

1.8. LEMMA. *Let E^{n+1} be an oriented $(n+1)$-Euclidean space. Let t_0^1 and t_0^2 be points on the sphere $S^n(c,r) \subset E^{n+1}$. Let $b_0^1 = c - t_0^1$ and $b_0^2 = c - t_0^2$. Let r_{12} be a rotation of E^{n+1} onto itself which preserves the orientation of E^{n+1} and for which*

$$r_{12}(b_0^1) = b_0^2.^5 \tag{1.4}$$

Then

$$r_{12}(\vec{E}_{t_0^1}) = \vec{E}_{t_0^2}, \tag{1.5}$$

where $E_{t_0^1}$ and $E_{t_0^2}$ are the oriented tangent spaces (see subsection 1.7) to $S^n(c,r)$ at the points t_0^1 and t_0^2 resp. Equivalently

$$r_{12}(\vec{T}_{t_0^1}) = \vec{T}_{t_0^2}, \tag{1.6}$$

where $T_{t_0^1}, T_{t_0^2}$ are the corresponding tangent planes.

For a proof see the Notes.[6]

1.9. LEMMA. *With the notation of Lemma 1.8, let $x_1 = t_0^1$ and $x_2 = t_0^2$. Let Ω_1 be an open bounded set in T_{x_1}. Then the degree*

$$d(r_{12}, \Omega_1, t_2) = +1 \tag{1.7}$$

for every $t_2 \in r_{12}(\Omega_1) \subset T_{x_2}$.

The proof is obvious from Lemma 1.8, from Definition 2.3 in Chapter 6 with $h = (r_{12})^{-1}$, and from remark (c) in subsection 1.7.

1.10. LEMMA. *For $i = 0,1$, let n_i be a point of $S^n(c,r)$, let s_i be the antipode of n_i, let T_i be the tangent plane at s_i, and let σ_i be the stereographic projection with pole n_i. Let U_0 be an open subset of $S^n(c,r)$ having a positive distance from n_0 and n_1 such that the subsets $\Omega_i = \sigma_i(U_0)$ of T_i are well defined. Let $\sigma_{01} = \sigma_1\sigma_0^{-1}$ and*

$$t_1 \in \sigma_1 U_0 = \Omega_1. \tag{1.8}$$

Then, assuming $n \geq 2$,

$$d((\sigma_{01})^{-1}, \Omega_1, t_0) = d(\sigma_{01}, \Omega_0, t_1) = +1, \tag{1.9}$$

where

$$t_0 \in \Omega_0 = \sigma_0 U_0. \tag{1.10}$$

The proof will be given in the Notes.[7]

1.11. For $i = 1, 2$, let E_i^{n+1} be an oriented $(n+1)$-space, and let $S^n(c_i, r_i) \subset E_i^{n+1}$ where c_i is a point in E_i^{n+1} and r_i a positive number. Let f be a continuous map $S_1^n = S^n(c_1, r_1)$ into $S_2^n = S_2^n(c_2, r_2)$. We want to define a degree $d(f, S_1^n, S_2^n)$.

If $n = 0$ the sphere $S^0(c_1, r_1)$ consists of the two points $x_1 = c_1 - r_1$, $x_2 = c_1 + r_1$ of the oriented 1-space (x-axis) E_1^1, and the sphere $S_2^0(c_2, r_2)$ consists of the points $y_1 = c_2 - r_2$, $y_2 = c_2 + r_2$ of the oriented "y-axis E_2^1." The degree $d(f, S_1^0(c_1, r_1), S_2^0(c_2, r_2))$ is defined as follows:

$$d(f, S_1^0(c_1, r_1), S_2^0(c_2, r_2)) = \begin{cases} +1 & \text{if } f(x_1) = y_1, \ f(x_2) = y_2, \\ -1 & \text{if } f(x_1) = y_2, \ f(x_2) = y_1, \\ 0 & \text{if } f(x_1) = f(x_2) = y_1, \\ 0 & \text{if } f(x_1) = f(x_2) = y_2. \end{cases}$$

Now let $n = 1$. If in addition $E_1^2 = E_2^2$, $c_1 = c_2 = 0$, $r_1 = r_2$, we define $d(f, S_1^n, S_1^n)$ as the winding number $u(f, S_1^n)$ discussed in subsection 1.6 of Chapter 4 and Note 1 to §4.1. It is obvious how to formulate the definition if the above additional conditions are not satisfied.

If $n \geq 2$ we will first define a degree $d(f, S_1^n(c_1, r_1), y_0)$ where y_0 is a point of $S_2^n(c_2, r_2)$ and prove later (see Lemma 1.12) that this degree is independent of y_0. We distinguish three cases: Case I: $y_0 \notin f(S_1^n(c_1, r_1))$; Case II: f maps $S^n(c_1, r_1)$ into a single point of $S_2^n(c_2, r_2)$; Case III: neither of the preceding two conditions holds. In Case I as well as in Case II we define

$$d(f, S_1^n(c_1, r_1), y_0) = 0. \tag{1.11}$$

In Case III we note first that there exists a point $n_1 \in S_1^n(c_1, r_1)$ such that

$$f(n_1) \neq y_0, \tag{1.12}$$

and the set

$$C_1 = f^{-1}(y_0) \tag{1.13}$$

is not empty. If moreover n_2 is a point of $S_2^n(c_2, r_2)$ different from y_0, then as will be proved in the Notes[8] there exists an open subset U_1 of $S_1^n(c_1, r_1)$ which has the following properties:

(a) $\text{dist}(n_1, U_1) > 0$;   (b) $C_1 \subset U_1$;   (c) $n_2 \notin \overline{f(U_1)}$.   (1.14)

For $i = 1, 2$, let s_i be the antipode of n_i, let T_i be the tangent plane at s_i, and let σ_i be the stereographic projection with pole n_i. Then $\Omega_1 = \sigma_1(U_1)$ is a bounded open set of T_1 and by (1.14c) the projection $\sigma_2(f(\overline{U_1})$ is well defined. Thus

$$f_1 = \sigma_2 f \sigma_1^{-1} \tag{1.15}$$

maps the subset $\overline{\Omega_1} = \overline{\sigma_1(U_1)}$ of T_1 continuously into T_2. Moreover $t_1 \in \Omega_1$ is a solution of

$$f_1(t_1) = \sigma_2(y_0) \tag{1.16}$$

if and only if $x = \sigma_1^{-1}(t_1) \in U_1$ is a solution of

$$f(x) = y_0, \tag{1.17}$$

i.e., if

$$x \in f^{-1}(y_0) = C_1. \tag{1.18}$$

Now it is easy to see that

$$\sigma_2(y_0) \notin f_1(\partial \Omega_1) \tag{1.19}$$

(see the Notes).[9] Therefore the degree $d(f_1, \Omega_1, \sigma_2(y_0))$ is defined. As will be shown in the Notes,[10] this degree is independent of the choice of n_1 and U_1 provided that $n_2 \neq y_0$, and that (1.12) and (1.14) hold. Therefore we may define

$$d(f, S_1^n, y_0) = d(f, U_1, y_0) = d(f_1, \Omega_1, \sigma_2(y_0)). \qquad (1.20)$$

1.12. LEMMA. *The degree $d(f, S_1^n(c_1, r_1), y_0)$ is independent of the choice of the point $y_0 \in S_2^n(c_2, r_2)$.*

PROOF. If f maps $S_1^n(c_1, r_1)$ into a single point of $S_2^n(c_2, r_2)$, i.e., in Case II of subsection 1.11, then, by definition, (1.11) holds for all $y_0 \in S_2(c_2, r_2)$. Suppose now that $f(S_1^n(c_1, r_1))$ is not a single point. Since the sphere is compact, it will be sufficient to prove that to every $y_0 \in S_2^n(c_2, r_2)$ there exists a neighborhood $V(y_0) \subset S_2(c_2, r_2)$ such that

$$d(f, S_1^n(c_1, r_1), y) = \text{const.} \quad \text{for } y \in V(y_0). \qquad (1.21)$$

Suppose first that for some $y_0 \in S_2^n(c_2, r_2)$ the equation

$$f(x) = y_0 \qquad (1.22)$$

has no solutions, i.e., there are some y_0 in Case I (see subsection 1.11). It is clear that such a y_0 has a neighborhood $V(y_0)$ such that the equation $f(x) = y$ for $y \in V(y_0)$ has no solution since $f(S_1^n(c_1, r_1))$ is closed and is therefore a positive distance from y_0. Consequently every $y \in V(y_0)$ is in Case I, and thus (1.21) holds for $y \in V(y_0)$, with the constant being zero.

Finally consider Case III. Then y_0 is an image point, and $d(f, S_1^n, y_0)$ is given by (1.20). If $x_0 \in U_1$ is a root of (1.22), then $t_1 = \sigma_1(x_0)$ is a root of $f_1(t_1) = \sigma_2(y_0)$ and $\sigma_2(y_0)$ satisfies (1.19). This implies that $\sigma_2(y_0) \in \Omega_2 = \sigma_2(U_1) \subset T_2$. Consequently there exists a neighborhood W_0 of $\sigma_2(y_0)$ which does not intersect $f_1(\partial\Omega_1)$. Thus $d(f_1, \Omega_1, w) = d(f_1, \Omega_1, \sigma_2(y_0))$ for all $w \in W_0$, and, by definition (1.20), the relation (1.21) will be true with $V = \sigma_2^{-1}(W_0)$.

Lemma 1.12 allows us to define the following.

1.13. DEFINITION. For $n \geq 2$ the degree $d(f, S_1^n(c_1, r_1), S_2^n(c_2, r_2))$ is defined by the equality $d(f, S_1^n(c_1, r_1), S_2^n(c_2, r_2)) = d(f, S_1^n(c_1, r_1), y_0)$ where y_0 is an arbitrary point of $S_2^n(r_2, c_2)$.

§2. Properties of the degree $d(f, S_1^n, S_2^n)$

We use the notations of the preceding section and assume $n \geq 2$. For $S_i^n(c_i, r_i)$ we will often simply write S_i^n.

2.1. LEMMA. *If $f(S_1^n)$ is a proper subset of S_2^n, then $d(f, S_1^n, S_2^n) = 0$.*

PROOF. By assumption there exists a $y_0 \in S_2^n$ such that $y_0 \notin f(S_1^n)$. Therefore, by definition, (1.11) holds. Our assertion follows now from Definition 1.13.

2.2. THEOREM. *If the continuous map $f: S_1^n \to S_2^n$ is one-to-one, then* (i) $d(f, S_1^n, S_2^n) = \pm 1$, *and* (ii) f *is onto.*

PROOF. Let $y_0 \in f(S_1^n)$. Since f is one-to-one, neither of the Cases I or II takes place (see subsection 1.11). We are thus in Case III. Therefore $d(f, S_1^n, S_2^n) = d(f, S_1^n, y_0)$ (Definition 1.13), and by (1.20), $d(f, S_1^n, y_0) = d(f_1, \Omega_1, \sigma_2(y_0))$ where f_1 is defined by (1.15). But the right member of this equality equals ± 1 since the "plane" mapping f_1 is obviously also one-to-one. This proves assertion (i). If assertion (ii) were not true then, by Lemma 2.1, $d(f, S_1^n, S_2^n) = 0$ which contradicts assertion (i).

2.3. On the index $j(f; x_0)$ of an isolated solution x_0 of (1.17). We recall first the second part of (1.20), where U_1 is an open subset of S_1^n which has the property (among others) that it contains all roots of (1.17) (see (1.14) and part (b) of (1.14)). Suppose now that there exists a point x_0 in S_1^n which is an isolated root of (1.17), i.e., there exists a neighborhood of x_0 in which x_0 is the only root. Such neighborhood contains an open set U_0 such that

$$x_0 \in U_0 \subset U_1, \qquad (2.1)$$

and therefore $\sigma_1(x_0) \in \sigma_1(U_0) \subset \sigma_1(U_1) = \Omega_1$. It is clear from the sum theorem for the "plane" map f_1 that $d(f_1, \sigma_1(U_0), \sigma_2(y_0))$ does not depend on the particular choice of U_0 provided that (2.1) is satisfied and that x_0 is the only root of (1.17) which lies in U_0. For such U_0 we define (cf. (1.20))

$$d(f, U_0, y_0) = d(f_1, \sigma_1(U_0), \sigma_2(y_0)). \qquad (2.2)$$

We denote this number by $j(f; x_0)$ and call it the index of x_0 as a solution of (1.17). If in particular this equation has only a finite number of roots, say $x_1, x_2, \ldots, x_r$, and if for $\rho = 1, 2, \ldots, r$ the open set U^ρ contains x_ρ and is contained in U_1, and if the $\overline{U^\rho}$ are disjoint, then

$$d(f_1, \sigma_1(U_1), \sigma_2(y_0)) = \sum_{\rho=1}^{r} d(f_1, \sigma_1(U^\rho), \sigma_2(y_0))$$

as follows from the sum theorem. Therefore by (1.20) and (2.2)

$$d(f, S_1^n, y_0) = \sum_{\rho=1}^{r} d(f, U^\rho, y_0) = \sum_{\rho=1}^{r} j(f; x_\rho). \qquad (2.3)$$

2.4. THEOREM. *Let $f_t(x) = f(x, t)$ be a continuous map $S_1^n \times [0, 1] \to S_2^n$. Then $d(f_t, S_1^n, S_2^n)$ is independent of t.*

The proof will be given in the Notes.[11]

2.5. THE PRODUCT THEOREM. *For $i = 1, 2, 3$, let E_i^{n+1} be an oriented $(n+1)$-space, and let $S_i^n = S_i(c_i, r_i)$ be the sphere in E_i^{n+1} with center c_i and radius r_i. Let f and g be continuous maps $S_1^n \to S_2^n$ and $S_2^n \to S_3^n$ resp. Then*

$$d(gf, S_1^n, S_3^n) = d(f, S_1^n, S_2^n) \cdot d(g, S_2^n, S_3^n). \qquad (2.4)$$

PROOF. Let n_1 be a point of S_1^n, and

$$n_2 = f(n_1), \qquad n_3 = g(n_2). \tag{2.5}$$

Suppose first that at least one of the following conditions holds: (a) $f(S_1^n) = n_2$; (b) $g(S_2^n) = n_3$. In either case $gf(S_1^n) = n_3$, and, therefore, the left member of (2.4) equals zero. But in case (a) the first factor and in case (b) the second factor of the right member of (2.4) equals zero. Thus (2.4) holds in either case.

Suppose now that neither of the conditions (a) and (b) holds. We next prove (2.4) for the special case that

$$gf(S_1^n) = n_3. \tag{2.6}$$

In this case the left member of (2.4) equals zero. Now we excluded case (b). Therefore $g^{-1}(n_3)$ is a proper subset of S_2^n, and since by (2.6), $f(S_1^n) \subset g^{-1}(n_3)$ the set $f(S_1^n)$ is a proper subset of S_2^n. Thus, by Lemma 2.1, $d(f, S_1^n, S_2^n) = 0$, and (2.4) is true, both of its members being zero.

Suppose finally that neither (a), (b), nor (2.6) holds. It is then easy to verify (see the Notes[12]) that there exists an $x_0 \in S_1^n$ of the following property: if

$$y_0 = f(x_0), \qquad z_0 = g(y_0), \tag{2.7}$$

then

$$\text{(a)} \quad f^{-1}(y_0) \not\ni n_1, \qquad \text{(b)} \quad g^{-1}(z_0) \not\ni n_2,$$
$$\text{(c)} \quad f(gf)^{-1}(z_0) \subset g^{-1}(z_0), \qquad \text{(d)} \quad f^{-1}(y_0) \subset (gf)^{-1}(z_0). \tag{2.8}$$

For $i = 1, 2, 3$, let s_i be the antipode to n_i, let T_i be the tangent plane to S_i^n at s_i, and let σ_i be the stereographic projection of S_i^n with pole n_i. It is then clear from the proof given in Note 8 for (1.14) (see also (1.13)) that a δ_2 of the following property can be chosen: if

$$U_2 = \bigcup_{y \in g^{-1}(z_0)} B(y, \delta_2), \tag{2.9}$$

then

$$\text{(a)} \quad \text{dist}(n_2, U_2) > 0; \qquad \text{(b)} \quad g^{-1}(z_0) \subset U_2; \qquad \text{(c)} \quad n_3 \notin \overline{g(U_2)}. \tag{2.10}$$

Moreover (cf. (1.20))

$$d(g, S_2^n, z_0) = d(g_1, \sigma_2(U_2), \sigma_3(z_0)), \tag{2.11}$$

where

$$g_1 = \sigma_3 g \sigma_2^{-1}. \tag{2.12}$$

In the same way we see that if δ_1 is a small enough positive number and

$$U_1 = \bigcup_{x \in (gf)^{-1}(z_0)} B(x_1, \delta_1), \tag{2.13}$$

then

$$\text{(a)} \quad \text{dist}(n_1, U_1) > 0; \qquad \text{(b)} \quad (gf)^{-1}(z_0) \in U_1; \qquad \text{(c)} \quad n_3 \notin \overline{gf(U_1)}. \tag{2.14}$$

Moreover

$$d(gf, S_1^n, z_0) = d((gf)_1, \sigma_1(U_1), \sigma_3(z_0)), \qquad (2.15)$$

where

$$(gf)_1 = \sigma_3 g f \sigma_1^{-1} = g_1 f_1 \qquad (2.16)$$

with

$$f_1 = \sigma_2 f \sigma_1^{-1}. \qquad (2.17)$$

But because of part (c) of (2.8) and the closedness of $g^{-1}(z_0)$, we see from (2.9) and (2.13) that δ_1 can be chosen in such a way that

$$\overline{f(U_1)} \subset U_2. \qquad (2.18)$$

Finally we note that the U_1 thus defined satisfies (1.14). Indeed, part (a) of (1.14) follows from part (a) of (2.14); part (b) follows from part (b) of (2.14) because of part (d) of (2.8) (and (1.13)); if part (c) of (1.14) were not true, then $n_2 = f(u_1)$ for some $u_1 \subset \overline{U}_1$. But then, by (2.5), $n_3 = g(n_2) = gf(u_1) \subset \overline{gf(U_1)}$ which contradicts part (c) of (2.14). But (1.14) thus proved implies that

$$d(f, S_1^n, y_0) = d(f_1, \sigma_1(U_1), \sigma_2(y_0)). \qquad (2.19)$$

It now follows from Definition 1.13 and from (2.11), (2.25), (2.19) that our assertion (2.4) is equivalent to

$$d(g_1 f_1, \sigma_1 U_1, \sigma_3(z_0)) = d(g_1, \sigma_2 U_2, \sigma_3(z_0)) \cdot d(f_1, \sigma_1 U_1, \sigma_2(y_0)). \qquad (2.20)$$

To prove this equality we want to apply the product theorem in subsection 2.12 of Chapter 6. We have to verify that the assumptions of that theorem are satisfied with $\Omega = \sigma_1 U_1$, $Y = \sigma_2 U_2$, and with f and g replaced by f_1 and g_1 resp., and with z_0 replaced by $\sigma_3(z_0)$, we have to show that

$$\text{(a))} \quad f_1(\sigma_1 U_1) \subset \sigma_2 U_2, \qquad \text{(b)} \quad \sigma_3(z_0) \notin g(\partial(\sigma_2 U_2)),$$

$$\text{(c)} \quad \sigma_3(z_0) \notin g_1 f_1(\partial \sigma_1 U_1).$$

Now (a) follows from (2.17) and (2.18). Suppose now (b) were not true. Then $\sigma_3(z_0) \in g_1(\partial(\sigma_2 U_2)) = g_1(\sigma_2(\partial U_2))$, and, by (2.12), $\sigma_3(z_0) \in \sigma_3(g(\partial U_2))$ which implies that $z_0 \in g(\partial U_2)$. This however contradicts part (b) of (2.10). Finally (c) follows similarly from (2.12), (2.17) and part (b) of (1.14) together with (1.13).

We are now in a position to apply Theorem 2.12 and obtain

$$d(g_1 f_1, \sigma_1 U_1, \sigma_3(z_0)) = \sum_\alpha d(g_1, Y_\alpha, \sigma_3(z_0)) \cdot d(f_1, \sigma_1 U_1, \sigma_2(y_\alpha)), \qquad (2.21)$$

where $\{Y_\alpha\}$ is the set of components of $\sigma_2(U_2) - f_1(\partial \sigma_1 U_1)$, where $\sigma_2(y_\alpha) \in Y_\alpha$, and where the summation is extended over those α for which $g_1(\eta) = \sigma_3(z_0)$ has a solution $\eta \in Y_\alpha$. Now by (1.20) and by Lemma (1.12)

$$d(f_1, \sigma_1 U_1, \sigma_2(y_\alpha)) = d(f, S_1^n, y_\alpha) = d(f, S_1^n, y_0) = d(f_1, \sigma_1 U_1, \sigma_2(y_0)). \qquad (2.22)$$

But by the sum theorem

$$\sum_\alpha d(g_1, Y_\alpha, \sigma_3(z_0)) = d(g_1, \sigma_2 U_2, \sigma_3(z_0)) = d(g, S_2^n, z_0). \qquad (2.23)$$

The assertion (2.20) now follows from (2.21), (2.23), and (2.22).

2.6. THEOREM. *Let f and g be continuous maps $S_1^n \to S_2^n$ and suppose that for no $x \in S_1^n$ are the points $f(x)$ and $g(x)$ of S_2^n antipodal. Then*

$$d(f, S_1^n, S_2^n) = d(g, S_1^n, S_2^n). \tag{2.24}$$

PROOF. Since $S_2^n = S_2^n(c_2, r_2)$, it is clear that for two nonantipodal points y_0 and y_1 of S_2^n

$$(1-t)y_0 + ty_1 - c_2 = (1-t)(y_0 - c_2) + t(y_1 - c_2) \neq 0 \quad \text{for } t \in [0,1].$$

Therefore

$$f(x,t) = \frac{(1-t)f(x) + tg(x) - c_2}{\|(1-t)f(x) + tg(x) - c_2\|} r_2 + c_2$$

is a continuous map defined on $S_1^n \times [0,1]$. Its range is on S_2^n since $\|f(x,t) - c_2\| = r_2$, and the assertion (2.24) follows from Theorem 2.4.

2.7. DEFINITION. Let c be a point of E^{n+1}, and let M and Σ be subsets of E^{n+1} not containing c. Suppose that each ray issuing from c (see §1.23) intersects Σ in exactly one point. Then the central projection $\pi = \pi(M, \Sigma, c)$ from c of M on Σ is defined as follows: if m is a point of M, then $\pi(m)$ is the intersection of the ray $\overrightarrow{cm}$ with Σ.

2.8. LEMMA. *Let E^{n+1} be oriented, and let S_0^n and S_1^n be positively oriented spheres in E^{n+1} with common center c and with radii r_0 and r_1 resp. Let π_1 be the central projection from c of S_0^n on S_1^n. Then*

$$d(\pi_1, S_0^n, S_1^n) = +1. \tag{2.25}$$

For a proof see the Notes.[13]

2.9. Let S_0^n and S_1^n be two positively oriented spheres in E^{n+1} of equal radius and with centers c_0 and c_1 respectively. Let $\tau \colon S_0^n \to S_1^n$ be given by $\tau(x) = x + c_1 - c_0$. Then $d(\tau, S_0^n, S_1^n) = +1$.

The rather obvious proof is omitted.

§3. The order of the image of a sphere with respect to a point

3.1. Let E_1^{n+1} and E_2^{n+1} be oriented spaces. Let S_1^n be a positively oriented sphere in E_1^{n+1}, and let c_1 be a point in E_2^{n+1}. Let f be a continuous map $S_1^n \to E_2^{n+1}$. In subsection 2.10 of Chapter 6 the winding number $u = u(f(S_1^n), c_1)$ was defined provided that

$$c_1 \notin f(S_1^n). \tag{3.1}$$

In the present chapter we will give another interpretation of the winding number. The assumption (3.1) implies the existence of a positive r_2 such that

$$\overline{B(c_1, r_2)} \cap f(S_1^n) = \emptyset. \tag{3.2}$$

Therefore the central projection $\pi_2 = \pi_2(c_1)$ with center c_1 of $f(S_1^n)$ on the sphere $S_2^n = \partial B(c_1, r_2)$ is defined. Thus $f_1 = \pi_2 f(S_1^n)$ is a continuous map $S_1^n \to S_2^n$ and its degree $d(f_1, S_1^n, S_2^n)$ is defined. Moreover it follows immediately from Lemma 2.8 in conjunction with the product theorem in subsection 2.5 that this degree is independent of the particular choice of an r_2 satisfying (3.2). Therefore the following definition is legitimate.

3.2. DEFINITION. With the assumption and notations of subsection 3.1 the order $v(f(S_1^n), c_1)$ of $f(S_1^n)$ with respect to c_1 is defined by

$$v(f(S_1^n), c_1) = d(\pi_2 f(S_1^n), S_2^n). \tag{3.3}$$

3.3. LEMMA. $v(f(S_1^n), c_1)$ *is constant if c_1 varies continuously in such a way that (3.1) remains true during the change.*

The proof will be given in the Notes.[14]
The theorem whose proof is the main object of this section is the following.

3.4. THEOREM. *With the above notations and assumptions*

$$v(f(S_1^n), c_1) = u(f(S_1^n), c_1). \tag{3.4}$$

By Definition 2.10 in Chapter 6 of the winding number u the above theorem is equivalent to the following one.

3.5. THEOREM. *Let B_1^{n+1} be the ball whose boundary is S_1^n. Let f be a continuous map of $\overline{B_1^{n+1}}$ into E_2^{n+1}. Then*

$$v(f(S_1^n), c_1) = d(f, B_1^{n+1}, c_1) \tag{3.5}$$

provided that (3.1) holds.

3.6. Our first step to proving Theorem 3.5 consists in formulating an equivalent theorem in which the ball B_1^{n+1} is replaced by an $(n+1)$-simplex β^{n+1} whose barycenter γ is the center of B_1^{n+1} and whose vertices lie on $S_1^{n+1} = \partial B_1^{n+1}$: let κ be the central projection from γ of S_1^n on $\partial \beta^{n+1}$ (see subsection 2.7). We extend the definition of κ from S_1^n to B_1^{n+1} as follows: we set $\kappa(\gamma) = \gamma$; if $x \in B_1^{n+1} - \{\gamma\}$ and if $\dot{x}$ is the point in which the ray $\overrightarrow{\gamma x}$ intersects S_1^n, we set

$$\kappa(x) = \gamma + \|\kappa(\dot{x}) - \gamma\| \frac{(x - \gamma)}{r_1}. \tag{3.6}$$

Then κ is a one-to-one continuous map of $\overline{B_1^{n+1}}$ onto β^{n+1}. We define a continuous map $g \colon \beta^{n+1} \to E_2^{n+1}$ by setting

$$g(x) = f(\kappa^{-1}(x)), \qquad x \in \beta^{n+1}, \ f \text{ as in Theorem 3.5.} \tag{3.7}$$

Then by part (i) of subsection 2.15 in Chapter 6

$$d(f, B_1^{n+1}, c_1) = d(g\kappa, B_1^{n+1}, c_1) = d(g, \beta^{n+1}, c_1) \cdot d(\kappa, B_1^{n+1}, \eta),$$

where η is an arbitrary point in $\kappa B_1^{n+1} = \beta^{n+1}$. Here the second factor at the right equals $+1$ (see the Notes).[15] Thus

$$d(f, B_1^{n+1}, c_1) = d(g, \beta^{n+1}, c_1), \tag{3.8}$$

and the assertion (3.5) is equivalent to

$$v(f(S_1^n), c_1) = d(g, \beta^{n+1}, c_1). \tag{3.9}$$

3.7. Proof of (3.9) **in a special case.** Since $g(\beta^{n+1})$ is bounded there exists a positive R_2 such that the ball $B(\theta, R_2) \in E_2^{n+1}$ contains $g(\beta^{n+1})$. Now let y_1 be a point in E_2^{n+1} for which

$$\|y_1\| > R_2. \tag{3.10}$$

We will prove (3.9) for $c_1 = y_1$ by showing that both members of that formula are zero. Since by definition of y_1 the equation $g(x) = y_1$ has no solution, it is clear that the right member equals zero. To show that the same is true for the left member, we choose an $r_1 > 0$ such that

$$\overline{B(y_1, r_1)} \cap \overline{B(\theta, R_2)} = \varnothing. \tag{3.11}$$

Then by Definition 3.2

$$v(f(S_1^n), y_1) = d(\pi f, S_1^n, S_2^n), \tag{3.12}$$

where $S_2^n = \partial B(y_1, r_1)$ and where π is the central projection from y_1 on S_2^n. Now $f(S_1^n) = g(\partial \beta^{n+1}) \subset B(\theta, R_2)$. Therefore

$$\pi f(S_1^n) \subset \pi B(\theta, R_2). \tag{3.13}$$

$$\pi B(\theta, R_2) = \text{ a proper subset of } S_2^n, \tag{3.14}$$

as follows from (3.11) and (3.10),[16] and by (3.13) the same is true of $\pi f(S_1^n)$. By Lemma 2.1 this implies that the right member of (3.12) equals zero, and we see that the left member of (3.9) (with $c_1 = y_1$) equals zero as asserted.

3.8. To prove (3.9) for an arbitrary $y_0 = c_1$ satisfying (3.1) or what is the same

$$y_0 \notin g(\partial \beta_1^{n+1}), \tag{3.15}$$

we consider the segment

$$y(t) = (1 - t)y_0 + ty_1, \qquad 0 \le t \le 1, \tag{3.16}$$

where y_1 is as in subsection 3.7. Since we proved (3.9) for $c_1 = y_1$, that equality will be proved for $c_1 = y_0$ once it is shown that both members of (3.9) with c_1 replaced by $y(t)$ undergo the same change as t varies from 0 to 1, i.e., that

$$v(f(S_1^n), y_0) - v(f(S_1^n), y_1) = d(g, \beta_1^{n+1}, y_0) - d(g, \beta_1^{n+1}, y_1). \tag{3.17}$$

For those $t \in [0, 1]$ for which

$$y(t) \notin g(\partial \beta_1^{n+1}), \tag{3.18}$$

we denote by π^t the central projection from $y(t)$ on a sphere $S_2^n(y(t))$ of center $y(t)$ (for which the ball whose boundary is $S_2^n(y(t))$ does not intersect $g(\partial \beta_1^{n+1})$). Now in a subinterval of $[0, 1]$ all of whose points t satisfy (3.18), both members of (3.9) with c_1 replaced by $y(t)$ remain constant (see the Notes).[17] Thus a change of these members can occur only if t crosses a point $\bar{t}$ for which $y(\bar{t}) \in g(\partial \beta_1^{n+1})$. If there are no such $\bar{t}$, then both members of (3.17) are zero and this equality is proved. We therefore assume from now on that there are such $\bar{t}$. The difficulty in this case is that there may be infinitely many such $\bar{t}$ and that each intersection of the segment $\overline{y_0 y_1}$ with $g(\partial \beta_1^{n+1})$ may be of a complicated nature. To avoid this difficulty we will introduce a simplicial approximation of g.

3.9. A simplicial approximation. We note first that there exists a positive ε of the following property: if h is a continuous map $\beta_1^{n+1} \to E_2^{n+1}$ which satisfies

$$|g(x) - h(x)| < \varepsilon \tag{3.19}$$

for all $x \in \beta_1^{n+1}$, then neither member of (3.9) changes its value if g is replaced by h and $f = g\kappa^{-1}$ (cf. subsection 3.7). For the right member this follows from Theorem 2.9 of Chapter 6, and for the left member from Lemma 3.3 and Theorem 2.6.

Let ε be such a number. We know from Theorem 5.1 of Chapter 6 that there exists a barycentric subdivision $K^{n+1} = B^r \beta_1^{n+1}$ of β_1^{n+1} and a map $h \colon |K^{n+1}| = |\beta_1^{n+1}| \to E_2^{n+1}$ which satisfies (3.19), is simplicial with respect to K^{n+1}, and is such that the images under h of the vertices of K^{n+1} together with the points y_0 and y_1 form a point set of E_2^{n+1} which is in general position. In other words, for the proof of (3.9) we may assume that g is simplicial with respect to the subdivision K^{n+1} of β_1^{n+1} and has the following property: if $u_0, u_1, \ldots, u_N$ are the vertices of K^{n+1} and if $v_i = g(u_i)$ for $i = 0, 1, \ldots, N$, then the points

$$y_0, y_1, v_0, v_1, \ldots, v_N \tag{3.20}$$

are in general position. This property has the following consequences C_1, C_2, C_3: if for some $\bar{t} \in [0, 1]$ the point $y(\bar{t})$ of the segment (3.16) lies on $g(\partial K^{n+1})$, then

(C_1) the equation

$$g(x) = y(\bar{t}) \tag{3.21}$$

has in ∂K^{n+1} one and only one solution;

(C_2) if this solution lies in the simplex σ^n of ∂K^{n+1} and if $\tau^n = g(\sigma^n)$, then $y(\bar{t})$ lies in the interior of τ^n;

(C_3) $y(\bar{t})$ is the only intersection of the ray $\overrightarrow{y_0 y_1}$ with τ^n.

For a proof of these three properties see the Notes.[18]

Since there are only a finite number of n-simplices in $g(\partial K^{n+1})$, it follows that there are only a finite number of t-values in $(0, 1)$ for which $y(t) \in g(\partial \beta_1^{n+1})$, say $t_1 < t_2 < \cdots < t_r$. Since we know that $t = 0$ and $t = 1$ do not belong to them, the intersection points of the segment $\overline{y_0 y_1}$ with $g(\partial \beta_1^{n+1})$ are

$$y(t_1), y(t_2), \ldots, y(t_r), \qquad 0 < t_1 < t_2 < \cdots < t_r < 1. \tag{3.22}$$

If, for t different from any of the t_α, $\alpha = 1, 2, \ldots, r$, π^t is defined as in the lines directly following (3.18), we write for simplicity's sake $l(y(t))$ and $r(y(t))$ for the left and right member of (3.9) (with c_1 replaced by $y(t)$ resp.). Now the discussion in subsection 3.8 showed that the left member of (3.17) equals the sum of the "jumps" of $l(y(t))$ as t crosses one of the t_α and that the right member equals the sum of the corresponding jumps of $r(y(t))$. Thus the assertion (3.17) will be proved once it is shown that for each t_α the jumps of $l(y(t))$ and $r(y(t))$

are the same. More precisely we have to show that for $\alpha = 1, 2, \ldots, r$

$$l(y(t_\alpha^+)) - l(y(t_\alpha^-)) = r(y(t_\alpha^+)) - r(y(t_\alpha^-)), \tag{3.23}$$

where the numbers t_α^- and t_α^+ are subject to the inequalities

$$t_{\alpha-1} < t_\alpha^- < t_\alpha < t_\alpha^+ < t_{\alpha+1}^- < t_{\alpha+1} \tag{3.24}$$

if we define $t_0 = 0$ and $t_{r+1} = 1$. Now let x_α be the unique solution in ∂K^{n+1} of the equation

$$g(x) = y(t_\alpha), \tag{3.25}$$

let σ_α^n be the unique n-simplex of ∂K^{n+1} which contains x_α, let σ_α^{n+1} be the positively oriented $(n+1)$-simplex of K^{n+1} for which

$$\sigma_\alpha^n \subset \partial \sigma_\alpha^{n+1}, \tag{3.26}$$

and let

$$\tau_\alpha^n = g(\sigma_\alpha^n), \qquad \tau_\alpha^{n+1} = g(\sigma_\alpha^{n+1}). \tag{3.27}$$

Then let α_0 be a fixed one of the integers $1, 2, \ldots, r$. For this α_0 the simplex $\tau_{\alpha_0}^{n+1}$ may be a positive or negative simplex of E_2^{n+1}. Since the proof of (3.23) is essentially the same in these two cases, we carry it out under the assumption that

$$\tau_{\alpha_0}^{n+1} \quad \text{is positively oriented.} \tag{3.28}$$

We denote by $w_1, w_2, \ldots, w_n$ the vertices of $\tau_{\alpha_0}^n$, and by $w_0, w_1, \ldots, w_n$ the vertices of $\tau_{\alpha_0}^{n+1}$. Moreover, if Π_0^n denotes the n-plane spanned by $\tau_{\alpha_0}^n$ we distinguish two cases:

(i) w_0 and $y_{t_{\alpha_0}^+}$ lie in the same half space of E_2^{n+1} with respect to Π_0^n;

(ii) w_0 and $t_{t_{\alpha_0}^+}$ lie in different half spaces of E_2^{n+1} with respect to Π_0^n.

$$\tag{3.29}$$

3.10. Computation of the right member of (3.23) for $\alpha = \alpha_0$. We will show that

$$r(y(t_{\alpha_0}^+)) - r(y(t_{\alpha_0}^-)) = \begin{cases} +1 & \text{in case (i) of (3.29),} \\ -1 & \text{in case (ii) of (3.29).} \end{cases} \tag{3.30}$$

The idea of the proof is to show that as t varies from $t_{\alpha_0}^-$ to $t_{\alpha_0}^+$, then as t crosses t_α the point $y(t)$ enters exactly one simplex of $g(\partial K^{n+1})$, namely, $\tau_{\alpha_0}^{n+1}$ in case (i), while it leaves that simplex in case (ii). This together with the assumption (3.28) will yield the assertion (3.30).

Precisely, with the notations employed in subsection 3.9, let $\tilde{K}^{n+1}$ be the simplicial set consisting of the $(n+1)$-simplices of K^{n+1} except for σ_α^{n+1}. It then follows from (C_1) and (C_2) that for $\alpha = \alpha_0$ the equation (3.25) has no root $x \in \partial\tilde{K}^{n+1}$. Thus every root $x \in \tilde{K}^{n+1}$ of that equation is an interior point of $\tilde{K}^{n+1}$. Consequently we can choose $t_{\alpha_0}^-$ and $t_{\alpha_0}^+$ so near to t_{α_0} that

$$d(g, \tilde{K}^{n+1}, y(t_{\alpha_0}^-)) = d(g, \tilde{K}^{n+1}, y(t_{\alpha_0}^+)) = d(g, \tilde{K}_{n+1}, y(t_\alpha)). \tag{3.31}$$

Thus, by the sum theorem,

$$d(g, K^{n+1}, y(t_{\alpha_0}^-)) = d(g, \tilde{K}^{n+1}, y(t_{\alpha_0}^-)) + d(g, \sigma_{\alpha_0}^{n+1}, y(t_{\alpha_0}^-)),$$

and this equality remains true if $y(t_{\alpha_0}^-)$ is replaced by $y(t_{\alpha_0}^+)$. It therefore follows from (3.31) and the definition of $r(y(t))$ that

$$r(y(t_{\alpha_0}^+)) - r(y(t_{\alpha_0}^-)) = d(g, \sigma_{\alpha_0}^{n+1}, y(t_{\alpha}^+)) - d(g, \sigma_{\alpha_0}^{n+1}, y(t_{\alpha_0}^-)). \qquad (3.32)$$

Now let H^+ and H^- be the two half spaces into which the plane Π_0^n spanned by $\tau_{\alpha_0}^n$ divides E_2^{n+1} with $y(t_{\alpha_0}^+) \in H^+$ and $y(t_{\alpha_0}^-) \in H^-$. Then in case (i) of (3.29) the interior of the simplex $\tau_{\alpha_0}^{n+1} = g(\sigma_{\alpha_0}^{n+1})$ lies in H^+. Since by assertion (C_2) stated in subsection 3.9 the point $y(t_{\alpha_0})$ lies in the interior of the simplex $\tau_0^n = g(\sigma_{\alpha_0}^n) \in \Pi^n$, we may assume that $y(t_{\alpha_0}^+) \in \tau_{\alpha_0}^{n+1}$. By assumption (3.28) this implies that the first term of the right member of (3.32) equals $+1$. On the other hand the second term of the right member of that equality equals zero since $y(t_{\alpha_0}^-) \in H^-$ while $g(\sigma_{\alpha_0}^{n+1}) = \tau_{\alpha_0}^{n+1} \in H^+$.

This proves the first part of (3.30). The second part is proved correspondingly.

3.11. *Computation of the left member of* (3.23)   Before giving a formal proof for the result (3.48) we make the following intuitive remark: if the ray $\overrightarrow{y_0 y_1}$ intersects at $y(t_{\alpha_0})$ the simplex $\tau_{\alpha_0}^n$ of $\partial g K^{n+1}$ and if (3.27) and (3.26) hold with $\alpha = \alpha_0$, then on account of the general position of the points (3.20) there are two possibilities as t increasing crosses t_{α_0}: (i) $y(t)$ enters the interior of $\tau_{\alpha_0}^{n+1}$ from the exterior of $g(K^{n+1})$; (ii) $y(t)$ leaves the interior of $\tau_{\alpha_0}^{n+1}$ and enters the exterior of $g(K^{n+1})$. Thus under assumption (3.28) the degree $d(g, \sigma_{\alpha_0}^{n+1}, y(t))$ changes from 0 to 1 in case (i) and from 1 to 0 in case (ii). It is then not hard to compute the related degree obtained by replacing g by $\pi^t g$ appearing in the right member of (3.46). Therefore our first goal in our formal proof of (3.48) will be to prove (3.46).

Again let $y(t)$ be given by (3.16), $0 \le t \le 1$. For any t value for which (3.18) holds or what is the same for which

$$y(t) \notin f(S_1^n), \quad \text{i.e., } t \ne t_\alpha, \quad \alpha = 1, 2, \ldots, r, \qquad (3.32)$$

holds, we denote by $S_2(y(t))$ a sphere with center $y(t)$ and radius so small that the closure of the ball whose boundary is $S_2(y(t))$ contains none of the $y(t_\alpha)$. Then by definition

$$v(f(S_1^n), y(t)) = d(\pi^t f, S_1^n, S_2(y(t))), \qquad (3.33)$$

where $\pi^t f(x)$ denotes the central projection of $f(x)$ on $S_2(y(t))$ with center $y(t)$. Now by Definition 1.13 the right member of (3.33) equals the degree $d(\pi^t f, S_1^n, y^*)$ if y^* is an arbitrary point of $S_2(y(t))$. Now if $t_{\alpha_0} < t < t_{\alpha_0+1}$ we choose for y^* the point $y(t^*)$ which is the intersection of $S_2^n(y(t))$ with the segment $\overline{y(t)y_1}$. Then

$$t_{\alpha_0} < t < t^* < t_{\alpha_0+1} \qquad (3.34)$$

(if the radius of $S_2(t)$ is chosen small enough), and by (3.33)

$$v(f(S_1^n), y(t)) = d(\pi^t f, S_1^n, y(t^*)). \qquad (3.35)$$

Now from our discussion of the root x_α of (3.25) and the relation (3.7) between g and f, we see that the equation

$$f(\xi) = y(t_\alpha) \tag{3.36}$$

has on S_1^n exactly one root $\xi_\alpha = \kappa^{-1} x_\alpha$, and if

$$\kappa^{-1} \sigma_\alpha^n = U_\alpha, \tag{3.37}$$

then

$$\xi_\alpha \in U_\alpha \subset S_1^n. \tag{3.38}$$

But from (3.34) we see that

$$\pi^t y(t_\alpha) = y(t^*) \tag{3.39}$$

if and only if $\alpha = \alpha_0 + 1$, $\alpha_0 + 2, \ldots, \alpha_r$. It follows that the roots on S_1^n of

$$\pi^t f(\xi) = y(t^*) \tag{3.40}$$

are $\xi_{\alpha_0+1}, \ldots, \xi_r$. Therefore we see from (3.35) and subsection 2.3 that

$$v(f(S_1^n), y(t)) = \sum_{\alpha=\alpha_0+1}^r j(\pi^t f; \xi_\alpha). \tag{3.41}$$

To express the right member in terms of g rather than f, we note that the simplices σ_α^n and $\tau_\alpha^n = g(\sigma_\alpha^n)$ are oriented (as boundary simplices of σ_α^{n+1} and τ_α^{n+1} resp.) They therefore induce an orientation in the n-planes $P_\alpha^n \subset E_1^{n+1}$ and $Q_\alpha^n \subset E_2^{n+1}$ in which they lie resp.) Now $y(t_\alpha)$ is an interior point of $\tau_\alpha^n = g(\sigma_\alpha^n)$ and the only intersection of $\overline{y_0 y_1}$ with τ_α^n. Let $\pi^t g$ be the central projection (from $y(t)$) on the tangent plane T_*^n to $S_2^n(y(t))$ at $y(t^*)$. Then the degree $d(\pi^t g, \sigma_\alpha^n, y(t^*))$ is defined and, as will be verified in the Notes,[19]

$$j(\pi^t f; \xi_\alpha) = d(\pi^t g, \sigma_\alpha^n, y(t^*)). \tag{3.42}$$

Since t satisfies (3.34) we may choose an arbitrary value satisfying $t_{\alpha_0} < t < t_{\alpha_0+1}$. Therefore, by (3.24) we may choose $t = t_{\alpha_0}^+$. Then by (3.41) and (3.42)

$$v(f(S_1^n), y(t_{\alpha_0}^+)) = \sum_{\alpha=\alpha_0+1}^r d(\pi^{t_{\alpha_0}^+} g, \sigma_\alpha^n, y(t_{\alpha_0}^+)^*). \tag{3.43}$$

In the same way one proves

$$v(f(S_1^n), y(t_{\alpha_0}^-)) = \sum_{\alpha=\alpha_0}^r d(\pi^{t_{\alpha_0}^-} g, \sigma_\alpha^n, y((t_{\alpha_0}^-)^*)), \tag{3.44}$$

where $t_{\alpha_0}^-$ satisfies (3.24) with α replaced by α_0 and where $y((t_{\alpha_0}^-)^*)$ is the intersection of $S_2^n(y(t_{\alpha_0}^-))$ with the segment $\overline{y(t_{\alpha_0}^-), y_1}$. But for $\alpha = \alpha_0 + 1, \ldots, r$

$$d(\pi^{t_{\alpha_0}^+} g, \sigma_\alpha^n, y((t_{\alpha_0}^+)^*)) = d(\pi^{t_{\alpha_0}^-} g, \sigma_\alpha^n, y((t_{\alpha_0}^-)^*)) \tag{3.45}$$

(see the Notes).[20] It therefore follows from (3.43), (3.44) and the definition of $l(y(t))$ (given in the paragraph directly following (3.22)) that

$$l(y(t^+)) - l(y(t^-)) = -d(\pi^{t_{\alpha_0}^-} g, \sigma_{\alpha_0}^n, y((t_{\alpha_0}^-)^*)). \tag{3.46}$$

Now since $t_{\alpha_0}^- < t_{\alpha_0}$ we can apply the assertion (77) of Note 20 with $t = t_\alpha^-$ and $\alpha = \alpha_0$. Since, moreover, $t_{\alpha_0} < t_{\alpha_0}^+$ we see that for $\alpha = \alpha_0$ the cases (i) and (ii) defined in (76) of Note 20 are identical with cases (i) and (ii) resp. (defined in (3.29)). Because of our assumption (3.28) we apply (77) with $\varepsilon_\alpha = +1$ to obtain

$$d(\pi^{t_{\alpha_0}^-} g, \sigma_{\alpha_0}^n, y((t_{\alpha_0}^-)^*)) = \begin{cases} -1 & \text{in case (i),} \\ +1 & \text{in case (ii).} \end{cases} \tag{3.47}$$

Therefore by (3.46)

$$l(y(t_{\alpha_0}^+)) - l(y(t_{\alpha_0}^-)) = \begin{cases} +1 & \text{in case (i),} \\ -1 & \text{in case (ii).} \end{cases} \tag{3.48}$$

Comparison of this formula with (3.30) proves (3.23). This finishes the proof of Theorems 3.5 and 3.4.

3.12. *Interpretation of order and degree as intersection numbers* (cf. subsections 2.1–2.6 in Chapter 4). We use the definitions and notations of the previous sections of the present chapter. In addition we state the following definition.

3.13. DEFINITION. A map $f \colon \overline{B_1^{n+1}} \to E_2^{n+1}$ as well as a map $f \colon \partial B_1^{n+1}$, $S_1^n \to E_2^{n+1}$ is called simplicial if f is given by (3.7) where g is simplicial with respect to the barycentric subdivision K^{n+1} of the simplex β_1^{n+1} (see subsection 3.9) and where the points (3.20) are in general position.

Now let f be a simplicial map $S_1^n \to E_2^{n+1}$. We consider the intersections of $f(S_1^n) = g(\partial K^{n+1})$ with the ray $\overrightarrow{y_0 y_1}$ given by

$$y(t) = (1-t)y_0 + ty_1, \qquad 0 \le t < \infty, \tag{3.49}$$

where y_0 and y_1 are as in subsection 3.8. If t_α is one of the at most finite number of positive t-values for which $y(t)$ intersects $f(S_1^n) = g(\partial K^{n+1})$, then for t_α^+ and t_α^- as in (3.24) the "jump" $v(f(S_1^n), y(t_\alpha^+)) - v(f(S_1^n), y(t_\alpha^-))$ given by (3.48) is different from zero, actually ± 1. However if $y(t)$ is not an intersection point with $f(S_1^n)$ and $t^- < t < t^+$ with $t^+ - t^-$ small enough, then $v(f(S_1^n), y(t^+)) - v(f(S_1^n), y(t^-))$ equals zero. Therefore for any $t \ge 0$ we define the intersection number $i(f(S_1^n), y(t))$ at $y(t)$ of $f(S_1^n)$ with our ray (3.49) by

$$i(f(S_1^n), y(t)) = v(f(S_1^n), y(t^+)) - v(f(S_1^n), y(t^-)). \tag{3.50}$$

The sum over t of all these intersection numbers is called the intersection number of $f(S_1^n)$ with the ray $\overrightarrow{y_0 y_1}$ and denoted by $i(f(S_1^n), \overrightarrow{y_0 y_1})$. Then

$$i(f(S_1^n), \overrightarrow{y_0 y_1}) = \sum_{\alpha=1}^r v(f(S_1^n), y(t_\alpha^+)) - v(f(S_1^n), y(t_\alpha^-)). \tag{3.51}$$

We recall that $v(f(S_1^n), y(t))$ is constant in every t interval which contains none of the t_α (see subsection 3.8). From this, from (3.24) and the fact that $v(f(S_1^n), y_1) = 0$ (see subsection 3.8), it follows that the right member of (3.51) equals $-v(f(S_1^n), y_0)$. Thus

$$v(f(S_1^n), y_0) = -i(f(S_1^n), \overrightarrow{y_0 y_1}). \tag{3.52}$$

This represents the order (for a simplicial f) as an intersection number. To obtain a corresponding representation for the degree we need the following definition: let τ^{n+1} be an oriented simplex in the oriented E^{n+1}, let $\overset{\circ}{\tau}{}^{n+1}$ be its interior, and let y_0 be a point in E^{n+1}. Then if ε_0 denotes one of the symbols $+$ and $-$, we define the intersection number $i(\varepsilon_0 y_0, \tau^{n+1})$ of the oriented zero simplex $\varepsilon_0 y_0$ (see subsection 1.16 in Chapter 6) with τ^{n+1} by setting

$$
i(\varepsilon_0 y_0, \tau^{n+1}) = \begin{cases}
\varepsilon_0 \cdot 1 & \text{if } y_0 \in \overset{\circ}{\tau}{}^{n+1} \text{ and if } \tau^{n+1} \\
& \text{is positively oriented,} \\
-\varepsilon_0 \cdot 1 & \text{if } y_0 \in \overset{\circ}{\tau}{}^{n+1} \text{ and if } \tau^{n+1} \\
& \text{is negatively oriented,} \\
0 & \text{if } y_0 \notin \tau^{n+1}.
\end{cases}
\tag{3.53}
$$

The intersection numbers are not defined if $y_0 \in \partial \tau^{n+1}$. But because of the general position of the points (3.20) the numbers

$$
i(\varepsilon_0 y_0, \tau_\alpha^{n+1}) \quad \text{where } \tau_\alpha^{n+1} = g(\sigma_\alpha^{n+1})
\tag{3.54}
$$

are defined for all $(n+1)$-simplices σ_α^{n+1} of K^{n+1}. Moreover the 0-face containing y_0 of the oriented 1-simplex (y_0, y_1) is $-y_0$ according to the definition given at the end of subsection 1.16 in Chapter 6. Thus $\varepsilon_0 = -1$. We now define the intersection number of $-y_0$ with $g(K_1^{n+1})$ as the sum of the numbers (3.54):

$$
i(-y_0, g(K^{n+1})) = \sum_\alpha i(-y_0, g(\sigma_\alpha^{n+1})),
\tag{3.55}
$$

where we sum over all $(n+1)$-simplices σ_α^{n+1} of K_1^{n+1}. But for each α the restriction of g to σ_α^{n+1} is affine. Therefore we see from (3.53) and the proof of assertion (iv) of subsection 3.7 in Chapter 6 that

$$
i(-y_0, g(\sigma_\alpha^{n+1})) = -d(g, \sigma_\alpha^{n+1}, y_0).
\tag{3.56}
$$

Consequently, by (3.55) and the sum theorem for the degree,

$$
i(-y_0, g(K^{n+1})) = -d(g, |K_1^{n+1}|, y_0).
\tag{3.57}
$$

But $|K_1^{n+1}| = \beta_1^{n+1}$ (see subsection 3.6) and, by (3.8), the right member of (3.57) equals $-d(f, B_1^{n+1}, y_0)$. Thus

$$
-d(f, B_1^{n+1}, y_0) = -d(g, \beta^{n+1}, y_0) = i(-y_0, g(\beta^{n+1})),
\tag{3.58}
$$

and we expressed the degree in terms of an intersection number.

Now in (3.52) as well as in (3.58) f was supposed to be a simplicial map. Now let f be an arbitrary continuous map satisfying only $y_0 \notin f(S_1^n)$. But we know from subsection 3.9 that for small enough ε the values of the numbers $v(f(S_1^n), y_0)$ and $d(f, B_1^{n+1}, y_0)$ do not change if f is replaced by a continuous f_1 satisfying $\|f - f_1\| < \varepsilon$, and that among such f_1 there are simplicial maps. Thus (3.52) and (3.58) hold for continuous f if the left member is replaced for such an approximation f_1. We therefore define for continuous maps f the intersection numbers $i(f(S_1^n), \overrightarrow{y_0 y_1})$ and $i(-y_0, f(B_1^{n+1}))$ by setting

$$i(f(S_1^n), \overrightarrow{y_0 y_1}) = -v(f(S_1^n), y_0), \qquad i(-y_0, f(B_1^{n+1})) = -d(f, B_1^{n+1}, y_0).$$
(3.59)

We finally note the "duality" formula

$$i(f(\partial B_1^{n+1}), \overrightarrow{y_0 y_1}) = i(-y_0, B_1^{n+1}),$$
(3.60)

which on account of (3.59) follows directly from Theorem 3.5 since $S_1^n = \partial B_1^{n+1}$.

§4. Two approximation lemmas

4.1. It is often important to approximate a given equation in such a way that the approximating equation has at most a finite number of roots. The lemmas of this chapter deal with such approximations. They will be used later on.

4.2. LEMMA. *Let β_1^{n+1} and β_2^{n+1} be $(n+1)$-simplices in E_1^{n+1} and E_2^{n+1} resp. We suppose $n \geq 1$. Let y_0 be a point in the interior of β_1^{n+1}, and let $z, \bar{z}$ be two points in $\partial \beta_2^{n+1}$. Let the map $\phi \colon \partial \beta_1^{n+1} \to \partial \beta_2^{n+1}$ be continuous, and let η be a positive number. Then there exists a continuous map $g \colon \partial \beta_1^{n+1} \to \partial \beta_2^{n+1}$ such that (i) for all $x \in \partial \beta_1^{n+1}$*

$$|\phi(x) - g(x)| < \eta,$$
(4.1)

and (ii) *each of the equations*

$$g(x) = z, \qquad g(x) = \bar{z}$$
(4.2)

has at most a finite number of solutions.

PROOF. Let R_1 be such a positive number that

$$\phi(\partial \beta_1^{n+1}) \subset B(y_0, R_1).$$
(4.3)

Let y_1 and $\bar{y}_1$ be points on the rays $\overrightarrow{y_0 z}$ and $\overrightarrow{y_0 \bar{z}}$ resp. for which

$$R_1 < \begin{cases} |y_1|, \\ |\bar{y}_1|. \end{cases}$$
(4.4)

Then if ε is a positive number, one sees by arguments quite similar to those used in subsection 3.9 that there is a barycentric subdivision K_1^{n+1} of β_1^{n+1} and a map $\gamma \colon \partial \beta_1^{n+1} \to E_2^{n+1}$, which is simplicial with respect to ∂K_1^{n+1}, which satisfies

$$|\phi(x) - \gamma(x)| < \varepsilon$$
(4.5)

for all $x \in \partial \beta_1^{n+1}$, and which has the following property: if $u_1, \ldots, u_N$ are the vertices of ∂K^{n+1} and if $v_1 = \gamma(u_1), \ldots, v_N = \gamma(u_N)$, then the points

$$y_0, y_1, \bar{y}_1, v_1, \ldots, v_N$$
(4.6)

are in general position. Moreover, if we choose

$$\varepsilon < \text{dist}(y_0, \partial \beta_2^{n+1}),$$
(4.7)

then

$$y_0 \notin \gamma(\beta_1^{n+1}), \tag{4.8}$$

and each of the rays $\overrightarrow{y_0 z} = \overrightarrow{y_0 y_1}$ and $\overrightarrow{y_0 z} = \overrightarrow{y_0 \overline{y}_1}$ intersects $\gamma(\partial K_1^{n+1})$ in at most a finite number of points. Finally, if $\tilde{y}$ is one of these intersection points, then the equation

$$\gamma(x) = \tilde{y} \tag{4.9}$$

has (because of the independence of the points (4.6)) exactly one solution x in ∂K_1^{n+1}, and this solution lies in the interior of one of the n-simplices of ∂K_1^{n+1} (cf. condition (C_1) on equation (3.21)).

It follows from all this that if π is the central projection from y_0 on $\partial \beta_2^{n+1}$ and if

$$g(x) = \pi(\gamma(x)), \tag{4.10}$$

then g maps $\partial \beta_1^{n+1}$ into $\partial \beta_2^{n+1}$ and is continuous since the convexity of β_2^{n+1} implies that $\pi(y)$ is continuous for $y \neq y_0$ (see the Notes).[21] Moreover, x is a solution of one of the equations (4.2) if and only if x is a solution of (4.9) with a $\tilde{y}$ as defined above. Thus equations (4.2) have at most a finite number of solutions.

It remains to show that for $\eta > 0$ given ε can be chosen also in such a way that the inequality (4.1) holds with g given by (4.10). Now since $\pi(y)$ is uniformly continuous on any closed bounded set not containing y_0, there is an ε such that for all $\dot{y} \in \partial \beta_2^{n+1}$

$$|\pi y - \pi \dot{y}| = |\pi y - \dot{y}| < \eta \tag{4.11}$$

if

$$|y - \dot{y}| < \varepsilon. \tag{4.12}$$

Since $\phi(x) \in \partial \beta_2^{n+1}$ we may apply this with $\dot{y} = \phi(x)$ and $y = \gamma(x)$ to see that for such ε the inequality (4.5) implies (4.1) with g given by (4.10).

4.3. LEMMA. *Let S_1^n and S_2^n be spheres in E_1^{n+1} and E_2^{n+1} resp. Let p and $\overline{p}$ be two points of S_2^n which are not antipodal. Let ψ be a continuous map $S_1^n \to S_2^n$ and $\varepsilon > 0$. Then there exists a continuous map $h \colon S_1^n \to S_2^n$ of the following properties:* (i)

$$\|\psi(x) - h(x)\| < \varepsilon \tag{4.13}$$

for all $x \in S_1^n$; (ii) *the equations*

$$h(x) = p, \qquad h(x) = \overline{p} \tag{4.14}$$

have at most a finite number of solutions.

This lemma is a corollary to the preceding one. Indeed, let the simplex β_1^{n+1} be as in subsection 3.6, let, as in that section, $\kappa_1 = \kappa$ be the projection from the center of S_1^n on $\partial \beta_1^{n+1}$, and let β_2^{n+1} and κ_2 be defined in the corresponding way with respect to S_2^n. Then from the continuity of κ_1 and κ_2, we see that $\phi = \kappa_2 \psi \kappa_1^{-1}$ is a continuous map $\partial \beta_1^{n+1} \to \partial \beta_2^{n+1}$ and that to a given $\varepsilon > 0$ there corresponds an $\eta > 0$ such that for all $x \in S_1^n$

$$\|\kappa_2^{-1} \phi \kappa_1(x) - \kappa_2^{-1} g \kappa(x)\| < \varepsilon \tag{4.15}$$

if g is a continuous map $\partial \beta_1^{n+1} \to \partial \beta_2^{n+1}$ satisfying for all $\xi \in \partial \beta_1^{n+1}$

$$\|\phi(\xi) - g(\xi)\| < \eta. \tag{4.16}$$

We now choose for g a map $\partial \beta_1^{n+1} \to \partial \beta_2^{n+1}$ which has the properties (i) and (ii) asserted in Lemma 4.2. Then (4.1), i.e., (4.16), and therefore (4.15) is satisfied. Thus (4.13) holds with $h = \kappa_2^{-1} g \kappa_1$. But also assertion (ii) of the present lemma is satisfied since x is a solution, e.g., of the first of the equations (4.14), if and only if $\xi = \kappa_1(x)$ is a solution of the first of the equations (4.2) with $z = \kappa_2 p$.

§5. Notes to Chapter 7

Notes to §1.

1. *Proof of Lemma 1.2.* Π^{s+1} is a translate of an $(s+1)$-dimensional subspace of E^{n+1} (Definition 1.5 in Chapter 6). It therefore follows from the well-known orthogonality properties of E^{n+1} that there exists a $c_0 \in \Pi^{s+1}$ such that $c_0 - c$ and $x - c_0$ are orthogonal for every $x \in \Pi^{s+1}$. Since $x - c = x - c_0 + c_0 - c$ we see that $\|x - c\|^2 = \|x - c_0\|^2 + \|c_0 - c\|^2$ for all $x \in \Pi^{s+1}$. If in particular $x \in \Pi^{s+1} \cap S(c, r)$, then $\|x - c\|^2 = r^2$. Thus

$$\|x - c_0\|^2 = r^2 - \|c_0 - c\|^2. \tag{1}$$

This equation for x has no solution if $r < \|c_0 - c\|$, and the unique solution $x = c_0$ if $r = \|c_0 - c\|$. If $r > \|c_0 - c\|$ we see from (1) that $\Pi^{s+1} \cap S^n(c, r) = \Pi^{s+1} \cap S^n(c_0, (r^2 - \|c_0 - c\|^2)^{1/2})$.

2. *Proof of Lemma 1.4.* By definition a great circle S^1 through N is the intersection of $S^n(c, r)$ with a two-dimensional plane Π^2 containing N and c. It therefore contains the antipode S of N. That no third point lies on the intersection of two (not identical) great circles follows from the fact that a Π^2 is uniquely determined by three independent points.

3. With the notations of subsection 1.6 let $x_0 = t_0 = n_0^*$, and let $t = \sigma(x)$ be the stereographic projection of x with pole n_0. Then the formulas (3) and (4) below hold, which obviously imply the asserted continuity:

$$\text{(a)} \quad t = n_0 + \mu(x - n_0), \qquad \text{(b)} \quad \mu = 4r^2/\|x - n_0\|^2, \tag{3}$$

$$\text{(a)} \quad x = n_0 + \lambda(t - n_0), \qquad \text{(b)} \quad \lambda = 4r^2/\|t - n_0\|^2, \tag{4}$$

Proof of (3) and (4). It is obvious that (3a) and (4a) hold for some real μ and λ. To prove that these numbers have the values given by (3b) and (4b), we note that

$$(n_0 - c, t - c) = (n_0 - c, t - x_0) + (n_0 - c, x_0 - c) = -r^2, \tag{5}$$

since $(n_0 - c, t - x_0) = 0$. Moreover, since $\|n_0 - c\|^2 = \|x - c\|^2 = r^2$, we see that

$$2(n_0 - c, n_0 - x) = 2(n_0 - c, n_0 - c) + 2(n_0 - c, c - x)$$
$$= \|n_0 - c\|^2 + 2(n_0 - c, c - x) + \|c - x\|^2.$$

Thus

$$2(n_0 - c, n_0 - x) = \|n_0 - c + c - x\|^2 = \|n_0 - x\|^2. \tag{6}$$

To prove (3b) we subtract c from both members of (3a) and then multiply by the scalar $n_0 - c$. We then see from (5) and (6) that

$$-r^2 = r^2 - (\mu/2)\|n_0 - x\|^2.$$

This proves (3b). To prove (4b) we note that by (3)

$$(x - n_0, t - n_0) = \mu\|x - n_0\|^2 = 4r^2.$$

Therefore scalar multiplication of (4a) by $t - n_0$ yields $4r^2 = \lambda\|t - n_0\|^2$. This proves (4b).

4. As in subsection 1.7 of Chapter 6 let $b_1, b_2, \ldots, b_n$ be a base for the tangent space E_x. Suppose, moreover, that $b_0 = c - t_0, b_1, \ldots, l_n$ form a positive base for E^{n+1} and that likewise $b'_1, b'_2, \ldots, b'_n$ is a base for E^n_x and that $b'_0 = c - t_0, b'_1, \ldots, b'_n$ form a positive base for E^{n+1}. We have to show that $b_1, \ldots, b_n$ and $b'_1, \ldots, b'_n$ determine the same orientation of E^n_x, i.e., (see subsection 1.22 of Chapter 6) that if

$$b'_i = \sum_{j=1}^{n} \beta_{ij} b_j, \qquad i = 1, 2, 3, \ldots, n, \tag{7}$$

then the determinant Δ^n of the β_{ij} is positive. If we enlarge (7) by adding $b'_0 = b_0$, then (again by subsection 1.22, Chapter 6) the determinant Δ^{n+1} of the enlarged system is positive since the b_i and b'_i, $i = 0, 1, \ldots, n$, both form a positive base. But obviously $\Delta^n = \Delta^{n+1}$.

5. It is easy to see that an r_{01} with the properties assumed in Lemma 1.8 exists: assuming $c = 0$ let E^2 be the 2-subspace of E^{n+1} containing b_0^1 and b_0^2, and let b_1^1 be a point of E^2 orthogonal to b_0^1. Then obviously there exists an orthogonal map (rotation) of E^2 onto itself which maps b_0^1 into b_0^2 and whose determinant with respect to the base b_0^1, b_1^1 is positive. If $n = 1$ we subject b_1^1 to the additional condition that b_0^1, b_1^1 form a positive base for $E^2 = E^{n+1}$. If $n \geq 2$ we choose points b_i^1, $i = 2, 3, \ldots, n$, such that the points $b_0^1, b_1^1, b_2^1, \ldots, b_n^1$ form an orthogonal base for E^{n+1} which is positive. We extend the map r_{12} defined on E^2 to E^{n+1} by setting

$$r_{12}(b_i^1) = b_i^1, \qquad i = 2, 3, \ldots, n. \tag{6}$$

The linear map determined by these additional conditions has the desired properties.

6. *Proof of Lemma 1.8.* With the notations of that lemma, let $b_0^1, b_1^1, \ldots, b_n^1$ be an orthogonal base for E^{n+1}, which is positive. Then $b_1^1, b_2^1, \ldots, b_n^1$, being orthogonal to $b_0^1 = c - t_0^1$, form a base for the tangent space $E_{t_0^1}$ of $S^n(c, r)$ at t_0^1.

Moreover by the definitions given in subsection 1.7 this base for $E_{t_0^1}$ is a positive one. On the other hand, since r_{12} is orthogonal and preserves the orientation, it follows that the $b_i^2 = r_{12}(b_i^1)$, $i = 0, 1, \ldots, n$, form a positive orthogonal base for E^{n+1}. This again implies that $r_{12}(b_1^1), r_{12}(b_2^1), \ldots, r_{12}(b_n^1)$ form a positive base for the tangent space $E_{t_0^2}$ to $S(c, r)$ at $t_0^2 = c - b_0^2$. This obviously proves the assertion (1.5). The equivalence of (1.6) with (1.5) follows from the definitions given in subsection 1.7.

7. *Proof of Lemma* 1.10. σ_{01} is a continuous one-to-one map of $\overline{\Omega}_0 \subset T_0$ onto $\overline{\Omega}_1 \subset T_1$, and if $\bar{t}_1 \in \Omega_1$, then the equation

$$\sigma_{01}(t_0) = \bar{t}_1 \tag{7}$$

has one and only one solution $t_0 = \bar{t}_0 \in \Omega_0$. So $d(\sigma_{01}, \Omega_0, \bar{t}_1) = d(\sigma_{01}, \Omega, \bar{t}_1)$ for any open set Ω satisfying $\bar{t}_0 \in \Omega \subset \Omega_0 \subset T_0$. Thus

$$d(\sigma_{01}, \Omega_0, \bar{t}_1) = d(\sigma_{01}, \sigma_0(U), \bar{t}_1) \tag{8}$$

for any open subset U of $S^n(c, r)$ for which

$$\bar{x}_0 = \sigma_0^{-1}(\bar{t}_0) \subset U \subset U_0. \tag{9}$$

Consequently to prove the assertion (1.9) it will be sufficient to show that

$$d(\sigma_{01}, \sigma_0(U), \bar{t}_1) = +1 \tag{10}$$

for an arbitrary U satisfying the above conditions. Now let Π^2 be a 2-plane containing c, n_0, and n_1 (and therefore s_0 and s_1). Let S^1 be the great circle $S^n(c, r) \cap \Pi^2$ (cf. subsection 1.3). We first consider the case that

$$\bar{s}_0 = \sigma_0^{-1}(\bar{t}_0) \notin S^1. \tag{11}$$

We then choose for U an open set which, in addition to (9), satisfies

$$\overline{U} \cap S^1 = \varnothing. \tag{12}$$

Let $S_{01}^1 : x = n(\theta)$, $0 \le \theta \le 1$, be one of the segments of S^1 which is bounded by n_0 and n_1 such that

$$n(0) = n_0, \qquad n_0(1) = n_1. \tag{13}$$

Let $s(\theta)$ be the antipode of $n(\theta)$, let T_θ be the tangent plane to $S^n(c, r)$ at $s(\theta)$, and let σ_θ be the stereographic projection with pole $n(\theta)$. Then by assumption (12) the stereographic projection $\sigma_\theta(x)$ is defined for every $x \in \overline{U}$ and $\theta \in [0, 1]$. We denote the open subset $\sigma_\theta(U)$ of T_θ by Ω_θ. For $\theta \in [0, 1]$ we set

$$\sigma_{\theta 1} = \sigma_1 \sigma_\theta^{-1}, \qquad \tau_{\theta 1} = (\sigma_{\theta 1})^{-1}. \tag{14}$$

Then

$$\sigma_{01}(t_0) = \sigma_1\sigma_0^{-1}(t_0) = t_1 \in T_1 \quad \text{for } t_0 \in \sigma_0(U),$$

$$\sigma_{11}(t_1) = t_1 \in T_1 \quad \text{for } t_1 \in \sigma_1(U), \quad (\tau_{\theta1})\Omega_1 = \Omega_\theta. \tag{15}$$

Now similar to the construction used in Note 5 we define a rotation r_θ of E^{n+1} with center c which keeps the orientation of E^{n+1}, which maps s_1 into s_θ and for which $r_\theta(\vec{T}_1) = \vec{T}_\theta$, while for an open set $V_1 \subset T_1$ and for an arbitrary point $t_\theta \in r_\theta(V_1) \subset T_1$

$$d(r_\theta, V_1, \bar{t}_\theta) = +1. \tag{16}$$

Now let $\phi_\theta = (r_\theta)^{-1}\tau_{\theta1}$ such that, by (15), $\phi_\theta(\Omega_1) = (r_\theta)^{-1}\Omega_\theta \subset T_1$. Since then $\tau_{\theta1} = r_\theta\phi_\theta$, we see from subsection 2.15 in Chapter 6 and (15) that for $t_\theta \in \Omega_\theta$

$$d(\tau_{\theta1}, \Omega_1, t_\theta) = d(r_\theta\phi_\theta, \Omega_1, t_\theta) = d(r_\theta, \phi_\theta\Omega_1, t_\theta) \cdot d(\phi_\theta, \Omega_1, t_1), \tag{17}$$

where t_1 is an arbitrary point of $\phi_\theta(\Omega_1) = (r_\theta)^{-1}\Omega_\theta \subset T_1$. We now see from (16) with $V_1 = \phi_\theta(\Omega_1)$ and from (17) that

$$d(\tau_{\theta1}, \Omega_1, t_\theta) = d(\phi_\theta, \Omega_1, t_1). \tag{18}$$

Now let $\bar{u}$ be an arbitrary point of the open subset U of $S^{n+1}(c, r)$ (cf. (12)), and let $t_1(\theta) = (r_\theta)^{-1}\sigma_\theta(\bar{u})$. Then $t_1(\theta) \in (r_\theta)^{-1}\Omega_\theta = (r_\theta)^{-1}\tau_{\theta1}(\Omega_1) = \phi_\theta(\Omega_1)$ by (15) and definition of ϕ_θ, which shows that the right member of (18) equals $d(\phi_\theta, \Omega_1, t_1(\theta))$ and also that $t_1(\theta) \notin \phi_\theta(\partial\Omega_1)$ since ϕ_θ is continuous and one-to-one while Ω_1 is open. Therefore, by Theorem 5.4 in Chapter 3, $d(\phi_\theta, \Omega_1, t_1(\theta))$ is independent of θ and has for all $\theta \in [0, 1]$ the value $+1$ since ϕ_1 is the identity on Ω_1. It now follows from (18) that $d(\tau_{01}, \Omega_1, t_0) = +1$. By (14) this proves the assertion (1.9) since the first part of (1.9) is obvious.

Suppose now the assumption (11) is not satisfied. In this case, using the notations of (7), we choose a positive number ρ_1 such that

$$d(\sigma_{01}, \Omega_0, \bar{t}_1) = d(\sigma_{01}, \Omega_0, t_1) \tag{20}$$

if

$$t_1 \in B(\bar{t}_1, \rho_1). \tag{21}$$

On the other hand $\bar{t}_0$ is the unique solution of (7). Therefore there exists a position ρ_0 such that

$$\sigma_{01}(t_0) \in B(t_1, \rho_1)$$

if

$$t_0 \in B(\bar{t}_0, \rho_0). \tag{22}$$

Thus (20) is true with $t_1 = \sigma_{01}(t_0) = \sigma_1\sigma_0^{-1}(t_0)$ if t_0 satisfies (22). But since $n \geq 2$ the condition (22) is satisfied for some $t_0 = \bar{\bar{t}}_0$ for which $\sigma_0^{-1}(\bar{\bar{t}}_0) \notin S'$, i.e., for which (11) is true with $\bar{t}_0$ replaced by $\bar{\bar{t}}_0$. Therefore

$$d(\sigma_{01}, \Omega_0, \bar{\bar{t}}_1) = 1 \tag{23}$$

for $\bar{\bar{t}}_1 = \sigma_{01}(\bar{\bar{t}}_0)$. The assertion (1.9) follows now from (20) and (23) since (20) holds for $t_1 = \bar{\bar{t}}_1$.

8. *Proof of the existence of a U_1 satisfying* (1.14). Since f is continuous the set $C_1 = f^{-1}(y_0)$ is closed and since, by (1.12), $n_1 \notin C_1$, there exists a positive δ_1 such that

$$\|n_1 - c\| > \delta_1 \quad \text{for all } c \in C_1. \tag{24}$$

Moreover, since $n_2 \neq y_0$ and since f is uniformly continuous on C_1, there is a positive δ_2 such that

$$\|f(x) - f(c)\| < \|n_2 - y_0\|/2 \tag{25}$$

for all couples x, c with $c \in C_1$, $x \in S_1^n(c_1, r_1)$ for which $\|x - c\| < \delta_2$. We claim that if $\delta = \min(\delta_1, \delta_2)$ and

$$U_1 = \bigcup_{c \in C_1} B(c, \delta), \tag{26}$$

then U_1 satisfies (1.14). Indeed part (a) of this assertion follows from (24), and part (b) is obvious from (26). To prove part (c) we note that for $x \in U_1$ we can, on account of (26), find a $c \in C_1$ for which $\|x - c\| < \delta < \delta_2$ and for which therefore (25) holds. Since $f(c) = y_0$ we see that

$$\|n_2 - f(x)\| \geq \|n_2 - f(c)\| - \|f(c) - f(x)\|$$
$$= \|n_2 - y_0\| - \|f(c) - f(x)\| \geq \|n_2 - y_0\|/2 > 0.$$

9. *Proof of* (1.19). Suppose that (1.19) is not true, i.e., that (1.16) holds for some $t_1 \in \partial\Omega_1 = \sigma_1(\partial U_1)$. Then we know that $x = \sigma_1^{-1}(t_1) \in \partial U_1$ satisfies (1.17) and, therefore, (1.18). Thus $x \in (\partial U_1) \cap C_1$. But this intersection is empty as follows from part (b) of (1.14) since U_1 is open and C_1 is closed.

10. *Proof of the independence of the degree* $d(f_1, \Omega_1, \sigma_2(y_0))$ *of the choice of* n_1 *and* U_1. Let n_1' be a point of $S_1^n(c_1, r_1)$ satisfying

$$f(n_1') \neq y_0. \tag{27}$$

Let n_2' be a point of $S_2^n(c_2, r_2)$ different from y_0, and let U_1' be an open subset of $S_1^n(c_1, r_1)$ which satisfies (1.14) with n_1, n_2, U_1 replaced by n_1', n_2', U_1' resp. For $i = 1, 2$ let s_i' be the antipode of n_i', let T_i' be the tangent plane to $S_i^n(c_i, r_i)$ at s_i', let σ_i' be the stereographic projection with pole n_i', and let $\Omega_1' = \sigma_1'(U_1')$. Finally let f_1' be the map $\overline{\Omega_1'} \to T_2'$ defined by

$$f_1'(t_1') = \sigma_2' f(\sigma_1')^{-1}(t_1'), \qquad t_1' \in \Omega_1' = \sigma_1'(U_1'). \tag{28}$$

We have to prove that

$$d(f_1', \sigma_1'(U_1'), \sigma_2'(y_0)) = d(f_1, \sigma_1(U_1), \sigma_2(y_0)). \tag{29}$$

Clearly the intersection $U_1'' = U_1 \cap U_1'$ satisfies (1.14) with U_1 replaced by U_1''. In particular part (b) of that relation together with (1.13) shows that

$$d(f_1, \sigma_1(U_1), \sigma_2(y_0)) = d(f_1, \sigma_1(U_1''), \sigma_2(y_0)). \tag{30}$$

In the same way one sees that

$$d(f_1', \sigma_1'(U_1'), \sigma_2'(y_0)) = d(f_1', \sigma_1'(U_1''), \sigma_2'(y_0)). \tag{31}$$

It follows from (30) and (31) that for the proof of (29) it will be sufficient to show that that relation holds with U_1' and U_1 replaced by U_1''. In other words we may assume that

$$U_1' = U_1. \tag{32}$$

Now

$$\sigma_1'(U_1) = (\sigma_1', \sigma_1^{-1})\sigma_1(U_1), \tag{33}$$

and by (28) and subsection 1.15

$$f_1' = (\sigma_2'\sigma_2^{-1})\sigma_2 f \sigma_1^{-1}\sigma_1(\sigma_1')^{-1} = \sigma_2'\sigma_2^{-1} f_1(\sigma_1(\sigma_1')^{-1}). \tag{34}$$

Now from (32)–(34) together with part (i) of subsection 2.15 in Chapter 6 we see that

$$d(f_1', \sigma_1'(U_1), \sigma_2'(y_0)) = d(\sigma_2'\sigma_2^{-1} f_1, \sigma_1(U_1), \sigma_2'(y_0)) \cdot d(\sigma_1(\sigma_1')^{-1}, \sigma_1'(U_1), t_1),$$

where $t_1 \in \sigma_1(U_1)$. But the second factor of the right member equals $+1$ by Lemma 1.10. Thus

$$d(f_1', \sigma_1'(U_1), \sigma_2'(y_0)) = d(\sigma_2'\sigma_2^{-1} f_1, \sigma_1(U_1), \sigma_2'(y_0)). \tag{35}$$

Using part (ii) of subsection 2.15 in Chapter 6 we see that

$$d(\sigma_2'\sigma_2^{-1} f_1, \sigma_1(U_1), \sigma_2'(y_0)) = d(\sigma_2'\sigma_2^{-1}, f_1\sigma_1 U_1, \sigma_2'(y_0)) \cdot d(f_1, \sigma_1 U_1, t_2), \tag{36}$$

where t_2 is an arbitrary point in $f_1\sigma_1 U_1 = \sigma_2 f(U)$. Since $y_0 \in f(U)$ we may choose $t_2 = \sigma_2(y_0)$. On the other hand, again by Lemma 1.10, the first factor at the right member of (36) equals $+1$ since $\sigma_2'(y_0) \in \sigma_2'f(U_1) = \sigma_2'\sigma_2^{-1}(f_1\sigma_1 U_1)$. This together with (36), (35), and (32) proves the assertion (29).

Notes to §2.

11. *Proof of Lemma 2.4.* By Definition 1.13 it will be sufficient to prove that $d(f_t, S_1^n, y_0)$ is independent of t for some $y_0 \in S_2^n$. For this again it will, by the Borel theorem, be sufficient to show that if $t_0 \in [0,1]$ then there exists a positive $\delta = \delta(t_0)$ such that

$$d(f_t, S_1^n, y_0) = d(f_{t_0}, S_1^n, y_0) \tag{37}$$

if

$$|t - t_0| < \delta, \qquad t \in [0,1]. \tag{38}$$

We consider the three cases I, II, III defined in subsection 1.11.

Case I. $y_0 \notin f_{t_0}(S_1^n)$ for some $y_0 \in S_2^n$. Then the right member of (37) equals zero (e.g. by Lemma 2.1), and y_0 has a positive distance 2ε from $f_{t_0}(S_1^n)$ since the latter set is closed. We now choose δ in such a way that for all $x \in S_1^n$, $\|f_t(x) - f_{t_0}(x)\| < \varepsilon$ if (38) is satisfied. For such t

$$\|y_0 - f_t(x)\| \geq \|y_0 - f_{t_0}(x)\| - \|f_{t_0}(x) - f_t(x)\| \geq \varepsilon > 0$$

for all $x \in S_1^n$. This shows that $f_t(S_1^n)$ is a proper subset of S_2^n. Therefore $d(f_t, S_1^n, y_0) = 0$ by Lemma 2.1. This proves (37), both members of that equality being zero.

Case II. $f_{t_0}(S_1^n)$ is a single point y_1 of S_2^n. Then, for any point $y_0 \in S_2^n$ different from y_1, we are in Case I.

Case III. In this case there exists an $n_1 \in S_1^n$ such that

$$f_{t_0}(n_1) \neq y_0. \tag{39}$$

Let n_2 be a point of S_2^n such that

$$n_2 \neq y_0, \tag{40}$$

and for $t \in [0, 1]$ let

$$C_t = \{x \in S_1^n \mid f_t(x) = y_0\}. \tag{41}$$

Then we know that there exists an open set $U_1 \subset S_1^n$ such that (1.14) holds with C_1 replaced by C_{t_0} and f by f_{t_0}, and that

$$d(f_{t_0}, S_1^n, y_0) = d(\sigma_2 f_{t_0} \sigma_1^{-1}, U_1, \sigma_2(y_0)) \tag{42}$$

(see (1.20) and (1.15)). On the other hand, we see from (39), (40), from part (c) of (1.14) (with f replaced by f_{t_0}), and from the uniform continuity of f_t on $S_1^n \times [0, 1]$ that there exists a $\delta_1 > 0$ such that

$$f_t(n_1) \neq y_0, \qquad n_2 \notin f_t(\overline{U}_1) \tag{43}$$

if

$$|t - t_0| < \delta_1. \tag{44}$$

Moreover, as will be proved below, δ_1 can be chosen in such a way that (44) implies the inclusion

$$C_t \subset U_1. \tag{45}$$

But from (43) and (45) we see that (1.14) is true if f is replaced by f_t and C_1 by C_t for t satisfying (44). It follows that for such t

$$d(f_t, S_1^n, y_0) = d(\sigma_2 f_t \sigma_1^{-1}, \sigma_1(U_1), \sigma_2(y_0)). \tag{46}$$

Now since $\sigma_2 f_t \sigma_1^{-1}$ for fixed t is a "plane" map, it follows from Theorem 2.9 of Chapter 6 that there exists a $\delta_2 > 0$ such that for $|t - t_0| < \delta_2$ the right members of (46) and (42) are equal. Thus these two equalities show that (38) with $\delta = \min(\delta_1, \delta_2)$ implies (37).

It remains to prove that δ_1 can be chosen in such a way that (44) implies (45). Since S_1^n is a compact subset of E^{n+1}, this assertion is a consequence of the following lemma: *let Σ be a compact subset of E_1^{n+1} and f a continuous map $\Sigma \rightarrow E_2^{n+1}$. Let y_0 be a point of E_2^{n+1} and $C(f) = \{x \in \Sigma \mid f(x) = y_0\}$. Let U_1 be an open bounded subset of E_1^{n+1} containing $C(f)$. Then there exists a positive η of the following property: if the map $g : \Sigma \rightarrow E_2^{n+1}$ satisfies*

$$\|f(x) - g(x)\| < \eta \qquad \text{for all } x \in \Sigma, \tag{47}$$

then

$$C(g) \subset U_1, \qquad (48)$$

where $C(g) = \{x \in S_1^n \mid g(x) = y_0\}$.

PROOF. Suppose there is no such η. Then there exists a sequence $\eta_1, \eta_2, \ldots$ of positive numbers converging to zero and continuous maps g_ν such that for all $x \in \Sigma$

$$\|g_\nu(x) - f(x)\| < \eta_\nu, \qquad (49)$$

while there exist points

$$x_\nu \in g_\nu^{-1}(y_0) \qquad (50)$$

for which

$$x_\nu \notin U_1. \qquad (51)$$

Since Σ is compact we may assume that

$$\lim_{\nu \to \infty} x_\nu = x_0, \qquad (52)$$

where $x_0 \in \Sigma$. Now $y_0 = g_\nu(x_\nu)$ by (50), and therefore

$$\|f(x_0) - y_0\| = \|f(x_0) - g_\nu(x_\nu)\| \leq \|f(x_0) - f(x_\nu)\| + \|f(x_\nu) - g_\nu(x_\nu)\|.$$

Here the two terms of the right member tend to zero as $\nu \to \infty$, the first one by (52) and the second one by (49). It follows that $f(x_0) = y_0$, i.e., that

$$x_0 \in C(f) \subset U_1. \qquad (53)$$

Now U_1 is open. Therefore (53) together with (52) implies that $x_\nu \in U_1$ for ν big enough. But this contradicts (51).

12. *Proof of* (2.8). Since (2.6) does not hold, there exists an $x_0 \in S_1^n$ such that

$$gf(x_0) \neq n_3. \qquad (54)$$

We now verify that for such x_0 the assertions (a)–(d) of (2.8) are true: if (a) were not true, then $f(n_1) = y_0$, and, by (2.5) and (2.7), $n_3 = gf(n_1) = g(y_0) = gf(x_0)$, which contradicts (54). Similarly, if (b) were not true, then $g(n_2) = z_0$ and $n_3 = gf(n_1) = z_0 = gf(x_0)$, which again contradicts (54). For the proof of (c) let $x \in (gf)^{-1}(z_0)$. Then $gf(x) = z_0$, i.e., $f(x) \in g^{-1}(z_0)$, which proves (c). Finally, for the proof of (d) let $x \in f^{-1}(y_0)$. Then $f(x) = y_0$ and, by (2.7),

$$gf(x) = g(y_0) = z_0, \quad \text{i.e., } x \in (gf)^{-1}(z_0).$$

13. *Proof of Lemma* 2.8. The sphere S_t^n is defined for $t = 0$ and $t = 1$. We extend this definition to all $t \in [0, 1]$ by denoting by S_t^n the sphere with center c and radius $r_t = (1 - t)r_0 + tr_1$. By π_t we denote the central projection from c of S_0^n on S_t^n. Now let a_0 be an arbitrary point on S_0^n, let $a_t = \pi_t(a_0) \in S_t^n$, and let n_t be its antipode. We will show that

$$d(\pi_t, S_0^n, a_t) = +1. \qquad (55)$$

For $t = 1$ this equality is, by Definition 1.13, equivalent to our assertion (2.25). In order to prove (55) we consider a positive number $\delta < \min(r_0, r_1)$. If $x_t = \pi_t(x)$ for $x \in S_0^n$, we see that $\|a_t - n_t\| = 2r_t > 2\delta$. Now if $U_0 = \{x \in S_0^n | \|x - a_0\| < \delta\}$, then

$$\|x - n_0\| \geq \|a_0 - n_0\| - \|x - a_0\| = 2r_0 - \|x - a_0\| > \delta \qquad (56)$$

for $x \in U_0$. We claim that then (1.14) is satisfied with $f = \pi_t$ and n_1, U_1 replaced by n_0, U_0 resp., i.e., that

(a) $\operatorname{dist}(n_0, U_0) > 0$;   (b) $\pi_t^{-1}(a_t) \subset U_0$;   (c) $n_t \notin \overline{\pi_t(U_0)}$.   (57)

Indeed, (a) follows directly from (56), and (b) from the fact that $\pi_t^{-1}(a_t) = a_0 \in U_0$. To prove (c) we note that $\pi_t(x)$ is given by $\pi_t(x) - c = (x - c)r_t(r_0)^{-1}$, and therefore $\pi_t(x) - n_t = \pi_t(x) - \pi_t(n_0) = (x - n_0)r_t(r_0)^{-1}$. Consequently, by (56), for $x \in U_0$

$$\|\pi_t(x) - n_t\| \geq \delta r_t(r_0)^{-1} \geq \delta \min(r_0, r_1)r_0^{-1}.$$

This proves (c). Now in subsection 1.11 we showed that (1.14) implies (1.20). Thus by (57)

$$d(\pi_t, S_0^n, a_t) = d(\sigma_t \pi_t \sigma_0^{-1}, \sigma_0 U_0, \sigma_t(a(t))), \qquad (58)$$

where, for each $t \in [0, 1]$, σ_t is the stereographic projection of S_t^n with pole n_t on the tangent plane T_t to S_t^n at the point a_t. But the degree at the right of (58) is—by remark (c) in subsection 1.7—equal to the degree of the corresponding mapping between the tangent spaces E_0 and E_t to S_0^n and S_t^n at a_0 and a_t resp. Now E_t is the translate of T_t which maps the point a_t into θ. Therefore taking into account the definition of the orientation of tangent spaces we see that $E_t = E_0$ (as oriented spaces) since the planes T_t are parallel to T_0. Thus

$$d(\sigma_t \pi_t \sigma_0^{-1}, \sigma_0 U_0, \sigma_t(a(t))) = d(\phi_t, \Omega_0, \theta), \qquad (59)$$

where $\Omega_0 = \{\xi \in E_0 | \xi + a_0 \in \sigma_0 U_0\}$ and where ϕ_t is the map $\overline{\Omega_0} \to E_0$ given by $\phi_t(\xi) - a_t = \sigma_t \pi_t \sigma_0^{-1}(\xi + a_0)$. But, by the continuity property of the degree, the right member of (59) is independent of t and therefore equals $+1$ since ϕ_0 is the identity on Ω_0. The assertion (55) now follows from (59) and (58).

Notes to §3.

14. *Proof of Lemma 3.3.* It will be sufficient to show that if c_1 is a given point in E_2^{n+1} satisfying (3.1), then there exists a $\delta > 0$ such that

$$v(f(S_1^n), c) = v(f(S_1^n), c_1) \qquad (60)$$

if

$$|c - c_1| < \delta. \qquad (61)$$

Now let r_2 be a positive number satisfying (3.2). Then obviously

$$B(c, r_2/2) \subset B(c_1, r_2) \qquad (62)$$

if c satisfies (61) with

$$0 < \delta < r_2/2. \tag{63}$$

Now it was already established in subsection 3.1 that the right member of (3.3) does not depend on the particular choice of the (small enough) radius of S_2^n. Therefore

$$v(f(S_1^n), c_1) = d(\pi_{c_1} f, S_1^n, S_2^n(c_1, r_2/2)), \tag{64}$$

where π_{c_1} is the central projection with center c_1 on $S_2^n(c_1, r_2/2)$. On the other hand,

$$v(f(S_1^n), c) = d(\pi_c f, S_1^n, S_2^n(c, r_2/2)). \tag{65}$$

We have to prove that for $\|c - c_1\|$ small enough the right members of (64) and (65) are equal. For this purpose we introduce the translation $\tau_c(y) = y + c_1 - c$. Then $\tau_c S_2^n(c, r_2/2) = S_2^n(c_1, r_2/2)$, and $\tau_c \pi_c f$ maps S_1^n on $S_2^n(c_1, r_2/2)$. From the product theorem of subsection 2.5 in conjunction with Lemma 2.9 we see that

$$d(\tau_c \pi_c f, S_1^n, S_2^n(c_1, r_2/2))$$
$$= d(\pi_c f, S_1^n, S_2^n(c, r_2/2)) \cdot d(\tau_c, S_2^n(c, r_2/2), S_2^n(c_1, r_2/2)) \tag{66}$$
$$= d(\pi_c f, S_1^n, S_2^n(c, r_2/2)).$$

This holds if δ and c satisfy (63) and (61) resp. We will show below that δ can be chosen in such a way that (61) implies that also

$$d(\tau_c \pi_c f, S_1^n, S_2^n(c_1, r_2/2)) = d(\pi_{c_1} f, S_1^n, S_2^n(c_1, r_2/2)). \tag{67}$$

But (66) and (67) together show that the right members of (64) and (65) are equal as we wanted to prove. Finally to verify (67) (for $|c - c_1|$ small enough) we note that $\tau_c \pi_c f$ and $\pi_{c_1} f$ both map S_1^n into $S_2^n(c_1, r_2/2)$. Therefore by Theorem 2.6 it will for the proof of (67) be sufficient to show that $\tau_c \pi_c f(x)$ tends to $\pi_{c_1} f(x)$ uniformly for $x \in S_1^n$ as $c \to c_1$. To prove this assertion we note that for $y \in f(S_1^n)$, $\pi_c(y) = c + (y - c)(2\|y - c\|)^{-1} r_2$, and therefore $\tau_c \pi_c(y) = c_1 + (y - c)(2\|y - c\|)^{-1} r_2$. Thus

$$\tau_c \pi_c(y) - \pi_{c_1}(y) = \left(\frac{y - c}{\|y - c\|} - \frac{y - c_1}{\|y - c_1\|} \right) \frac{r_2}{2}.$$

This proves our assertion since $(f(x) - c) \cdot \|f(x) - c\|^{-1}$, on account of (3.2), is uniformly continuous on $S_1^n \times \{|c - c_1| \leq r_2/2\}$.

15. *Proof that*

$$d(\kappa, B_1^{n+1}, \eta) = +1 \tag{68}$$

(cf. subsection 3.6). It is sufficient to prove (68) with $\eta = \gamma$. For $t \in [0, 1]$ and $x \in \overline{B_1^{n+1}}$, let $\kappa_t(x)$ be defined by $\kappa_t(x) - \gamma = (1 - t)(\kappa(x) - \gamma) + t(x - \gamma)$. Then by (3.6):

$$\kappa_t(x) - \gamma = \left[(1 - t) \frac{\|\kappa(\dot{x}) - \gamma\|}{r_1} + 1 \cdot t \right] (x - \gamma). \tag{69}$$

If $\delta = \text{dist}(\partial \beta_1^{n+1}, \gamma)$, then $\|\kappa(\dot{x}) - \gamma\| \geq \delta$ and the bracket in (69) is not smaller than $\min(\delta r_1^{-1}, 1)$, and it follows from (69) that $\kappa_t(x) \neq \gamma$ for $x \in S_1^{n+1} = \partial B_1^{n+1}$. Therefore $d(\kappa_t, B_1^{n+1}, \gamma)$ is independent of t and thus equals $d(\kappa_1, B_1^{n+1}, \gamma) = +1$ since, by (69), $\kappa_1(x) = x$.

16. *Proof of* (3.14). Since $\pi B(\theta, R_2) = \pi \partial B(\theta, R_2)$ (as point sets), the assertion is a consequence of the following lemma: *for $i = 1, 2$ let Σ_i be a sphere in E^{n+1} of radius r_i and center y_i. Suppose that*

$$r_1 + r_2 < \|y_1 - y_2\| \tag{70}$$

(i.e., the closures of the balls bounded by Σ_1^n and Σ_2^n are disjoint). Let π be the central projection from y_1 of Σ_2^n on Σ_1^n. Then $\pi \Sigma_2^n$ is a proper subset of Σ_1^n.

PROOF. Let P^n be the n-plane through y_1 which is orthogonal to $y_1 - y_2$. Let H_1 and H_2 denote the two open half spaces into which P^n decomposes E^{n+1} (see subsection 1.24 of Chapter 6). By (70) the sphere Σ_2^n lies in one of them, say in H_2. Since $\overline{H_2}$ is convex, all rays $\overrightarrow{y_1 x}$, $x \in \Sigma_2^n$, lie in $\overline{H_2}$. Thus no point of $\pi \Sigma_2$ lies in H_1. But obviously Σ_1^n contains points in H_1.

17. That the right member of (3.9) (with c_1 replaced by $y(t)$) is constant in a t-interval whose elements satisfy (3.20) follows from Theorem 2.9 of Chapter 6. The corresponding assertion for the left member follows from Lemma 3.3, since $f(S_1^n) = g(\partial \beta^{n+1})$.

18. *Proof of the properties* (C_1)–(C_3) *in subsection* 3.9. The proof will be based on the following properties (C_4) and (C_5).

(C_4) Let σ_1^n and σ_2^n be two different n-simplices of ∂K^{n+1}, and let $\tau_1^n = g(\sigma_1^n)$ and $\tau_2^n = g(\sigma_2^n)$. Suppose that

$$\tau_{12}^n = \tau_1^n \cap \tau_2^n \tag{71}$$

is not empty. Then

$$\dim \tau_{12}^n \leq n - 1, \tag{72}$$

where $\dim \tau_{12}^n$ denotes the dimension of the plane spanned by τ_{12}^n.

(C_5) $y(\bar{t})$ does not lie in a simplex τ^m of $g(\partial K^{n+1})$ of dimension $m \leq n - 1$ if $\bar{t}$ is as in the sentence following (3.20).

Proof of (C_4). Let $b_0, b_1, \ldots, b_n$ be the vertices of σ_1^n, and $c_0, c_1, \ldots, c_n$ be those of σ_2^n. Since σ_1^n and σ_2^n are different, at least one of the vertices of σ_2^n, say c_0, is different from all vertices of σ_1^n. Then the $n + 2$ points $\gamma_0 = g(c_0)$, $\beta_0 = g(b_0), \ldots, \beta_n = g(b_n)$ are in general position since they form a subset of the points (3.20). They therefore span the $(n+1)$-space E_2^{n+1}. We thus see that if Π_1 and Π_2 are the planes spanned by τ_1^n and τ_2^n resp., then their hull H (see Definition 1.7 of Chapter 6) equals E_2^{n+1}. Suppose now the assertion (72) not to be true. Then $\dim(\tau_1^n \cap \tau_2^n) = n$, and therefore $\dim(\Pi_1 \cap \Pi_2) = n$. Thus

$$\dim(\Pi_1 \cap \Pi_2) + \dim H = 2n + 1 > 2n = \dim \Pi_1 + \dim \Pi_2$$

in contradiction to Lemma 1.8 of Chapter 6. This proves (C_4).

Proof of (C_5). Suppose the assertion to be wrong and $y(\bar{t}) \in \tau^m$ with $m \le n - 1$. Then the plane Π^r spanned by τ^m and the ray $\overrightarrow{y_0 y_1}$ is of dimension

$$r \le m + 1 \le n. \tag{73}$$

On the other hand, if $v_0, v_1, \ldots, v_m$ are the vertices of τ^m, then the $m+3 \le n+2$ points $y_0, y_1, v_0, \ldots, v_m$ are in general position, being a subset of the points (3.20). Therefore they do not lie in a plane of dimension $< m + 2$. But these points lie in the plane Π^r whose dimension, by (73), is $\le m + 1 < m + 2$. This contradiction proves (C_5).

We now turn to the proofs of assertions (C_1)–(C_3) of the text.

Proof of (C_1). Suppose the equation (3.21) has two roots $x_1 \neq x_2$ in ∂K^{n+1}. For $i = 1, 2$ let $x_i \in \sigma_i^n$, where σ_i^n is an n-simplex of ∂K^{n+1}. Since the restriction of g to a single simplex of ∂K^{n+1} is one-to-one, the simplices σ_1^n and σ_2^n are different. Since x_1 and x_2 satisfy (3.21), the point $y(\bar{t})$ lies in the intersection of $\tau_1 = g(\sigma_1^n)$ and $\tau_2 = g(\sigma_2^n)$. Thus this intersection is not empty and therefore (72) holds by (C_4). But this contradicts (C_5) since $y(\bar{t})$ is a point of $\tau_1^n \cap \tau_2^n$.

Proof of (C_2). Let $y(\bar{t}) \in \tau^n = g(\sigma^n)$ where σ^n is an n-simplex of ∂K^{n+1}. Suppose now (C_2) not to be true. Then $y(\bar{t}) \in \partial \tau^n$, and $y(\bar{t}) \in \tau^m$, where τ^m is an m-simplex of $g(\partial K^{n+1})$ of dimension $m \le n - 1$. This again contradicts (C_5).

Proof of (C_3). Suppose $\overrightarrow{y_0 y_1}$ has two intersections $y(t_1)$ and $y(t_2)$, $t_1 \neq t_2$, with τ^n. Then $\overrightarrow{y_0 y_1}$ lies in the n-plane Π^n spanned by τ^n. Thus if $v_0, v_1, \ldots, v_n$ are the vertices of τ^n, then the $n + 2$ points $y_0, v_0, \ldots, v_n$ lie in the plane Π^n of dimension $n < n + 1$. This contradicts the fact that these points as a subset of the points (3.20) are in general position.

19. *Proof of* (3.42). Let $U_\alpha = \kappa^{-1} \sigma_\alpha^n$. Then U_α is a neighborhood of ξ_α which contains no other root of (3.40). Therefore

$$j(\pi^t f, \xi_\alpha) = d(\pi^t f, U_\alpha, y(t^*)). \tag{73a}$$

Now if T_1^n is the tangent plane at ξ_α to S_1^n and s_α the central projection of S_1^n from its center on T_1^n, then

$$d(\pi^t f, U_\alpha, y(t^*)) = d(\pi^t f s_\alpha^{-1}, \Omega_\alpha, y(t^*)),$$

where $\Omega_\alpha = s_\alpha U_\alpha$. But $f s_\alpha^{-1} = g \kappa s_\alpha^{-1}$. So $j(\pi^t f, \xi_\alpha) = d(\pi^t g \kappa s_\alpha^{-1}, \Omega_\alpha, y(t^*))$. Since $\kappa s_\alpha^{-1} \Omega_\alpha = K U_\alpha = \sigma_\alpha^n$, we see from the product theorem that $j(\pi^t f, \xi_\alpha) = d(\pi^t g, \sigma_\alpha^n, y(t^*)) \cdot d(\kappa s_\alpha^{-1}, \Omega_\alpha, x)$, where x is a fixed point of σ_α^n. This proves (3.42) since $d(\kappa s_\alpha^{-1}, \Omega_\alpha, x) = +1$. (Note that by (2.2) the right member of (73a) is defined as the degree of a map $\Omega_\alpha \to T_*^n$.)

20. *Proof of* (3.45). Since $t_{\alpha_0}^-$ and $t_{\alpha_0}^+$ are both less than t_α for $\alpha \ge \alpha_0 + 1$ (cf. (3.24)), the equality (3.45) is a consequence of the following assertion: with the notations used in subsection 3.11 the degree

$$d(\pi^t g, \sigma_\alpha^n, y(t^*)) \tag{74}$$

is independent of t if

$$0 < t < t_\alpha. \tag{75}$$

This assertion in turn is a consequence of the following one: let Q_α^n be the n-plane in E_2^{n+1} which contains the simplex $\tau_\alpha^n = g(\sigma_\alpha^n) = (w_1, w_2, \ldots, w_{n+1})$. Let $\tau_\alpha^{n+1} = g(\sigma_\alpha^{n+1}) = (w_0, w_1, \ldots, w_{n+1})$. Let H^+ and H^- be the two halfspaces of E_2^{n+1} with respect to Q_α^n, where we suppose that $y(t^*)$ (and therefore $y(t)$) lies in H_α^-. As in subsection 3.11 let T_*^n be the tangent plane to $S_2^n(y(t))$ at $y(t^*)$. Finally let ε_α equal $+1$ or -1 according to whether τ_α^{n+1} is a positive or negative simplex. We distinguish two cases:

$$\text{(i)} \quad w_0 \in H^+, \qquad \text{(ii)} \quad w_0 \in H^-, \tag{76}$$

and assert that

$$d(\pi^t g, \sigma_\alpha^n, y(t^*)) = \begin{cases} -\varepsilon_\alpha & \text{in case (i),} \\ +\varepsilon_\alpha & \text{in case (ii),} \end{cases} \tag{77}$$

if t satisfies (75).

We present the proof of (77) for case (i) with $\varepsilon_\alpha = +1$: we note that the simplex $(y(t), \pi^t \tau_\alpha^n) = (y(t), \pi^t w_1, \ldots, \pi^t w_{n+1})$ has the same orientation as the simplex $(y(t), \tau_\alpha^n) = (y(t), w_1, \ldots, w_{n+1})$. But since $y(t) \in H^-$ and $w_0 \in H^+$, it follows from subsection 1.25 of Chapter 6 that the simplices $(y(t), \tau_\alpha^n)$ and $(w_0, \tau_\alpha^n) = \tau_\alpha^{n+1}$ have different orientations, i.e., by the assumption $\varepsilon_\alpha = +1$, it follows that the simplex $(y(t), \tau_\alpha^n)$, and therefore the simplex $(y(t), \pi^t \tau_\alpha^n)$, is negative. But by subsection 1.7 the negativity of $(y(t), \pi^t \tau_\alpha^n)$ implies that the simplex $\pi^t \tau_\alpha^n$ is a negative simplex of the tangent plane T_*^n. Thus $d(\pi^t, \tau_\alpha^n, y(t^*)) = -1$. Since $\tau_\alpha^n = g(\sigma_\alpha^n)$, this proves (77) under the assumptions considered. The remaining assertions follow similarly.

Notes to §4.

21. THEOREM. *Let V be a bounded open convex set in a Banach space E, let y_0 be a point in V, and let y be a point in E different from y_0. Then (i) the ray $\overrightarrow{y_0 y}$ intersects the boundary ∂V of V in exactly one point $\overline{y} = \overline{y}(y)$, and (ii) $\overline{y}(y)$ is continuous at every point $y = y_1 \neq y_0$.*

PROOF. Assertion (i) was already proved in Note 2 to Chapter 4. For the proof of (ii) we note that for $y \neq y_0$

$$\overline{y} = y_0 + (y - y_0) \frac{\|\overline{y} - y_0\|}{\|y - y_0\|}. \tag{78}$$

If we set

$$t(z) = \frac{\|z\|}{\|y_0 + z - y_0\|}, \qquad z \neq 0, \tag{79}$$

then by (78)

$$\bar{y} = y_0 + \frac{y - y_0}{t(y - y_0)} \tag{80}$$

To prove the continuity of $\bar{y}(y)$ at $y = y_1 \neq y_0$ we note that if $\|y_1 - y_0\| = 2\delta_1$, then

$$\|y - y_0\| \geq \delta_1 \tag{81}$$

if

$$\|y - y_1\| \leq \delta_1. \tag{82}$$

Now $t(y - y_0)$ is bounded away from zero for y satisfying (82). Indeed, for such y, by (81)

$$t(y - y_0) = \frac{\|y - y_0\|}{\|\bar{y} - y_0\|} \geq \delta_1/r_0,$$

where r_0 is the positive distance of y_0 from ∂V.

It now follows from (80) that for our continuity proof it will be sufficient to show that $t(y - y_0)$ is continuous at $y = y_1$. We claim next that this continuity is a consequence of the following "triangular inequality" to be proved below:

$$t(y_2 - y_0) \leq t(y_2 - y_1) + t(y_1 - y_0), \tag{83}$$

where y_2 denotes a point y satisfying (82). Indeed from (83), from the inequality obtained from (83) by interchanging y_1 and y_2, and from the definition (79) of $t(z)$, we see that

$$|t(y_2 - y_0) - t(y_1 - y_0)| \leq t(y_2 - y_1) = \frac{\|y_2 - y_1\|}{\|y_2 - y_1 + y_0 - y_0\|}. \tag{84}$$

But $\overline{y_2 - y_1 + y_0} \in \partial V$ since we obviously may assume $y_2 \neq y_1$. Thus the denominator of the right member of (84) is not smaller than $r_0 = \text{dist}(y_0, \partial V)$, and (84) implies the continuity of $t(y - y_0)$ at $y = y_1$.

It remains to prove the inequality (83). We follow the proof given by Minkowski in [40, pp. 12, 13]. We note first that for $\alpha > 0$

$$t(\alpha(y - y_0)) = \alpha t(y - y_0), \qquad y \neq y_0, \tag{85}$$

since obviously $\overline{y_0 + \alpha(y - y_0)} = \overline{y_0 + (y - y_0)}$. Suppose now that (83) is not true for some $y = y_2$ satisfying (82). Then there exists a number τ for which

$$t(y_2 - y_0) > \tau > t(y_2 - y_1) + t(y_1 - y_0). \tag{86}$$

To arrive at a contradiction, we consider points y_3, y_4, y_5, y_6 defined by

$$y_3 = y_0 + (y_1 - y_0)/\tau, \qquad y_4 = y_3 + (y_2 - y_1)/\tau,$$
$$y_5 = y_0 + (y_3 - y_0)/(1 - \gamma), \qquad y_6 = y_0 + (y_4 - y_3)/\gamma,$$

where

$$\gamma = \frac{t(y_4 - y_3)}{t(y_3 - y_0) + t(y_4 - y_3)}.$$

From these definitions, from (86), and from (85) it follows that

$$t(y_3 - y_0) = t(y_1 - y_0)/\tau < 1, \qquad t(y_4 - y_0) = t(y_2 - y_0)/\tau > 1,$$

$$t(y_5 - y_0) = t(y_3 - y_0) + t(y_4 - y_3) = (1/\tau)[t(y_1 - y_0) + t(y_2 - y_1)] < 1,$$

$$t(y_6 - y_0) = t(y_3 - y_0) + t(y_4 - y_3) = (1/\tau)[t(y_1 - y_0) + t(y_2 - y_1)] < 1.$$

Now, by definition, $t(y - y_0) = \|y - y_0\| \cdot (\|\bar{y} - y_0\|)^{-1}$. Thus $t(y - y_0) > 1$ if and only if y is exterior to V, and $t(y - y_0) < 1$ if and only if y is a point of the open convex set V. In particular we see from the above inequalities that y_5 and y_6 belong to V and y_4 does not. This however contradicts the convexity of V since y_4 is a convex combination of y_5 and y_6, viz., $y_4 = (1 - \gamma)y_5 + \gamma y_6$ with $0 < \gamma < 1$.

CHAPTER 8

Some Extension and Homotopy Theorems

§1. An extension and a homotopy theorem

1.1. Let E be a Banach space, let Ω be a bounded open connected set in E, let y_0 be a point of E, and let f be a g.L.-S. map (see subsection 6.2 of Chapter 3) which maps $\partial\Omega$ in $E - y_0$. Then by Definition 1.5 in Chapter 4 of the winding number

$$u(f(\partial\Omega), y_0) = d(\tilde{f}, \Omega, y_0) \tag{1.1}$$

for every g.-L.S. extension $\tilde{f}$ of f to $\overline{\Omega}$. Consequently if $u(f(\partial\Omega), y_0) \neq 0$, then the equation

$$\tilde{f}(x) = y_0 \tag{1.2}$$

has at least one solution $x \in \Omega$. Thus for the existence of a g.L.-S. extension of f mapping $\overline{\Omega}$ into $E - y_0$, the condition

$$u(f(\partial\Omega), y_0) = 0 \tag{1.3}$$

is necessary. The following theorem asserts that this condition is also sufficient.

1.2. THEOREM. *Let E, Ω, y_0, and f be as in subsection 1.1, and suppose that (1.3) is satisfied. Then* (i) *there exists a g.L.-S. extension of f which maps $\overline{\Omega}$ into $E - y_0$;* (ii) *if f is L.-S., then there exists an L.-S. extension of the above property.*

In the finite-dimensional case this theorem is a special case of H. Hopf's classical extension theorem (see [**2**, p. 499ff.] and the Hopf papers quoted on p. 621 of [**2**], cf. also [**29**, p. 142ff.]). For this reason we refer to Theorem 1.2 as the Hopf extension theorem.

1.3. Let f and g be g.L.-S. maps $\partial\Omega \to E - y_0$. We know from Theorem 1.7 of Chapter 4 that if f and g are g.L.-S. homotopic, then

$$u(f(\partial\Omega), y_0) = u(g(\partial\Omega), y_0). \tag{1.4}$$

Thus this equality is necessary for f and g to be g.L.-S. homotopic. The following theorem states that this equality is also sufficient for the g.L.-S. homotopy of f and g if Ω is convex.

1.4. THEOREM. *Let Ω be an open bounded convex set of the Banach space E. Let f and g be g.L.-S. maps $\partial\Omega \to E - y_0$. Suppose that (1.4) holds. Then f and g are g.L.-S. homotopic in $E - y_0$.*

We refer to this theorem as the Hopf-Krasnoselskiĭ homotopy theorem since in the finite-dimensional case it is a special case of the Hopf theorem quoted in subsection 1.2 and since as already mentioned in the introduction Krasnoselskiĭ was the first to prove the theorem for L.-S. maps in arbitrary Banach spaces.

1.5. We start the proof of Theorems 1.2 and 1.4 by considering the finite-dimensional case. For $i = 1, 2$ let E_i^{n+1} be a Banach space of finite dimension $n + 1$, let θ_i be a point of E_i^{n+1}, let B_i^{n+1} be a ball with center θ_i, and let $S_i^n = \partial B_i^{n+1}$. Let f be a continuous map $S_1^n \to S_2^n$ satisfying

$$u(f(S_1^n), \theta_2) = 0. \tag{1.5}$$

Under this assumption we will prove the following assertions h^n and e^{n+1} for $n = 0, 1, \dots$.

1.6. *Assertion h^n.* f is homotopic to a continuous map f_1 which maps S_1^n into a single point of S_2^n.

1.7. *Assertion e^{n+1}.* f can be extended to a continuous map $\overline{B_1^{n+1}} \to S_2^n$.

Moreover we will prove

1.8. *Assertion e_*^{n+1}.* If f is a continuous map $S_1^n \to E_2^{n+1} - \theta_2$ which satisfies (1.5), then f can be extended to a continuous map $\overline{B_1^{n+1}} \to E_2^{n+1} - \theta_2$.

For the proof of these assertions the following lemma is basic.

1.9. LEMMA. *Let $n \geq 0$. Then (a) h^n and e^{n+1} are equivalent, and (b) e^{n+1} implies e_*^{n+1}.*

PROOF. (a) Suppose h^n is true. Then there exists a continuous $f(x, t): S_1^n \times [0, 1] \to S_2^n$ such that $f(x, 0) = f(x)$ and such that $f(x, 1)$ maps S_1^n into a single point y_2 of S_2^n. We now define $\overline{f}: B_1^{n+1} \to S_2^n$ as follows: we set $f(\theta_1) = y_2$; if $\xi \in B_1^{n+1} - \theta_1$, then the ray $\overrightarrow{\theta_1 \xi}$ intersects S_1^n in a unique point $x = x(\xi)$ and

$$\xi = (1 - t)x + t\theta_1, \tag{1.6}$$

where $t = t(\xi)$ is a unique value in $[0, 1]$. We then set $\overline{f}(\xi) = f(x, t)$ to obtain the desired extension.

Now suppose e^{n+1} is true. Then there exists a continuous map $\overline{f}: \overline{B_1^{n+1}} \to S_2^n$ such that $\overline{f}(x) = f(x)$ for $x \in S_1^n$. We define a map $f(x, t): S_1^n \times [0, 1] \to S_2^n$ as follows: we set $f(x, 1) = \overline{f}(\theta_1)$ for all $x \in S_1^n$. For $0 \leq t < 1$ and $x \in S_1^n$ the point (1.6) lies in B_1^{n+1} but is different from θ_1. We then set $f(x, t) = \overline{f}((1-t)x + t\theta_1)$ to obtain the desired homotopy.

Proof of (b). Since by assumption $\theta_2 \notin f(S_1^n)$ there exists a ball B_2^{n+1} with center θ_2 whose closure does not intersect $f(S_1^n)$. Let k be the central projection from θ_2 of $f(S_1^n)$ on $S_2^n = \partial B_2^{n+1}$, and let $\phi(x) = kf(x)$. Then ϕ is a continuous map $S_1^n \to S_2^n$, and by Definition 3.2 and Theorem 3.4, both in Chapter 7,

$$u(f(S_1^n), \theta_2) = v(f(S_1^n), \theta_2) = d(\phi, S_1^n, S_2^n) = v(\phi(S_1^n), \theta_2) = u(\phi(S_1^n), \theta_2).$$

Thus $u(\phi(S_1^n), \theta_2) = 0$ by assumption (1.5), and the assumed validity of e^{n+1} implies the existence of a continuous extension $\bar{\phi}$ of ϕ which maps $\overline{B_1^{n+1}}$ into S_2^n.

To obtain an extension $\bar{f}$ of f of the desired properties, we set

$$r(x) = \|f(x) - \theta_2\| \quad \text{for } x \in S_1^n.$$

Since for $f(x) \neq \theta_2$ for these x there exists α, β such that $0 < \alpha \leq r(x) \leq \beta$ and since the interval $[\alpha, \beta]$ is convex, it follows from §1.31 that $r(x)$ can be extended to a continuous function $\bar{r}(x)$ which is defined in all of E_1^{n+1} and satisfies the inequality $0 < \alpha \leq \bar{r}(x) \leq \beta$. We now define $\bar{f}(x)$ for each $x \in B_1^{n+1}$ as that point on the ray $\overrightarrow{\theta_2\bar{\phi}(x)}$ whose distance from θ_2 has the positive value $\bar{r}(x)$.

1.10. We turn to the proof of assertions h^n, e^{n+1}, e_*^{n+1}, $n \geq 0$.

Proof of h^0. From the definition of the winding number and by definition of the degree $d(f, S_1^0, S_2^0)$ given in subsection 1.11 of Chapter 7 it follows that $u(f(S_1^0), \theta_2) = d(f, S_1^0, S_2^0)$ and that the right member of this equality equals zero if and only if f maps S_1^0 into one of the two points of which S_2^0 consists. Thus in this case (1.5) implies that f maps S_1^0 into a single point of S_2^0.

Proof of h^1. We refer to subsection 1.6 of Chapter 4 and Note 1 to Chapter 4. If r_1 and r_2 are the radii of S_1^1 and S_2^1 resp. and if ϕ_1 and ϕ_2 are polar angles with the poles θ_1 and θ_2 resp., then the argument given in Step II of that note shows that the map $f: S_1^1 \to S_2^1$ is homotopic to a map $f: (r_1, \phi_1) \to (r_2, \phi_2)$ where $\phi_2 = \phi_1$ multiplied by an integer whose value, by (1.13) of Chapter 4, equals $u(f(S_1^n), \theta_2)$. Thus, by assumption (1.5), f_0 maps S_1^n into the single point $(r_2, 0)$ of S_2^n.

Proof of h^n for $n > 1$. Suppose h^{k-1} has been proved for some integer k in the interval $2 \leq k \leq n$. We will show that then h^k is true. This will prove h^n since h^1 has been established. Let f be a continuous map $S_1^k \to S_2^k$ which satisfies (1.5) with n replaced by k. We have to show that f is homotopic to a map f_1 which maps S_1^k into a single point of S_2^k. We note first that it is sufficient to show that f is homotopic to a continuous map f_1 which maps S_1^k into a proper subset of S_2^k. Indeed, for such map there exists a point y_0 of S_2^k which is not contained in $f_1(S_1^k)$. Let n_0 be the antipode to y_0, and let f_2 be the map which maps S_1^n into n_0. Then for no $x \in S_1^k$ are the points $f_1(x)$ and $f_2(x)$ antipodal. But this implies that f_1 and f_2 are homotopic (cf. the proof of Theorem 2.6 in Chapter 7). It follows that f is homotopic to f_2 if it is homotopic to f_1.

We note next that if in Lemma 4.3 of Chapter 7 with $\psi = f$ and $n = k$ we choose ε less than the diameter of S_2^k, then (4.13) in Chapter 7 shows that for no $x \in S_1^k$ are the points $f(x)$ and $h(x)$ antipodal. As pointed out above, this implies that h is homotopic to f. Therefore for our proof it is sufficient to assume that f has the property (ii) asserted for h in that lemma. In other words, using different notation, we may assume that there are two different but not antipodal points y_0 and n_2 on S_2^k which have the property that each of the equations

$$\text{(a)} \quad f(x) = y_0, \qquad \text{(b)} \quad f(x) = n_2 \tag{1.7}$$

has at most a finite number of solutions.

We will show that f is homotopic to a map f_1 for which at least one of the points y_0, n_2 does not lie in $f(S_1^k)$, thus proving that the latter set is a proper subset of S_2^k. Now if one of the equations (1.7) has no roots, this is obvious; we simply have to take $f_1 = f$.

Suppose now that each of the equations has roots. In this case we will show that f is homotopic to a map $f_1 \colon S_1^k \to S_2^k$ for which

$$y_0 \notin f_1(S_1^k). \tag{1.8}$$

Then let $x_1, x_2, \ldots, x_r$ be the points of $f^{-1}(y_0)$, and $n_1^1, n_1^2, \ldots, n_1^s$ be the points of $f^{-1}(n_2)$. Let U_1 be an open subset of S_1^k which contains the finite set $f^{-1}(y_0)$ and whose closure is disjoint from the finite set $f^{-1}(n_2)$. Such U_1 satisfies (1.14) in Chapter 7 with $n_1 = n_1^1$ and $C_1 = f^{-1}(y_0)$. Therefore if the tangent planes T_1, T_2, the stereographic projections σ_1, σ_2, and Ω_1 are defined as in the lines directly following (1.14) in Chapter 7, we see from Definition 1.13 and from (1.20), both also in Chapter 7, that

$$d(f, S_1^k, S_2^k) = d(f, S_1^k, y_0) = d(f^*, \Omega_1, y_0), \tag{1.9}$$

where

$$f^*(\xi) = \sigma_2 f \sigma_1^{-1}(\xi), \quad \Omega_1 = \sigma_1(U_1), \quad \xi \in \overline{\Omega_1}. \tag{1.10}$$

If $\eta_0 = \sigma_2(y_0)$, then the roots of the equation

$$f^*(\xi) = \eta_0 \tag{1.11}$$

are the points $\xi_\rho = \sigma_1(x_\rho)$, $\rho = 1, 2, \ldots, r$, which obviously lie in Ω_1. Now let $a > 0$ be such that

$$B_1^k = B(\xi_1, a) \subset \Omega_1. \tag{1.12}$$

As will be shown in the Notes,[1] there exists a homotopy

$$f_t^*(\xi) = f^*(x, t) \colon \overline{\Omega_1} \times [0, 1] \to T_2$$

which has the following properties:

(i) $f_0^*(\xi) = f^*(\xi)$;
(ii) $f_t^*(\xi) = f^*(\xi)$ for $\xi \in \partial \Omega_1$, $t \in [0, 1]$;
(iii) all roots of the equation

$$f_1^*(\xi) = \eta_0 \tag{1.13}$$

lie in B_1^n. Then

$$d(f_1^*, \Omega_1, \eta_0) = d(f^*, \Omega_1, \eta_0) \tag{1.14}$$

as follows, e.g., from subsection 2.9 of Chapter 6 and (ii). Moreover from (iii) and (1.12) together with the sum theorem we see that

$$d(f_1^*, \Omega_1, \eta_0) = d(f_1^*, B_1^k, \eta_0). \tag{1.15}$$

Now since f_1^* is homotopic to f^*, it follows that the maps $f_1 = \sigma_2^{-1} f_1^* \sigma_1$ and $f = \sigma_2^{-1} f^* \sigma_1$ (cf. 1.10) are homotopic maps $S_1^k \to S_2^k$. Therefore

$$d(f_1^*, \Omega_1, \eta_0) = d(f_1, S_1^k, S_2^k) = d(f, S_1^k, S_2^k). \tag{1.16}$$

But by Definition 3.2 and Theorem 3.4, both in Chapter 7, the right member of this equality equals $u(f(S_1^k), \theta_2)$ and thus equals zero by assumption (1.5) with n replaced by k. We thus see from (1.16) and (1.15) that $d(f_1^*, B_1^k, \eta_0) = 0$, or what, by definition of the winding number, is the same that

$$u(f_1^*(\partial B_1^k), \eta_0) = 0. \tag{1.17}$$

We now recall that h^{k-1} is true by our induction assumption. But by Lemma 1.9, h^{k-1} implies e^k. Therefore (1.17) implies the existence of a continuous map $\overline{f_1^*} : \overline{B_1^k} \to T_2 - \eta_0$ which satisfies

$$\overline{f_1^*}(\xi) = f_1^*(\xi) \quad \text{for } \xi \in \partial B_1^k. \tag{1.18}$$

We extend the definition of $\overline{f_1^*}$ to $\overline{\Omega_1}$ by setting

$$\overline{f_1^*}(\xi) = f_1^*(\xi) \quad \text{for } \xi \in \overline{\Omega_1} - B_1^k.$$

Because of (1.18) this is a continuous map. Moreover by property (iii) of f_1^*, our extension $\overline{f_1^*}$ maps $\overline{\Omega_1} \to T_2 - \eta_0$. Finally we see that $\overline{f_1^*}(\xi) = f^*(\xi)$ for $x \in \partial\Omega_1$. Thus $\overline{f_1^*}$ is homotopic to f^* (cf. the proof of Theorem 1.3 in Chapter 4). Denoting the extension $\overline{f_1^*}$ of f_1^* simply by f_1^*, we see that $f_1 = \sigma_2^{-1} f_1^* \sigma_1$ is a continuous map of the subset $\overline{U}_1$ of S_1^k into $S_2^k - y_0$ which agrees with f on ∂U_1. If we extend the definition of f_1 to all of S_1^n by setting $f_1(x) = f(x)$ for $x \in S_1^n - \overline{U}_1$, we obviously obtain a continuous map $f_1 : S_1^n \to S_2^n - y_0$ which is homotopic to f.

1.11. We proved h^n for $n \geq 0$ and therefore by Lemma 1.9 also assertions e^{n+1} and e_*^{n+1}. Our next goal is to prove the Hopf extension theorem in subsection 1.2. This entails two steps: (I) to show that e_*^{n+1} is still true if B_1^{n+1} is replaced by an arbitrary bounded open connected subset Ω^{n+1} of E_1^{n+1} and S_1^n by $\partial\Omega_1^{n+1}$; (II) to show that the theorem thus obtained remains true if the finite-dimensional spaces E_1^{n+1}, E_2^{n+1} are replaced by an arbitrary Banach space E provided the maps involved are g.L.-S. For both of these steps we will make use of the following theorem.

1.12. THE HOMOTOPY EXTENSION THEOREM. *Let C_1 be a bounded closed subset of the Banach space E, let C_0 be a closed subset of C_1, and let y_0 be a point of E. Let g_0 and f_0 be two g.L.-S. maps $C_0 \to E - y_0$. Assume that (i) there exists a g.L.-S. extension g of g_0 which maps C_1 into $E - y_0$, and (ii) g_0 and f_0 are g.L.-S. homotopic in $E - y_0$.*

Then there exists a g.L.-S. extension f of f_0 which maps C_1 into $E - y_0$. In the finite-dimensional case the theorem remains true if domain and range spaces are different spaces of the same dimension.

The proof will be given in the Notes.[2] We now turn to the generalization of assertion e_*^{n+1} stated as Step I of subsection 1.11. We will need the following lemma.

1.13. LEMMA. *Let Ω^{n+1} be a bounded open set in E_1^{n+1}, let y_0 be a point in E_2^{n+1}, and let f be a continuous map $\overline{\Omega^{n+1}} \to E_2^{n+1}$ for which*

$$y_0 \notin f(\partial\Omega^{n+1}). \tag{1.19}$$

Then there exists a map $g \in C'(\overline{\Omega^{n+1}})$ which on $\partial\Omega^{n+1}$ is (i) homotopic to f in $E_2^{n+1} - y_0$, and for which (ii) the equation

$$g(x) = y_0 \tag{1.20}$$

has at most a finite number of roots.

PROOF. By (1.19)

$$\|f(x) - y_0\| > 4\varepsilon \quad \text{for } x \in \partial\Omega^{n+1} \tag{1.21}$$

for some positive ε. Let g_1 be a C^1-map: $\overline{\Omega^{n+1}} \to E_2^{n+1}$ for which

$$\|g_1(x) - f(x)\| < 2\varepsilon \quad \text{for } x \in \overline{\Omega^{n+1}}. \tag{1.22}$$

Such g_1 exists by §1.32. By Sard's theorem the ball $B(y_0, \varepsilon)$ contains a point $\overline{y}_0$ which is a regular value for g_1. If $g(x) = g_1(x) + y_0 - \overline{y}_0$, then equation (1.20) is equivalent to the equation $g_1(x) = \overline{y}_0$, and the differential of g equals the one of g_1. Consequently y_0 is a regular value for g which implies that (1.20) has at most a finite number of roots. Moreover by (1.22)

$$\|g(x) - f(x)\| \leq \|g(x) - g_1(x)\| + \|g_1(x) - f(x)\| \leq \|y_0 - \overline{y}_0\| + 2\varepsilon < 3\varepsilon$$

for all $x \in \overline{\Omega}$. From this inequality and from (1.21) we see that for $x \in \partial\Omega$ and $t \in [0, 1]$

$$\|(1 - t)f(x) + tg(x) - y_0\| \geq \|f(x) - y_0\| - t\|g(x) - f(x)\| \geq 4\varepsilon - 3\varepsilon.$$

This proves that the homotopy $(1 - t)f + tg$ between f and g has the asserted properties.

1.14. We are now ready to prove the assertion stated as step I in subsection 1.11. We have to show that if f is as in subsection 1.13 and if in addition (1.3) is satisfied with $\Omega = \Omega^{n+1}$, then there exists a continuous extension $\overline{f}$ to $\overline{\Omega^{n+1}}$ of the restriction f_0 of f to $\partial\Omega^{n+1}$ such that

$$y_0 \notin \overline{f}(\overline{\Omega^{n+1}}). \tag{1.23}$$

By property (i) of Lemma 1.13 the restriction g_0 of g to $\partial\Omega^{n+1}$ is homotopic to f_0 in $E_2^{n+1} - y_0$. Suppose now we can prove that g_0 has an extension $\overline{g}$ to $\overline{\Omega^{n+1}}$ such that $y_0 \notin \overline{g}(\overline{\Omega^{n+1}})$. Then application of the homotopy extension theorem of subsection 1.12 with $C_0 = \partial\Omega^{n+1}$ and $C_1 = \overline{\Omega^{n+1}}$ shows also that f_0 has an extension on $\overline{f}$ satisfying (1.23). Thus we have to consider only a map g with the properties (i) and (ii) of Lemma 1.13. In other words it is sufficient for our proof to assume that the equation

$$f(x) = y_0 \tag{1.24}$$

has at most a finite number of solutions. But under this assumption we know
from Lemma 1 of Note 1 that if B_1^{n+1} is a ball in Ω having a positive distance
from $\partial\Omega^{n+1}$, then there exists a continuous map $f_1: \overline{B_1^{n+1}} \to E_2^{n+1}$ such that all
roots of the equation

$$f_1(x) = y_0 \tag{1.25}$$

lie in B_1^{n+1} and such that

$$f_1(x) = f(x) \quad \text{for } x \in \partial\Omega^{n+1}. \tag{1.26}$$

The first of these properties implies that

$$d(f_1, \Omega^{n+1}, y_0) = d(f_1, B_1^{n+1}, y_0), \tag{1.27}$$

and the second one together with assumption (1.3) that

$$u(f_1(\partial\Omega^{n+1}), y_0) = 0. \tag{1.28}$$

But by definition of the winding number u, the left member of (1.28) equals
the left member of (1.27). Therefore $0 = d(f_1, B_1^{n+1}, y_0) = u(f_1(\partial B_1^{n+1}), y_0)$.
But the latter equality implies by e_*^{n+1} (see subsection 1.8) the existence of
a continuous map $f_2: \overline{B_1^{n+1}} \to E_2^{n+1} - y_0$ which agrees with f_1 on ∂B_1^{n+1}.
Therefore if we set $f_2(x) = f_1(x)$ for $x \in \overline{\Omega^{n+1}} - B_1^{n+1}$ we obtain a continuous
map $\overline{\Omega^{n+1}} \to E_2^{n+1} - y_0$. Moreover $f_2 = f_1 = f$ on $\partial\Omega$. Thus f_2 is a continuous
extension of the restriction of f to $\partial\Omega$. This finishes the proof of the Hopf
extension theorem of subsection 1.2 in the finite-dimensional case.

1.15. Proof of Theorem 1.2 for the case that the g.L.-S. map $f(x) =$
$\lambda(x)x - F(x)$ is a finite layer map. In this case there exists a subspace E^{n+1}
of E of finite dimension $n + 1$ such that not only $F(\partial\Omega) \in E^{n+1}$ but also by
$y_0 \in E^{n+1}$. Let $\Omega_{n+1} = \Omega \cap E^{n+1}$, and let f_{n+1} be the restriction of f to
$\partial\Omega_{n+1} = E^{n+1} \cap \partial\Omega$. Then

$$u(f(\partial\Omega), y_0) = u(f_{n+1}(\partial\Omega_{n+1}), y_0). \tag{1.29}$$

Indeed by Definitions 3.4 and 6.4 of Chapter 3

$$d(\tilde{f}, \Omega, y_0) = d(\tilde{f}_{n+1}, \Omega_{n+1}, y_0) \tag{1.30}$$

if $\tilde{f}$ is a g.L.-S. extension of f to $\overline{\Omega}$ and f_{n+1} is its restriction to $\overline{\Omega_{n+1}}$. But by
definition of the winding number u, the left and right members of (1.29) equal
the left and right members of (1.30) resp.

Now by assumption (1.3) the equality (1.29) implies that $u(f_{n+1}(\Omega_{n+1}), y_0) =$
0. But by subsection 1.14 this condition implies the existence of a continu-
ous extension $\overline{f}_{n+1}$ of f_{n+1} which maps $\overline{\Omega_{n+1}}$ into $E - y_0$. We define a map
$\overline{F}_{n+1}: \overline{\Omega_{n+1}} \to E$ by the equality

$$\overline{f}_{n+1}(x) = \overline{\lambda}(x)x - \overline{F}_{n+1}(x), \tag{1.31}$$

where $\overline{\lambda}$ is an extension of λ to E which is positive and bounded away from zero
(cf. §31 of Chapter 1). $\overline{F}_{n+1}$ is continuous and therefore completely continuous

on the closed bounded subset $\overline{\Omega_{n+1}}$ of E^{n+1}. We thus obtain a completely continuous map $F_1: C = \overline{\Omega_{n+1}} \cup \partial\Omega \to E^{n+1}$ by defining

$$F_1(x) = \begin{cases} \overline{F}_{n+1}(x), & x \in \overline{\Omega_{n+1}}, \\ F(x), & x \in \partial\Omega. \end{cases} \tag{1.32}$$

Now C is a closed bounded set in E and $F(x)$ lies in the convex set E^{n+1}. Therefore by §1.31 there exists a completely continuous extension of F_1 which maps E into E^{n+1}. We set

$$\overline{f}(x) = \overline{\lambda}(x)x - \overline{F}_1(x), \tag{1.33}$$

where $\overline{\lambda}$ and $\overline{F}_1$ are extensions of λ and F_1 with the above properties. Then $\overline{f}$ is a g.L.-S. map $\overline{\Omega} \to E$, and our theorem will be proved if we can show that

$$\overline{f}(x) \neq y_0 \quad \text{for all } x \in \overline{\Omega}. \tag{1.34}$$

Now if $x \in \overline{\Omega_{n+1}}$, then (cf. (1.32)) $\overline{f}(x)$ equals $\overline{f}_{n+1}(x)$ (see (1.31)) and $\overline{f}_{n+1}(x) \neq y_0$. But if the point x of $\overline{\Omega}$ does not lie in $\overline{\Omega_{n+1}}$, then x and $\overline{\lambda}(x)x$ do not lie in E^{n+1}, and since $\overline{F}(x) \in E^{n+1}$ it follows from (1.33) that $\overline{f}(x) \notin E^{n+1}$. However, since $y_0 \in E^{n+1}$ the inequality (1.34) follows again.

1.16. Conclusion of the proof of the Hopf extension theorem 1.2. Since by assumption f is a g.L.-S. map $\partial\Omega \to E - y_0$, there exists a positive ε such that

$$\|f(x) - y_0\| \geq 2\varepsilon \tag{1.35}$$

for $x \in \partial\Omega$. On the other hand there exists a finite layer map g defined on $\partial\Omega$ such that

$$\|f(x) - g(x)\| < \varepsilon. \tag{1.36}$$

It follows from the last two inequalities that

$$\|g(x) - y_0\| \geq \varepsilon > 0, \tag{1.37}$$

and from the Rouché theorem (subsection 1.9 in Chapter 4) that $u(g(\partial\Omega), y_0) = u(f(\partial\Omega), y_0)$. Consequently $u(g(\partial\Omega), y_0) = 0$ by assumption (1.3). We now see from subsection 1.15 that g can be extended to a g.L.-S. map $\overline{g}: \overline{\Omega} \to E - y_0$. But on $\partial\Omega$ the maps f and g are homotopic in $E - y_0$ since the map $(1 - t)f(x) + tg(x) \neq y_0$ for $t \in [0, 1]$ and $x \in \partial\Omega$ as follows from (1.35) and (1.36) (cf. the last inequality in subsection 1.13). Therefore application of the homotopy extension theorem shows that with g also f can be extended to a g.L.-S. map $\overline{f}: \overline{\Omega} \to E - y_0$.

This proves part (i) of Theorem 1.2.

To prove (ii) we note that an L.-S. map f may be considered as a g.L.-S. map $f(x) = \lambda(x)x - F(x)$ with $\lambda(x) = 1$. Now in the proof of (i) we showed that the map $\overline{f}$ given by (1.33) maps $\overline{\Omega} \to E - y_0$ where the conditions on the extension $\overline{\lambda}$ of λ are that λ is continuous, bounded, nonnegative, and bounded away from zero. But if $\lambda = 1$, then $\overline{\lambda} = 1$ satisfies all these conditions. This finishes the proof of the Hopf extension theorem.

1.17. For the proof of the Hopf-Krasnoselskiĭ homotopy theorem 1.4 we will show that it is a consequence of Theorem 1.2 just proved. For this we need the following lemma: *let V_1 be a bounded open convex subset of the Banach space E, let a be a point of V_1, let θ be a positive number less than unity, and let*

$$V_0 = \{x \in E \mid a + (x - a)\theta^{-1} \in V_1\}. \tag{1.38}$$

Then (i) V_0 *is an open convex subset of V_1;* (ii) *if ε is a given positive number, then there exists a positive number δ such that $\|\dot{x}_1 - \dot{x}_0\| < \varepsilon$ if $0 < 1 - \theta < \delta$, where $\dot{x}_0$ and $\dot{x}_1$ are points of ∂V_0 and ∂V_1 resp. which lie on the same ray issuing from a.*

The proof will be given in the Notes.[3] Here we only remark that if V_1 is a ball with center a and radius r_1, then the V_0 defined above is a concentric ball with center a and radius $r_0 = \theta r_1$. Thus the assertions of the lemma are quite trivial in this case.

1.18. Proof of the Hopf-Krasnoselskiĭ homotopy theorem 1.4. Let $\overline{f}$ be a g.L.-S. extension of f to $\overline{\Omega}$. We assert first that it is possible to choose a positive number $\theta < 1$ with the following property: if $V_1 = \Omega$ and if V_0 is the set defined by (1.38), then $\Omega_0 = \Omega - \overline{V}_0$ contains no root of the equation

$$\overline{f}(x) = y_0. \tag{1.39}$$

This assertion is trivial if (1.39) has no roots in $\overline{V}_1 = \overline{\Omega}$, i.e., if $(\overline{f})^{-1}(y_0)$ is empty. Then let $(\overline{f})^{-1}(y_0)$ be not empty. This set is disjoint from $\partial V_1 = \partial \Omega$ since $\overline{f}(\partial \Omega) = f(\partial \Omega)$ is contained in $E - y_0$ by assumption, and it is compact by part (iii) of §1.10 and subsection 6.3 of Chapter 3. Therefore $(\overline{f})^{-1}(y_0)$ has a positive distance from ∂V_1, and our assertion follows from part (ii) of Lemma 1.17.

If V_0 has the above properties, it follows that

$$d(\overline{f}, V_1, y_0) = d(\overline{f}, V_0, y_0),$$

and we see from the definition of the winding number u and the fact that $f = \overline{f}$ on ∂V_1 that

$$u(\overline{f}(\partial V_0), y_0) = u(f(\partial V_1), y_0). \tag{1.40}$$

f and g are defined on $\partial V_1 = \partial \Omega$. We now "transfer" f to ∂V_0, i.e., we define a g.L.-S. map $h \colon \partial V_0 \to E - y_0$ as follows: if (as in the proof of Lemma 1.17) $\dot{x}_0$ and $\dot{x}_i$ denote points on ∂V_0 and ∂V_1 resp. which lie on the same ray issuing from a, we define

$$h(\dot{x}_0) = f(\dot{x}_1), \tag{1.41}$$

i.e.,

$$h(\dot{x}_0) = f(a + (\dot{x}_0 - a)\theta^{-1}). \tag{1.42}$$

We thus obtain a g.L.-S. map $b \colon \partial \Omega_0 = \partial V_1 \cup \partial V_0 \to E - y_0$ by setting

$$b(x) = \begin{cases} g(x) & \text{for } x \in \partial V_1, \\ h(x) & \text{for } x \in \partial V_0. \end{cases} \tag{1.43}$$

Then, as will be proved below, the assumption (1.4) of our theorem implies that

$$u(b(\partial \Omega_0), y_0) = 0. \tag{1.44}$$

Therefore we see from the Hopf extension theorem 1.2 that a g.L.-S. map $\phi: \overline{\Omega}_0 \to E - y_0$ exists for which

$$\phi(x) = \begin{cases} g(x) & \text{for } x = \dot{x}_1 \in \partial V_1 = \partial \Omega, \\ h(x) & \text{for } x = \dot{x}_0 \in \partial V_0, \end{cases} \tag{1.45}$$

Now if $x \in \Omega_0$ there are unique points $\dot{x}_0 \in \partial V_0$ and $\dot{x}_1 \in \partial V_1$ such that the points $\dot{x}_0, x, \dot{x}_1$ lie on the same ray issuing from a. Then

$$x = (1-t)\dot{x}_0 + t\dot{x}_1 = (1-t)(a + (\dot{x}_1 - a)\theta) + t\dot{x}_1, \qquad 0 \leq t \leq 1.$$

Denoting the right member here by $x_t(\dot{x}_1)$, we set $\psi_t(\dot{x}_1) = \phi(x_t(\dot{x}_1))$ (cf. (1.45)). Since $\phi(x) \neq y_0$ for $x \in \overline{\Omega}_0$, we see that

$$\psi_t(\dot{x}_1) \neq y_0 \quad \text{for all } \dot{x}_1 \in \partial \Omega \text{ and } t \in [0,1].$$

Moreover since ϕ is g.L.-S. and since the factor of $\dot{x}_1$ in $x_t(\dot{x}_1)$ equals $(1-t)\theta + t$ and thus lies in the interval $[\theta, 1]$ and therefore is bounded away from zero, we see that $\psi_t(\dot{x}_1)$ is for each $t \in [0,1]$ a g.L.-S. map. Finally we see from (1.45) and (1.41) that

$$\psi_0(\dot{x}_1) = \phi(x_0(\dot{x}_1)) = \phi(\dot{x}_0) = h(\dot{x}_0) = f(\dot{x}_1),$$
$$\psi_1(\dot{x}_1) = \phi(\dot{x}_1) = g(\dot{x}_1).$$

We thus see that $\psi_t(x)$, $x \in \partial \Omega$, $t \in [0,1]$, is a homotopy between $f(\dot{x}_1)$ and $g(\dot{x}_1)$ with the desired properties, and for the proof of our theorem it remains to verify (1.44). For this purpose we set

$$h_t(\dot{x}_0) = \overline{f}\left(a + \frac{\dot{x}_0 - a}{(1-t)\theta + t}\right), \qquad \dot{x}_0 \in \partial V_0, \ t \in [0,1]. \tag{1.46}$$

Then by (1.41)

$$h_0(\dot{x}_0) = \overline{f}(\dot{x}_1) = f(\dot{x}_1) = h(\dot{x}_0), \qquad h_1(\dot{x}_0) = \overline{f}(\dot{x}_0). \tag{1.47}$$

Moreover

$$h_t(\dot{x}_0) \neq y_0 \qquad \text{for } t \in [0,1]. \tag{1.48}$$

Indeed,

$$x(t) = a + \frac{\dot{x}_0 - a}{(1-t)\theta + t} \in \overline{V}_1 - V_0 = \overline{\Omega}_0.$$

Therefore $x(t)$ is not a root of (1.39). This together with (1.46) proves (1.48). But the latter inequality together with (1.47) shows that

$$u(h(\partial V_0), y_0) = u(h_0(\partial V_0), y_0) = u(\overline{f}(\partial V_0), y_0). \tag{1.49}$$

But by (1.40) and assumption 1.4 $u(\overline{f}(\partial V_0), y_0) = u(g(\partial V_1), y_0)$. Therefore by (1.49)

$$u(h(\partial V_0), y_0) = u(g(\partial V_1), y_0). \tag{1.50}$$

To prove (1.44) with b given by (1.43) we define a g.L.-S. map $\overline{b}: \overline{V}_1 \to E$ which on V_0 is an extension of the map h given on ∂V_0 and which on $\Omega_0 = V_1 - V_0$ is an extension of the b given on $\partial \Omega_0$. Then by the sum theorem

$$d(\overline{b}, V_1, y_0) = d(\overline{b}, V_0, y_0) + d(\overline{b}, \Omega_0, y_0).$$

By definition (1.43) of b and by the definition of winding numbers, we see from this equality that

$$u(g(\partial V_1), y_0) = u(h(\partial V_0), y_0) + u(b(\partial \Omega_0), y_0).$$

This equality together with (1.50) proves the assertion (1.44). This finishes the proof of the Hopf-Krasnoselskiĭ homotopy theorem.

§2. Two further extension theorems

2.1. In the finite-dimensional version of the Hopf extension theorem (see subsections 1.5 and 1.7) the domain and range space were of the same dimension $n+1$ and the condition (1.4) was necessary for the desired extension. But, as was also shown by Hopf (see [**2**, §1.5]), this condition is not necessary if the dimension of the domain space is less than that of the range space. For a more precise statement see Theorem 2.2 below. This theorem will be used to prove Theorem 2.3 which deals with "odd" mappings and which will be useful in Chapter 9.

2.2. THEOREM. *Let E_1^m and E_2^{n+1} be Banach spaces of finite dimension m and $n+1$ resp. with $m \leq n$. Let Ω_1 be a bounded open domain in E_1^m, and let y_0 be a point in E_2^{n+1}. Let f be a continuous map $\partial \Omega_1 \to E_2^{n+1}$ satisfying*

$$y_0 \notin f(\partial \Omega_1). \tag{2.1}$$

Then there exists a continuous extension $\overline{f}$ of f mapping

$$\overline{\Omega_1} \to E^{n+1} - y_0. \tag{2.2}$$

PROOF. The argument used in the last paragraph of the proof for part (b) of Lemma 1.9 shows that it will be sufficient to prove

Assertion A. If assumption (2.1) is replaced by

$$f: \partial \Omega_1 \to S_2^n = \partial B_2^{n+1}, \tag{2.3}$$

where B_2^{n+1} is a ball with center y_0, then f can be extended to a continuous map

$$\overline{f}: \overline{\Omega_1} \to S_2^n. \tag{2.4}$$

Now let σ_2^{n+1} be an $(n+1)$-simplex which is inscribed in S_2^n and which contains y_0 as an interior point, and let κ be the central projection from y_0 of S_2^n on $\partial \sigma^{n+1}$. Then $\overline{f}$ is a continuous map satisfying (2.4) if and only if κf is a continuous map mapping $\overline{\Omega_1}$ into $\partial \sigma_2^{n+1}$. It is therefore clear that assertion A is equivalent to

Assertion B. Let

$$f: \partial \Omega_1 \to \partial \sigma_2^{n+1} \tag{2.5}$$

be continuous, and suppose that

$$m \leq n, \tag{2.6}$$

where m and $n+1$ are the dimensions defined in the statement of our theorem. Then there exists a continuous extension

$$\bar{f}:\overline{\Omega_1} \to \partial\sigma_2^{n+1}. \tag{2.7}$$

Proof of assertion B. Let Q be a closed cube in E_1^n containing $\overline{\Omega_1}$ in its interior. By §1.30 there exists a continuous extension g of f:

$$g:Q \to E_2^{n+1}, \tag{2.8}$$

$$g(x) = f(x) \quad \text{for } x \in \partial\Omega_1. \tag{2.9}$$

g is uniformly continuous on Q. Therefore if r is the distance of y_0 from $\partial\sigma_2^{n+1}$ there exists a $\delta > 0$ such that

$$\|g(x') - g(x'')\| < r/2 \tag{2.10}$$

if

$$\|x' - x''\| < \delta, \qquad x', x'' \in Q. \tag{2.11}$$

Now by subsections 4.6 and 4.5 of Chapter 6 we can subdivide Q simplicially in such a way that the diameter of each simplex of the subdivision is less than δ. We denote by K^m the set of those simplices which have a point in common with $\overline{\Omega_1}$, and by $\dot{K}^m$ the set of those simplices which have a point in common with $\partial\Omega_1$. Thus if

$$x \in \dot{K}^m, \tag{2.12}$$

then x lies in some simplex τ^m of $\dot{K}^m$ which contains a point $\dot{x}$ of $\partial\Omega_1$. Therefore for such $\dot{x}$

$$\|x - \dot{x}\| \leq \operatorname{diam} \tau^n < \delta.$$

Thus (2.11) holds with $x' = x$, $x'' = \dot{x}$. Consequently $\|g(x) - g(\dot{x})\| < r/2$ by (2.10). Moreover since $g(\dot{x}) = f(\dot{x})$ by (2.9) we see that $\|g(x) - f(\dot{x})\| < r/2$. But $f(\dot{x}) \in \partial\sigma_2^{n+1}$ by (2.5), and therefore by definition of r, $\|f(\dot{x}) - y_0\| \geq r$. Consequently $\|g(x) - y_0\| \geq \|f(\dot{x}) - y_0\| - \|f(\dot{x}) - g(x)\| \geq r - r/2 > 0$ for all x satisfying (2.12). Therefore for such x we can project $g(x)$ centrally from y_0 on $\partial\sigma_2^{n+1}$ to obtain a continuous map

$$h: \dot{K}^m \to \partial\sigma_2^{n+1}. \tag{2.13}$$

Now $g(x) \in \partial\sigma_2^{n+1}$ for $x \in \partial\Omega_1$ by (2.9) and (2.5). Therefore $h(x) = g(x) = f(x)$ for these x. Thus

$$h(x) = f(x) \quad \text{for } x \in \partial\Omega_1. \tag{2.14}$$

Now $K^m \supset \overline{\Omega_1}$. We therefore see from (2.14) that assertion B will follow from

Assertion C. The map (2.13) can be extended to a continuous map

$$\bar{h}: K^m \to \partial\sigma_2^{n+1}. \tag{2.15}$$

For the proof of C we note first that this assertion is nearly trivial if the map (2.13) is simplicial with respect to $(\dot{K}^m, \partial\sigma_2^{n+1})$ in the sense of part (c) of Definition 3.5 in Chapter 6. For if we arbitrarily assign to every vertex v of K^m

which is not a vertex of $\dot{K}^m$ a vertex $w = \overline{h}(v)$ of $\partial \sigma_2^{n+1}$ and set $\overline{h}(v) = h(v)$ if v is a vertex of $\dot{K}^m$, we obtain by the construction given in subsection 3.8 of Chapter 6 an extension $\overline{h}$ of h which maps K^m into $\partial \sigma_2^{n+1}$ and is simplicial with respect to $(K^m, \partial \sigma_2^{n+1})$ and is therefore continuous.

The following assertion D reduces the case of a continuous h to the case of a simplicial map.

Assertion D. There exists a barycentric subdivision K_1^m of K^m of the following property: if $\dot{K}_1^m$ consists of those simplices of K_1^m which belong to a subdivision of $\dot{K}^m$, then the continuous map (2.13) is homotopic to a map $h_1 \colon \dot{K}_1^m \to \partial \sigma_2^{n+1}$ which is simplicial with respect to $(K_1^m, \partial \sigma_2^{n+1})$.

Before proving this assertion we show that it implies assertion C. We note first that by D there exists a continuous map $k(x,t) \colon K_1^m \times [0,1] \to \partial \sigma_2^{n+1}$ such that $h(x,0) = h(x)$ and $h(x,1) = h_1(x)$. Now as we just saw, the fact that h_1 is simplicial with respect to $(\dot{K}_1^m, \partial \sigma_2^{n+1})$ implies the existence of an extension $\overline{h}_1 \colon K_1^m \to \partial \sigma_2^{n+1}$ of h_1 which is simplicial with respect to $(K_1^m, \partial \sigma_2^{n+1})$ and is therefore a continuous map $K_1^m \to E_2^{n+1} - y_0$ since $y_0 \notin \partial \sigma_2^{n+1}$. It now follows from the homotopy extension theorem 1.12 that $h(x) = h(x,0)$ can be extended to a continuous map $\tilde{h} \colon K_1^m \to E_2^{n+1} - y_0$. It is then clear that $\overline{h}(x) = \kappa \tilde{h}(x)$ is a continuous extension of h mapping $K_1^m \to \partial \sigma_2^{n+1}$ if κ is the central projection from y_0 of $\tilde{h}(x)$ on $\partial \sigma_2^{n+1}$.

It remains to prove assertion D. Let $w_0, w_1, \ldots, w_{n+1}$ be the vertices of σ_2^{n+1}. Then the sets $W_i = \mathrm{st}(w_i, \partial \sigma_2^{n+1})$, $i = 0, 1, \ldots, n+1$, defined in subsection 3.9 of Chapter 6 form an open covering of $\partial \sigma_2^{n+1}$. This implies—as was already pointed out in that section—that there exists a positive ε such that every subset S of $\partial \sigma_2^{n+1}$ of diameter $< \varepsilon$ is contained in at least one of the W_i. Since h is continuous, we can choose a $\delta > 0$ such that

$$\|h(x') - h(x'')\| < \varepsilon \tag{2.16}$$

if

$$\|x' - x''\| < \delta. \tag{2.17}$$

Now by subsection 4.5 of Chapter 6 there exists a barycentric subdivision K_1^m of K^m such that the diameter of each simplex of K_1^m is less than $\delta/2$. If $\dot{K}_1^m$ consists of those simplices of K_1^m which are contained in a simplex of $\dot{K}^m$ and if v_i is a vertex of $\dot{K}_1^m$ and x', x'' are points in $\mathrm{st}(v_i, \dot{K}_1^m)$, then the inequality (2.17) holds since $\|x' - v_i\|$ and $\|x'' - v_i\|$ are both less than $\delta/2$ and $\|x' - x''\| \leq \|x' - v_i\| + \|x'' - v_i\|$. It follows that (2.16) is satisfied and that therefore $h(x')$ and $h(x'')$ lie in the same star of $\partial \sigma_2^{n+1}$, i.e., there is a vertex w_j of $\partial \sigma_2^{n+1}$ such that $h(x')$ and $h(x'')$ lie in $\mathrm{st}(w_j, \partial \sigma_2^{n+1})$. We thus see that if we set

$$w_j = h_0(v_i), \tag{2.18}$$

then

$$h(\mathrm{st}(v_i, \dot{K}_1^m)) \subset \mathrm{st}(h_0(v_i), \partial \sigma_2^{n+1}) \tag{2.19}$$

for all vertices v_i of $\dot{K}_1^m$. Using the construction described in subsection 3.8 of Chapter 6, we extend the map (2.18) to a map denoted again by h_0 which maps $\dot{K}_1^m$ into $\partial\sigma_2^{n+1}$ and which is simplicial with respect to $(\dot{K}_1^m, \partial\sigma_2^{n+1})$.

We now show that the simplicial map h_0 is homotopic (on $\partial\sigma_2^{n+1}$) to the continuous map h. Let x be a point of $\dot{K}_1^m$, let σ_1^μ be the simplex of smallest dimension of $\dot{K}_1^m$ which contains x, and let $v_0, v_1, \ldots, v_\mu$ be its vertices. Then $x \in \sigma_1^\mu \subset \mathrm{st}(v_i, \dot{K}_1^m)$ for $i = 0, 1, \ldots, \mu$. Therefore for these i by (2.19)

$$h(x) \subset h(\mathrm{st}(v_i, \dot{K}_1^m)) \subset \mathrm{st}(h_0(v_i), \partial\sigma_2^{n+1}). \tag{2.20}$$

Now the points $h_0(v_0), h_0(v_1), \ldots, h_0(v_\mu)$ are vertices of $\partial\sigma_2^{n+1}$ but not necessarily different ones. Therefore they are vertices of a ν-face σ_2^ν of $\partial\sigma_2^{n+1}$, $\nu \leq \mu$, and we see from (1.20) that

$$h(x) = \bigcap_{i=\sigma}^{\mu} \mathrm{st}(h_0(v_i), \partial\sigma_2^{n+1}) = \sigma_2^\nu. \tag{2.21}$$

But also $h_0(x) \in \sigma_2^\nu$ since h_0 is simplicial and the $h_0(v_i)$ are vertices of σ_2^ν. From this, from (2.21), and from the fact that σ_2^ν is convex, it follows that

$$(1-t)h_0(x) + th(x) \in \sigma_2^\nu \subset \sigma_2^{n+1}$$

for $0 \leq t \leq 1$. This establishes the desired homotopy between h and h_0.

2.3. THEOREM. *Let Ω^m be a bounded open subset of E^m whose closure does not contain the zero point θ_1 of E^m (e.g., a "ring" bounded by two spheres with center θ_1). We assume moreover that Ω^m is symmetric with respect to θ_1, i.e., that it contains the point $-x$ if it contains x. Let $n \geq m$ and let f be a continuous map $\partial\Omega^m \to E^{n+1} - \theta_2$, where θ_2 is the zero point of E^{n+1}.*

We assume that f is odd with respect to θ_1, i.e., that

$$f(-x) = -f(x) \tag{2.22}$$

for all $x \in \partial\Omega^m$.

Then f can be extended to a continuous odd map $\overline{\Omega}^m \to E^{n+1} - \theta_2$.

PROOF. For fixed n we prove the theorem by induction for $m = 1, 2, \ldots, n$. Let $m = 1$. By our assumption on Ω^1 we may write $\Omega^1 = \Omega_+^1 \cup \Omega_-^1$, where $\Omega_+^1 = \{x \in \Omega^1 \mid x > 0\}$ and $\Omega_-^1 = \{x \in \Omega^1 \mid x < 0\}$. Since $n + 1 > 1$ and $\theta_2 \notin f(\partial\Omega_+^1)$, application of Theorem 2.2 to Ω_+^1 shows that the restriction of f to $\partial\Omega_+^1$ can be extended to a continuous map $\overline{f}: \overline{\Omega}_+^1 \to E_2^{n+1} - \theta_2$. Now if $x \in \overline{\Omega}_-$, then $-x \in \overline{\Omega}_+$. Therefore we obtain an odd continuous extension which maps $\overline{\Omega}^1 \to E_2^{n+1} - \theta_2$ by setting $\overline{f}(x) = -\overline{f}(-x)$ for $x \in \Omega_-^1$.

Now suppose our assertion has been proved for an integer $m - 1 \leq n - 1$. We want to prove it for m. Let E_1^m, Ω^m, and f satisfy our assumptions, let $x_1, x_2, \ldots, x_m$ be coordinates for E^m, let E^{m-1} be the subspace of E^m given by $x_m = 0$, and let f_{m-1} be the restriction of f to $\partial\Omega^m \cap E^{m-1}$. By induction assumption f_{m-1} can be extended to an odd continuous map

$$\tilde{f}_{m-1}: \overline{\Omega}^m \cap E^{m-1} \to E_2^{n+1} - \theta_2. \tag{2.23}$$

We extend this definition by setting

$$\overline{\overline{f}}_{m-1}(x) = \begin{cases} \tilde{f}_{m-1}(x) & \text{for } x \in \overline{\Omega}^m \cap E^{m-1}, \\ f(x) & \text{for } x \in \partial\Omega^m. \end{cases} \tag{2.24}$$

Then $\overline{\overline{f}}$ is a continuous odd map whose range lies in $E_2^{n+1} - \theta_2$. Now let E_+^m and E_-^m be the two half spaces of E^m characterized by $x_m > 0$ and $x_m < 0$ resp., and let $\Omega_+^m = \Omega^m \cap E_+^n$, $\Omega_-^m = \Omega^m \cap E_-^m$. Then the restriction of $\overline{\overline{f}}$ to $\partial\Omega_+^m$ maps the latter set into $E_2^{n+1} - \theta_2$. Since $m \leq n$ we see from Theorem 2.2 that this map can be extended to a map $\overline{f}: \overline{\Omega}_+^m \to E_2^{n+1} - \theta_2$. Since Ω^m is symmetric with respect to θ, we simply have to define $\overline{f}(x) = -f(-x)$ for $x \in \partial\Omega_-^m$ to obtain a map defined on $\overline{\Omega}^m$ with the asserted properties.

§3. Notes to Chapter 8

1. The existence of a homotopy with the properties asserted in the lines between (1.12) and (1.13) is an obvious consequence of the following lemma.

LEMMA 1. *Let E_1^k and E_2^k be Banach spaces of finite dimension k. Let Ω be a bounded connected open set in E_1^k, let f be a continuous map $\overline{\Omega} \to E_2^k$, and let y_0 be a point of E_2^k. We assume that*

$$y_0 \notin f(\partial\Omega) \tag{1}$$

and that the number r of solutions of

$$f(x) = y_0 \tag{2}$$

is finite. Let B be a ball whose closure lies in Ω. Then there exists a continuous map $g: \overline{\Omega} \to E_2^k$ of the following properties: (i) all solutions of the equation

$$g(x) = y_0 \tag{3}$$

lie in B and their number is r; (ii)

$$g(x) = f(x) \quad \text{for } x \in \partial\Omega. \tag{4}$$

(Note that condition (ii) implies the existence of a homotopy connecting f and g whose values on $\partial\Omega$ are the same as those of f. This is clear from the proof given for Theorem 1.3 in Chapter 4.)

The proof of this lemma will be based on Lemmas 2 and 3 below.

LEMMA 2. *Let c be a point of E_1^k, and let a be a positive number. Let f be a continuous map $\overline{B(c,a)} \to E_2^k$. Let the point y_0 of E_2^k satisfy*

$$y_0 \notin f(\partial B(c,a)), \tag{5}$$

and let c_1 be an arbitrary point in $B(c,a)$. Suppose that the equation (2) has exactly one solution c_0 in $B(c,a)$.

Then there exists a continuous map $\phi \colon \overline{B}(c, a) \to E_2^k$ such that
(i) *c_1 is the only root of the equation*

$$\phi(x) = y_0; \tag{6}$$

(ii)

$$\phi(x) = f(x) \quad \text{for } x \in \partial B(c, a). \tag{7}$$

PROOF. We set

$$\phi(c_1) = y_0 \tag{8}$$

and define ϕ on $\partial B(c, a)$ by (7). Now let x be a point of $B(c, a)$ different from c_1. Then the ray $\overrightarrow{c_1 x}$ intersects $\partial B(c, a)$ in exactly one point $\overline{x} = \overline{x}(x)$; this point depends continuously on x, and

$$x = (1 - t)c_1 + t\overline{x}, \tag{9}$$

where t is a uniquely determined number in the interval $(0, 1)$. We then define $\phi(x)$ by setting

$$\phi(x) = (1 - t)y_0 + tf(\overline{x}). \tag{10}$$

It is clear from the above definitions that $\phi \colon \overline{B}(c, a) \to E_2^k$ is continuous, that (7) is satisfied, and that c_1 is a root of (6). It remains to verify that c_1 is the only root of (8). To do this we note that by (10) a root $x \in B(c, a)$ of (6) satisfies

$$\theta = \phi(x) - y_0 = t(f(\overline{x}) - y_0). \tag{11}$$

But since $\overline{x} \in \partial B(c, a)$ we see from (5) and (7) that $f(\overline{x}) - y_0 \neq \theta$. Thus, $t = 0$ by (11), and therefore $x = c_1$ by (9).

LEMMA 3. *We use the notation and assumptions of Lemma 1. Let x_1, $x_2, \ldots, x_r$ be the roots of equation (2), and let $\overline{x}_1$ be a point of Ω different from these roots. Then there exists a map $\psi \colon \overline{\Omega} \to E_2^k$ and an open subset Ω_1 of Ω having the following properties:*
(i) *x_1 and $\overline{x}_1$ lie in Ω_1;*
(ii) *Ω_1 has a positive distance from $\partial\Omega$ as well as from the points $x_2, x_3, \ldots, x_r$;*
(iii) *the roots of the equation*

$$\psi(x) = y_0 \tag{12}$$

are the points $\overline{x}_1, x_2, x_3, \ldots, x_r$;
(iv) *$\psi(x) = f(x)$ for $x \in \overline{\Omega} - \Omega_1$.*

PROOF. Since Ω is open and connected there exists a simple polygon $Q \subset \Omega$ connecting x_1 and $\overline{x}_1$ (see §§1.24 and 1.25) and having a positive distance $2\delta_1$ from $\partial\Omega$. Obviously Q can be chosen in such a way that it has a positive distance $2\delta_2$ from the points $x_2, x_3, \ldots, x_r$. Let $\delta = \min(\delta_1, \delta_2)$. We consider the balls $B(q, \delta)$ as q varies over Q. Since the polygon Q is compact, it contains a finite number of points $q_1 = x_1, q_2, q_3, \ldots, q_{N-1}, q_N = \overline{x}_1$ (monotone with respect to the arc length on Q counted from x_1) such that the intersections $I_\nu = Q \cap B(q_\nu, \delta)$, $\nu = 1, 2, \ldots, N$, cover Q and such that the intersections $I_\nu \cap I_{\nu+1}$, $\nu = 1, 2, \ldots, \nu - 1$, are not empty. Then the set

$$\Omega_1 = \bigcup_{\nu=1}^{N} B(q_\nu, \delta) \tag{13}$$

is an open subset of Ω which contains x_1 and $\bar{x}_1$ and whose distance from $\partial \Omega$ and the points $x_2, x_3, \ldots, x_r$ is not smaller than δ. It thus satisfies assertions (i) and (ii). It follows that $x_1 = q_1$ is the only root of (2) which lies in $B(x_1, \delta)$. Now let x_{12} be a point of $I_1 \cap I_2$. Then $x_{12} \in I_1 \cap B(x_1, \delta)$. Therefore by Lemma 2 there exists a continuous map $\psi_{12} : \overline{B(x_1, \delta)} \to E_2^k$ such that x_{12} is the only root of the equation

$$\psi_{12}(x) = y_0 \tag{14}$$

and such that $\psi_{12}(x) = f(x)$ for $x \in \partial B(x_1, \delta)$. We therefore obtain a continuous map $\psi_{12} : \overline{\Omega} \to E_2^k$ if we define

$$\psi_{12}(x) = f(x) \tag{15}$$

for $x \in \overline{\Omega} - B(x_1, \delta)$. Since $B(x_1, \delta) = B(q_1, \delta) \subset \Omega_1$ by (13), the equality (15) holds for $x \in \overline{\Omega} - \Omega_1$, and the roots in $\overline{\Omega}$ of (14) are $x_{12}, x_2, x_3, \ldots, x_r$. Moreover $x_{12} \in \Omega_1$. Thus conditions (i)–(iv) are satisfied with $\bar{x}_1$ replaced by x_{12} and ψ by ψ_{12}.

We repeat this process by replacing the point $x_1 = q_1 \in I_1$ by the point $x_{12} \in I_1 \cap I_2 \subset I_2$ and choosing a point $x_{23} \subset I_2 \cap I_3 \subset I_3$. Going on this way, we obtain a point $x_{N-1,N} \in I_{N-1} \cap I_N$ and a map $\psi_{N-1,N} : \overline{\Omega} \to E_2^k$ such that conditions (i)–(iv) are satisfied with $\bar{x}_1$ replaced by $x_{N-1,N}$ and ψ by $\psi_{N-1,N}$. Now $x_{N-1,N}$ and $\bar{x}_1$ both lie in I_N and thus in $B(x_{N-1,N}, \delta)$. Therefore by again employing Lemma 2 we obtain from $\psi_{N-1,N}$ a map ψ satisfying assertions (i)–(iv) by using the argument used above to obtain ψ_{12} from f.

Proof of Lemma 1. Let $x_1, x_2, \ldots, x_r$ be the roots of equation (2). Let $B \subset \Omega$ be a ball which has a positive distance from $\partial \Omega$, and let $\bar{x}_1, \bar{x}_2, \ldots, \bar{x}_r$ be points in B. Then by Lemma 2 there exists a continuous map $\psi_1 : \overline{\Omega} \to E_2^k$ which agrees with f on $\partial \Omega$ and is such that the roots of $\psi_1(x) = y_0$ are $\bar{x}_1, x_2, \ldots, x_r$. Again by Lemma 2 there exists a continuous map $\psi_2 : \overline{\Omega} \to E_2^k$ which on $\partial \Omega$ agrees with ψ_1, and therefore with f, and is such that the roots of $\psi_2(x) = y_0$ are $\bar{x}_1, \bar{x}_2, x_3, \ldots, x_r$. Going on this way, we obtain a map $g = \psi_r$ satisfying the assertion of Lemma 1.

2. *Proof of the homotopy extension theorem* 1.12. Let $f_0(x) = \lambda_0(x)x - F_0(x)$ and $g_0(x) = \mu_0(x)x - G_0(x)$, where F_0 and G_0 are completely continuous maps $C_0 \to E$ and where λ_0 and μ_0 are real-valued continuous functions defined on

C_0 and satisfying

$$0 < m_0 \leq \begin{cases} \lambda_0(x) \\ \mu_0(x) \end{cases} \leq M_0, \qquad x \in C_0, \tag{16}$$

for some constants m_0 and M_0. By assumption, there is a completely continuous map $H_0(x,t)$ and a continuous map $\nu(x,t)$ which maps $C_0 \times [0,1]$ into E and the reals resp. These two maps have the following properties (17)–(20).

$$H_0(x,0) = G_0(x), \qquad H_0(x,1) = F_0(x), \qquad x \in C_0, \tag{17}$$

$$\nu(x,0) = \mu(x), \qquad \nu(x,1) = \lambda(x), \qquad x \in C_0, \tag{18}$$

$$0 < m \leq \nu(x,t) \leq M \tag{19}$$

for some constants m, M.

$$h_0(x,t) = \nu(x,t)x - H_0(x,t) \neq y_0, \qquad (x,t) \in C_0 \times [0,1]. \tag{20}$$

Also by assumption there exists a completely continuous extension $\overline{G}_0$ of G_0 to C_1, and a continuous extension $\overline{\mu}$ of μ to C such that

$$0 < \overline{m} \leq \overline{\mu}(x) \leq \overline{M}, \qquad x \in C_1, \tag{21}$$

for some constants $\overline{m}$ and $\overline{M}$ and such that

$$\overline{g}_0(x) = \overline{\mu}(x)x - \overline{G}_0(x) \neq y_0, \qquad x \in C_1. \tag{22}$$

We now consider the set

$$Z = C_1 \times [0,1] \subset E \times [0,1]$$

and its subset

$$Z_0 = (C_1 \times 0) \cup (C_0 \times [0,1]).$$

(If as in subsection 1.4, and also in subsection 1.14, $C_1 = \overline{\Omega}$, $C_0 = \partial\Omega$, we may think of Z as a cylinder over $\overline{\Omega}$, with $C_1 \times 0$ as its base and $C_0 \times [0,1]$ as its "lateral surface.")

We define a completely continuous map $H_1(x,t)\colon Z_0 \to E$ by setting

$$H_1(x,0) = \overline{G}_0(x) \quad \text{for } x \in C_1 \times 0,$$
$$H_1(x,t) = H_0(x,t) \quad \text{for } (x,t) \in C_0 \times [0,1], \tag{23}$$

and define a real-valued continuous function $\mu_1(x,t)$ on Z_0 by setting

$$\mu_1(x,0) = \overline{\mu}(x), \qquad x \in C_1,$$
$$\mu_1(x,t) = \nu(x,t) \quad \text{for } (x,t) \in C_0 \times [0,1]. \tag{24}$$

Then by (21) and (19)

$$0 < m_1 \leq \mu_1(x,t) \leq M_1, \qquad (x,t) \in Z_0, \tag{25}$$

for some constants m_1 and M_1. If we define

$$h_1(x,t) = \mu_1(x,t)x - H_1(x,t), \qquad (x,t) \in Z_0, \tag{26}$$

we see from (22)–(24) and (20) that

$$h_1(x,0) = \bar{g}_0(x) \quad \text{for } x \in C_1,$$
$$h_1(x,t) = h_0(x,t) \quad \text{for } (x,t) \in C_0 \times [0,1], \tag{27}$$

and from (20) and (22) that

$$h_1(x,t) \neq y_0 \quad \text{for } (x,t) \in Z_0. \tag{28}$$

Now Z_0 is a closed set. Therefore, by §1.31, $H_1(x,t)$ has a completely continuous extension to Z. We denote one such extension again by $H_1(x,t)$. Again by §1.31, $\mu_1(x,t)$ has a continuous extension to Z satisfying (25) since the interval $[m_1, M_1]$ is convex. Calling one such extension again $\mu_1(x,t)$, we obtain a g.L.-S. map $h_1(x,t) = \mu_1(x,t)x - H_1(x,t)$ defined on all of Z. Now by (27), (20), (18), and (17)

$$h_1(x,1) = \lambda(x)x - F_0(x) = f_0(x) \tag{29}$$

for $x \in C_0$. Thus $h_1(x,1)$, $x \in C_1$, is an extension of f_0. Therefore if

$$h_1(x,1) \neq y_0 \quad \text{for } x \in C_1, \tag{30}$$

then $h(x,1)$ is an extension of f_0 of the desired property. In general, however, (30) will not be true for all $x \in C_1$. We therefore construct a g.L.-S. map $f(x,t) : Z \to E$ which agrees with $h_1(x,t)$ on Z_0 but which satisfies the condition

$$f(x,t) \neq y_0 \quad \text{for all } (x,t) \in Z. \tag{31}$$

For this purpose we consider the set

$$X_0 = \{x \in C_1 \mid h_1(x,t) = y_0 \text{ for some } t \in [0,1]\}.$$

Now by (28) the closed sets X_0 and Z_0 are disjoint. As is well known, this implies the existence of a real-valued continuous function ρ with the properties $\rho(x) = 1$ for $x \in Z_0$, $\rho(x) = 0$ for $x \in X_0$, and $0 \leq \rho(x) \leq 1$ (see e.g. [18, p. 15]). We now set for $(x,t) \in Z$

$$f(x,t) = \mu_1(x,t\rho(x))x - H_1(x,t\rho(x)).$$

Then (31) holds. Indeed suppose $f(x_0,t_0) = y_0$ for some $(x_0,t_0) \in Z$. Now $x_0 \in X_0$ and therefore $\rho(x_0) = 0$. Thus

$$y_0 = f(x_0,t_0) = \mu_1(x_0,0)x_0 - H_1(x_0,0) = h_1(x_0,0).$$

But this contradicts (28) since $(x_0,0) \in C_1 \times 0 \in Z_0$.

3. *Proof of the lemma in subsection* 1.17. That V_0 is an open subset of V_1 is obvious. To prove the convexity of V_0 we have to show that if x_0^0 and x_0^1 are points of V_0 and if

$$x_0^t = (1-t)x_0^0 + tx_0^1,$$

then

$$x_0^t \in V_0 \quad \text{for } t \in [0,1], \tag{32}$$

i.e., by the definition of V_0 that $a + (x_0^t - a)\theta^{-1} \in V_1$. Now elementary computation shows that

$$a + (x_0^t - a)\theta^{-1} = a + [(1-t)x_0^0 + tx_0^1 - a]\theta^{-1}$$
$$= (1-t)[a + (x_0^0 - a)\theta^{-1}] + t[a + (x_0^1 - a)\theta^{-1}].$$

But since x_0^0 and x_0^1 are points of V_0, it follows that $a + (x_0^i - a)\theta^{-1} \in V_1$ for $i = 0, 1$, and since V_1 is convex the right member of the last equality lies in V_1. Therefore so does the left member. But this implies (32). This proves assertion (i). To prove assertion (ii), let $\dot{x}_0$ and $\dot{x}_1$ be points on ∂V_0 and ∂V_1 resp. which lie on the same ray issuing from a. Then $\dot{x}_0 = a + \theta(\dot{x}_1 - a)$ or $\dot{x}_0 - \dot{x}_1 = (1 - \theta)(a - \dot{x}_1)$. This obviously proves (ii) since a and $\dot{x}_1$ are points of the bounded set $\overline{V}_1$.

CHAPTER 9

The Borsuk Theorem and Some of Its Consequences

§1. The Borsuk theorem

1.1. THEOREM.[1] *Let E be a Banach space and Ω a bounded open subset of E which contains the zero point θ of E and which is symmetric with respect to θ. Let $f: \partial\Omega \to E$ be an L.-S. map which is odd with respect to θ. We suppose moreover that*

$$\theta \notin f(\partial\Omega). \tag{1.1}$$

Then the winding number $u(f(\partial\Omega), \theta)$ is odd. (For the meaning of the terms symmetric and odd with respect to θ, see the statement of Theorem 2.3 in Chapter 8.)

The following theorem is equivalent to Theorem 1.1.

1.2. THEOREM. *Let E and Ω be as in Theorem 1.1. Let f be an L.-S. map $\overline{\Omega} \to E$ which is odd with respect to $\overline{\Omega}$, i.e., for which*

$$f(-x) = -f(x) \tag{1.2}$$

for all $x \in \overline{\Omega}$. We suppose moreover that (1.1) is satisfied. Then the degree $d(f, \Omega, \theta)$ is an odd number.

1.3. Proof of the equivalence of the two preceding theorems. That Theorem 1.1 implies Theorem 1.2 is obvious, since by definition

$$d(f, \Omega, \theta) = u(f(\partial\Omega), \theta) \tag{1.3}$$

for any L.-S. map f defined on $\overline{\Omega}$ and satisfying (1.1). To prove the converse implication, we note that an f satisfying the assumptions of Theorem 1.1 has an L.-S. extension f_1 to $\overline{\Omega}$ which is odd. Indeed if f_2 is an arbitrary L.-S. extension to $\overline{\Omega}$, we only have to set $f_1(x) = (f_2(x) - f_2(-x))/2$. Then (1.3) holds with f replaced by f_1 on the left, while $d(f_1, \Omega, \theta)$ is odd by Theorem 1.2.

1.4. It is instructive to note that if in addition to the assumptions of Theorem 1.2 we suppose that $f \in C'(\overline{\Omega})$ and that θ is a regular value for f, then that theorem is a nearly immediate consequence of the fact that the differential of an odd map is even (see §1.22b). To see this we note first that θ is a root of the equation

$$f(x) = \theta, \tag{1.4}$$

since $f(\theta) = f(-\theta) = -f(\theta)$ by (1.2). Now if θ is the only root of (1.4), then, θ being a regular value for f, $d(f,\Omega,\theta) = j(f,\theta) = \pm 1$ by subsection 2.1 of Chapter 2. Suppose now that θ is not the only root of (1.4). We know that the regularity of θ implies that the number of roots is finite, and (1.2) implies that, along with x, $-x$ also is a root of (1.4). Thus we may denote the roots of that equation by $\theta, x_1, -x_1, \ldots, x_r, -x_r$, where r is a positive integer. Then by Definition 2.2 in Chapter 2 of the degree

$$d(f,\Omega,\theta) = j(Df(\theta;\cdot)) + \sum_{\rho=1}^{r}[j(Df(x_\rho;\cdot)) + j(Df(-x_\rho;\cdot))]. \tag{1.5}$$

But, as already mentioned, (1.2) implies that $Df(-x_\rho;\cdot) = Df(x_\rho;\cdot)$. This shows that the right member of (1.5) is an odd integer since each index $j(Df(x_\rho;\cdot)) = \pm 1$.[2]

1.5. Turning to the proof of Theorem 1.2 without the additional assumptions made in subsection 1.4, we consider first the finite-dimensional case: $E = E^{n+1}$, a space of finite dimension $n+1$. Following J. T. Schwartz (see [51, III.33]) we start with a special case in which the assertion of our theorem is rather obvious.

1.6.   THEOREM. *Let $\Omega = \Omega^{n+1}$ be a bounded open subset of E^{n+1} which contains θ and is symmetric with respect to θ. Let g be a continuous map $\overline{\Omega} \to E^{n+1}$ which satisfies the relations (1.1) and (1.2) with f replaced by g. Moreover we assume that there exists a ball B^{n+1} with center θ whose closure is contained in Ω^{n+1} and for which*

$$g(x) = x \tag{1.6}$$

for $x \in B^{n+1}$. Finally we assume the existence of a subspace E^n of E^{n+1} such that

$$g(x) \neq \theta \quad \text{for } x \in E^n \cap (\overline{\Omega^{n+1}} - B^{n+1}). \tag{1.7}$$

Under these assumptions we assert that $d(g,\Omega^{n+1},\theta)$ is odd.

PROOF. Since (1.1) holds with $f = g$ we see from (1.6) and the sum theorem that

$$d(g,\Omega^{n+1},\theta) = d(g,B^{n+1},\theta) + d(g,\Omega^{n+1} - \overline{B^{n+1}},\theta) = 1 + d(g,\Omega^{n+1} - \overline{B^{n+1}},\theta).$$

It remains to show that here the second term of the right member is an even integer. Now if H_+^{n+1} and $H_-^{n+1}$are the half spaces into which E^n divides E^{n+1}, then the assumption (1.7) allows us by application of the sum theorem to obtain the relation

$$d(g,\Omega^{n+1} - \overline{B^{n+1}},\theta)$$
$$= d(g, H_+^{n+1} \cap (\Omega^{n+1} - \overline{B^{n+1}}), \theta) + d(g, H_-^{n+1} \cap (\Omega^{n+1} - \overline{B^{n+1}}), \theta).$$

But by Theorem 2.9 in Chapter 5 the two degrees on the right are equal, since the transformation $x \to -x$ maps

$$H_-^{n+1} \cap (\Omega^{n+1} - \overline{B^{n+1}}) \quad \text{onto} \quad H_+^{n+1} \cap (\Omega^{n+1} - \overline{B^{n+1}})$$

and since $-g(-x) = g(x)$.

1.7. To prove Theorem 1.2 with $E = E^{n+1}$ without the additional assumptions of Theorem 1.6 we recall that if f and g are continuous maps $\overline{\Omega^{n+1}} \to E^{n+1}$, if (1.1) is satisfied with $\Omega = \Omega^{n+1}$, and if

$$g(x) = f(x) \quad \text{for } x \in \partial\Omega^{n+1}, \tag{1.8}$$

then $d(g, \Omega^{n+1}, \theta) = d(f, \Omega^{n+1}, \theta)$. It is therefore clear from Theorem 1.6 that Theorem 1.2 (with $E = E^{n+1}$) is a consequence of the following theorem.

1.8. THEOREM. *Let $\Omega = \Omega^{n+1}$ be a subset of $E = E^{n+1}$ satisfying the assumptions made in Theorem 1.2. Let f be a continuous map $\partial\Omega^{n+1} \to E^{n+1} - \theta$ which is odd with respect to θ. Then there exists an extension g of f to $\overline{\Omega^{n+1}}$ which satisfies the assumptions of Theorem 1.6.*

PROOF. For $x \in \partial\Omega^{n+1}$ we define $g(x)$ by (1.8), and if B^{n+1} is as in Theorem 1.6 we set

$$g(x) = x \quad \text{for } x \in \overline{B^{n+1}}. \tag{1.9}$$

Then $g(x)$ is continuous and odd.

Moreover if

$$U^{n+1} = \Omega^{n+1} - \overline{B^{n+1}}, \tag{1.10}$$

then

$$g(x) \neq \theta \quad \text{for } x \in \partial U^{n+1}. \tag{1.11}$$

If, furthermore, E^n is an n-dimensional subspace of E^{n+1} and $U^n = U^{n+1} \cap E^n$ and if for a moment g_n denotes the restriction of g to ∂U^n, then the assumptions of Theorem 2.3 in Chapter 8 are satisfied if in that theorem we set $m = n$, $\Omega^m = U^n$, and replace f by g_n. It follows that g_n can be extended to a continuous odd map $\overline{U^n} \to E^{n+1} - \theta$. We denote this extension again by g. Now let $U_+^{n+1} = U^{n+1} \cap H_+^{n+1}$ and $U_-^{n+1} = U^{n+1} \cap H_-^{n+1}$ where the half spaces H_+^{n+1} and H_-^{n+1} are as in the proof of Theorem 1.6. Then ∂U_+^{n+1} belongs to the domain of g and $g(\partial U_+^{n+1}) \neq \theta$. Therefore g can be extended to a continuous map $\overline{U_+^{n+1}} \to E^{n+1}$. If we now define $g(x) = -g(-x)$ for $x \in U_-^{n+1}$, then g is continuous and odd on $\overline{U^{n+1}}$ while $g(x) = f(x)$ on $\partial\Omega^{n+1}$ and $g(x) = x$ on ∂B^{n+1}. Taking into account definition (1.9) we see that g is an extension of f to $\overline{\Omega^{n+1}}$ of the asserted properties.

This finishes the proof of Theorem 1.2 in the finite-dimensional case. The following lemma will allow us to reduce the proof of this theorem for the general Banach space case to the finite-dimensional one.

1.9. LEMMA. *Let Ω and $f(x) = x - F(x)$ satisfy the assumptions of Theorem 1.2. Then to every $\varepsilon > 0$ there exists a subspace E^n of E of finite dimension n and a layer map*

$$\lambda(x) = x - \Lambda(x) \tag{1.12}$$

with respect to E^n which, in addition to satisfying the inequality

$$\|f(x) - \lambda(x)\| = \|F(x) - \Lambda(x)\| < \varepsilon \tag{1.13}$$

for all $x \in \overline{\Omega}$, is an odd map.[3]

PROOF. Since F is completely continuous there exists according to the second Leray-Schauder Lemma 4.3 in Chapter 3 an E^n and a layer map $\mu(x) = x - M(x)$ with respect to E^n such that

$$\|F(x) - M(x)\| < \varepsilon \tag{1.14}$$

for all $x \in \overline{\Omega}$. If we set

$$\Lambda(x) = \tfrac{1}{2}(M(x) - M(-x)),$$

then the map λ defined by (1.12) is odd. Moreover, since F is odd by assumption,

$$F(x) - \Lambda(x) = \tfrac{1}{2}[F(x) - M(x) - (F(-x) - M(-x))].$$

This equality together with (1.14) proves (1.13).

1.10. *Conclusion of the proof of Theorem 1.2.* By (1.1) θ has a positive distance from $f(\partial\Omega)$. Let ε be a positive number less than this distance. With such an ε let E^n and λ be as in Lemma 1.9, and let λ^n be the restriction of λ to $\Omega^n = \Omega \cap E^n$. Then by subsections 5.5 and 5.6 of Chapter 3

$$d(f, \Omega, \theta) = d(\lambda^n, \Omega^n, \theta).$$

Here the right member is an odd number since λ^n is odd, and our theorem is proved in the finite-dimensional case.

§2. Some consequences of the Borsuk theorem

In this section Ω always denotes a subset of a Banach space E which is bounded, open, and symmetric with respect to the zero point θ of E which it contains.

2.1. THEOREM. *Let f be an L.-S. map $\overline{\Omega} \to E$ which satisfies (1.1). If, moreover, (1.2) holds for $x \in \partial\Omega$, then the equation*

$$f(x) = \theta \tag{2.1}$$

has at least one root $x \in \Omega$.

PROOF. By definition of the winding number

$$d(f, \Omega, \theta) = u(f(\partial\Omega), \theta). \tag{2.2}$$

Here the right member is an odd number by Theorem 1.1. Thus $d(f, \Omega, \theta) \neq 0$ which implies the existence of a root of (2.1).

2.2. LEMMA. *Let ψ be an L.-S. map $\partial\Omega \to E - \theta$. Suppose that for no $x \in \partial\Omega$ do the vectors $\psi(x)$ and $\psi(-x)$ have the same direction, i.e., that*

$$\psi(x) \neq \lambda\psi(-x) \tag{2.3}$$

for all $x \in \partial\Omega$ and all $\lambda > 0$. Then $u(\psi(\partial\Omega), \theta)$ is odd.

PROOF. $g(x) = (\psi(x) - \psi(-x))/2$ is odd. Moreover $g(x) \neq 0$ for all $x \in \partial\Omega$ since otherwise $\psi(x) = \psi(-x)$ for some $x \in \partial\Omega$ which contradicts the assumption (2.3). Thus by Theorem 1.2, $u(g(\partial\Omega), \theta)$ is odd, and our assertion will be proved once it is shown that

$$u(\psi(\partial\Omega), \theta) = u(g(\partial\Omega), \theta). \tag{2.4}$$

To prove this equality we set

$$h(x, t) = (1 - t)g(x) + t\psi(x), \qquad t \in [0, 1], \tag{2.5}$$

and show that

$$h(x, t) \neq \theta \tag{2.6}$$

for all $x \in \partial\Omega$ and $t \in [0, 1]$. This relation is obvious for $t = 0$ since $h(x, 0) = g(x)$. Suppose now that $h(x, t) = \theta$ for some $x \in \partial\Omega$ and some t in the interval $0 < t \leq 1$. We then see from (2.5) and from the definition of g by elementary calculations that

$$\psi(x) = \frac{1 - t}{1 + t}\psi(-x).$$

Now this relation contradicts (2.3) if $0 < t < 1$, and if $t = 1$ it contradicts the assumption that ψ maps $\partial\Omega$ into $E - \theta$. We thus proved (2.6). But by Theorem 1.8 in Chapter 4 the relation (2.6) implies (2.4).

2.3. THEOREM. *Let ψ be an L.-S. map $\bar{\Omega} \to E$ which on $\partial\Omega$ satisfies the assumption of Lemma 2.2. Then the equation $\psi(x) = \theta$ has at least one solution $x \in \Omega$.*

PROOF. The theorem follows from Lemma 2.2 the same way Theorem 2.1 followed from Theorem 1.1.

2.4. THEOREM. *Suppose that E_1 is a subspace of E of codimension 1, i.e., that there exists an element $e_0 \neq 0$ in E such that every $x \in E$ has the unique representation*

$$x = \lambda e_0 + e_1, \tag{2.7}$$

where $e_1 \in E_1$ and where λ is a real number. Let f be an odd L.-S. map $\partial\Omega \to E_1$. Then the equation

$$f(x) = \theta \tag{2.8}$$

has at least one solution $x \in \partial\Omega$.

PROOF. Suppose the assertion to be wrong. Then

$$f(\partial\Omega) \subset E_1 - \theta \subset E - \theta, \tag{2.9}$$

and we see from Theorem 1.1 that

$$u(f(\partial\Omega), \theta) \neq 0. \tag{2.10}$$

Thus by Theorem 2.2 in Chapter 4 every ray issuing from θ intersects $f(\partial\Omega)$. But the ray $x = \lambda e_0$, $\lambda \geq 0$, does not intersect $f(\partial\Omega)$ as is seen from (2.9). This contradiction proves the theorem.

2.5. BORSUK-ULAM THEOREM. [4] *Let E_1 be as in subsection 2.4. Let $g(x) = x - G(x)$ be an L.-S. map $\partial\Omega \to E_1$. Then the equation*

$$g(x) = g(-x) \tag{2.11}$$

has at least one solution $x \in \partial\Omega$. (Note: the theorem shows that E_1 contains no one-to-one L.-S. image of $\partial\Omega$.)

PROOF. Let

$$f(x) = \frac{g(x) - g(-x)}{2} = x - \frac{G(x) - G(-x)}{2}.$$

Then f satisfies the assumptions of Theorem 2.4. Therefore, the equation (2.8) has a solution. But this equation is equivalent to equation (2.11).

2.6. COROLLARY TO THEOREM 2.5. *Let U be an arbitrary open set in E. Then E_1 contains no one-to-one L.-S. image of U.*

PROOF. Obviously we may suppose that $\theta \in U$. Assume then that there exists a one-to-one L.-S. map g such that $g(U) \in E_1$. Let Ω be a bounded open subset of U which contains θ and is odd with respect to θ (e.g., a ball with center θ and with a small enough radius). Then the restriction of g to $\partial\Omega$ satisfies the assumption of Theorem 2.5. Therefore (2.11) holds for some $x \in \partial\Omega$. This contradicts our assumption that g is one-to-one.

2.7. THEOREM. *Let f be an L.-S. map $E \to E$. We assume that for every bounded set $\beta \in E$ the inverse $f^{-1}(\beta) = \{x \in E \mid f(x) \subset \beta\}$ is bounded. We assume moreover that there exists a monotone sequence of positive numbers $r_1, r_2, \ldots$ tending to ∞ such that if S_i is the sphere with center θ and radius r_i, then for $i = 1, 2, \ldots$ one of the following conditions (a) and (b) is satisfied:*
(a) f is odd on S_i, (b) f satisfies (2.3) with $\psi = f$ on S_i.
Then for every $y_0 \in E$ the equation

$$f(x) = y_0 \tag{2.12}$$

has a solution.

PROOF. Since the set β_0 consisting of the single point θ is bounded, it follows from our assumptions that $f^{-1}(\theta)$ is bounded. Therefore there exists an i_0 such that

$$\theta \notin f(S_i) \quad \text{for } i \geq i_0,$$

and we see from Theorem 1.1 in case (a) and from Lemma 2.2 in case (b) that $u(f(S_i), \theta)$ is odd for $i \geq i_0$. Therefore in either case

$$d(f, B(r_i, \theta), \theta) = u(f(S_i), \theta) \neq 0 \quad \text{for } i \geq i_0. \tag{2.13}$$

Now let y_0 be a given point in E. Then the set $\beta = \{y \in E \mid y = ty_0, 0 \leq t \leq 1\}$ is bounded. Consequently there exists an $i_1 \geq i_0$ such that $f^{-1}(\beta)$ lies in the open ball $B(\theta, r_{i_1})$ and thus contains no point of $S_{i_1} = \partial B(\theta, r_{i_1})$. In other words, $ty_0 \not\subset f(S_{i_1})$ for $0 \leq t \leq 1$. Therefore $d(f, B(r_{i_1}, \theta), y_0) = d(f, B(r_{i_1}, \theta), \theta)$. Since (2.13) holds for $i = i_1$, we see that $d(f, B(r_{i_1}, \theta), y_0) \neq 0$. But this implies that (1.12) has a root $x \in B(r_{i_1}, \theta)$.

§3. Notes to Chapter 9

Notes to §1.

1. Theorem 1.1 (and the equivalent Theorem 1.2) is referred to as Borsuk's theorem. In his original paper [7] Borsuk dealt with the case that $\Omega = B^{n+1}$, a ball in E^{n+1}, and proved without the use of degree theory that under condition (1.2) a continuous map $f: S^n = \overline{\partial B^{n+1}} \to E^{n+1} - \theta$ cannot be extended to a continuous map $\overline{B^{n+1}} \to E^{n+1} - \theta$ (Borsuk's "Antipodensatz z"). Obviously (see subsection 1.1 in Chapter 8) this theorem is a consequence of Theorem 1.1 with $\Omega = B^{n+1}$, proved by H. Hopf (see [2, p. 483]), since it implies that $u(f(S^n), \theta) \neq 0$. The generalization of the latter theorem to Banach spaces (i.e., Theorem 1.10) was given by Krasnoselskiĭ [31, p. 124], while the generalization to Banach spaces of the original Antipodensatz (and of related theorems) was given by Granas [24, Chapter IV].

2. For the finite-dimensional case the argument of subsection 1.4 was used by Eisenack-Fenske [19, p. 106].

3. Lemma 1.9 is taken from [24, p. 41].

4. In the finite-dimensional case with Ω a ball, Theorem 2.5 was conjectured by Ulam and proved by Borsuk [7, p. 178]. (See also [2, p. 486]). The generalization to the Banach space case with Ω a ball was given by Granas [24, p. 45].

The Linear Homotopy Theorem

§1. Motivation for the theorem and the method of proof

1.1. Let E be a Banach space, and let

$$l(x) = x - L(x) \tag{1.1}$$

be a nonsingular L.-S. map, $E \to E$. We want to define an index $j(l)$ which indicates how often E is covered by $l(E)$. Since l is a nonsingular L.-S. map, it maps, according to §§1.11 and 1.12, E onto E in a one-to-one fashion. Therefore we should define $j(l)$ as a number of absolute value 1. In order to decide for which l to define $j(l)$ as $+1$ and for which as -1, we consider the finite-dimensional case.

Let $E = E^n$ be an oriented space of finite dimension n (see subsections 1.21 and 1.22 in Chapter 6 for the concept of orientation). It is then natural to define $j(l)$ as $+1$ if l preserves the orientation (as does, e.g., the identity map I) and as -1 otherwise. By the sections just quoted, this is equivalent to defining

$$j(l) = \begin{cases} +1 & \text{if } \det l > 0, \\ -1 & \text{if } \det l < 0, \end{cases} \tag{1.2}$$

where $\det l$ denotes the determinant of l. (Note that $\det l \neq 0$ since l is supposed to be nonsingular.) Now definition (1.2) in the form stated cannot be used in general Banach spaces since it depends on determinants; however, it can be reformulated in a way to become meaningful for Banach spaces. This is possible on account of the following facts (a) and (b) which are well known in linear algebra, being a consequence of the existence of the Jordan normal form for $n \times n$ matrices (see, e.g., [**26**, §58 and §77]): (a) if L has real eigenvalues greater than 1, say $\lambda_1 > \lambda_2 > \cdots > \lambda_r > 1$, and if for $\rho = 1, \ldots, r$, $\mu_\rho = \mu(\lambda_\rho)$ is the generalized multiplicity of λ_ρ, i.e., the dimension of the subspace E^ρ of E^n which consists of those $x \in E^n$ for which

$$(\lambda_\rho I - L)^m x = \theta$$

for some positive integer m, then

$$\text{sign} \det l = (-1)^{\sum_{\rho=1}^{r} \mu_\rho}; \tag{1.3}$$

(b) if L has no real eigenvalues greater than 1, then

$$\text{sign} \det l = +1. \tag{1.4}$$

It follows from (1.3) and (1.4) that definition (1.2) is equivalent to

$$j(l) = \begin{cases} (-1)^{\sum_{\rho=1}^{r} \mu_\rho} & \text{in case (a)}, \\ +1 & \text{in case (b)}. \end{cases} \tag{1.5}$$

As is well known and as will be seen in the following sections, the right member of (1.5) makes sense for an arbitrary Banach space E if l is a nonsingular linear L.-S. map $E \to E$. We note that Definition 1.6 in Chapter 2 appears as a theorem in the original Leray-Schauder theory (see [**36**, II. 11]).

In §2 of this appendix we recall background material from the spectral theory of linear operators M, referring for proofs mainly to the presentation given in [**18**, Chapter VII]. Since in general the spectrum of M contains complex numbers, we deal here with a complex Banach space Z. In order to apply the theory to the given real Banach space E, we have to "complexify" E, i.e., embed it into a complex Banach space Z. This is done in §3. §4 discusses the index $j(l)$ in E as defined by (1.5) and also its definition for the complexification Z of E. Finally, in §5, the linear homotopy theorem (cf. subsection 1.3 in Chapter 2) is proved.

§2. Background material from the spectral theory in a complex Banach space Z

2.1. DEFINITION. Let Z be a complex Banach space, let M be a linear bounded operator on Z, and let I denote the identity map on Z. Then the set of complex numbers λ for which the inverse $R_\lambda(M)$ of $\lambda I - M$ exists as a bounded operator $Z \to Z$ is called the resolvent set of M and will be denoted by $\rho(M)$. The complement $\sigma(M)$ of $\rho(M)$ in the λ-plane is called the spectrum of M. $\sigma(M)$ is closed and bounded. A subset $\sigma_0(M)$ of $\sigma(M)$ is called a spectral set of M if it is open and closed with respect to $\sigma(M)$ [**18**, pp. 566, 567, 572].

2.2. For fixed M the domain of $R_\lambda(M)$ as a function of λ is the set $\rho(M)$, and its range is a subset of the family of bounded linear operators on Z. This family is a Banach space with the linear operations defined in the obvious way and with the norm defined [**18**, pp. 475, 477] by

$$\|M\| = \sup_{\|z\| \leq 1} \|M(z)\|. \tag{2.1}$$

This Banach space will be denoted by Z_1, and for M fixed we consider $R_\lambda(M)$ as a map $\rho(M) \to Z_1$. The topology of Z_1 induced by the norm (2.1) is called the uniform operator topology.

2.3. An example for a spectral subset of the spectrum $\sigma(M)$ is $\sigma(M)$ itself, and so is every isolated point of $\sigma(M)$.

2.4. LEMMA. $\rho(M)$ *is open, and* $\sigma(M)$ *is a nonempty closed bounded set.* $R_\lambda(M)$ *as a function of* λ *is analytic at each* $\lambda \in \rho(M)$, *i.e., at such* λ *the* *derivative of $R_\lambda(M)$ with respect to λ exists. Obviously*

$$\sum_{n=0}^{\infty} \|M\|^n |\lambda|^{-(n+1)}$$

converges if

$$|\lambda| > \|M\|. \tag{2.2}$$

Moreover for such λ

$$R_\lambda(M) = \sum_{n=0}^{\infty} \frac{M^n}{\lambda^{n+1}}. \tag{2.3}$$

For a proof of this lemma see [18, pp. 566, 567].

2.5. LEMMA. *Let Δ be a closed not necessarily bounded subset of $\rho(M)$. Then $R_\lambda(M)$ is bounded on Δ.*

PROOF. On the bounded closed set $\{\lambda \mid |\lambda| \le \|M\| + 1\}$ the boundedness of $R_\lambda(M)$ follows from its analyticity. But on the set $\{\lambda \mid |\lambda| > \|M\| + 1\}$ it follows from (2.3).

2.6. DEFINITION. $\mathcal{F}(M)$ denotes the family of all complex-valued functions f which are defined and analytic in some open neighborhood U of $\sigma(M)$. (U may depend on f.)

2.7. Let σ_0 be a spectral set of M, and let V_0 be an open set in the complex λ-plane containing σ_0 but no point of $\sigma - \sigma_0$ where $\sigma = \sigma(M)$. Let $f \in \mathcal{F}(M)$ and let U_f be the domain of definition of f. Let Γ_0 and Γ_1 be two rectifiable Jordan curves lying in $V_0 \cap U_f$ oriented in the counterclockwise sense and such that σ_0 lies in the open bounded sets whose boundaries are Γ_0 and Γ_1. Then

$$\int_{\Gamma_0} f(\mu) R_\mu(M)\, d\mu = \int_{\Gamma_1} f(\mu) R_\mu(M)\, d\mu. \tag{2.4}$$

We note that the integrands in (2.4) are operator-valued. For the definition of such integrals and for the proof of the "Cauchy theorem" (2.4), we refer the reader to [18, pp. 224–227].

2.8. DEFINITION. For $f \in \mathcal{F}(M)$ we define $f(M)$ by setting

$$f(M) = \frac{1}{2\pi i} \int_\Gamma f(\mu) R_\mu(M)\, d\mu, \tag{2.5}$$

where Γ is a rectifiable Jordan curve of the following properties: (i) $\sigma(M)$ lies in the bounded open domain whose boundary is Γ; (ii) Γ is contained in the domain U_f of definition for f; (iii) Γ is oriented in the counterclockwise sense. The uniqueness of Definition 2.5 follows from Lemma 2.7.

2.9. LEMMA. *Let Γ be a rectifiable Jordan curve satisfying conditions (i), (ii) and (iii) of subsection 2.8. Then*

$$M = \frac{1}{2\pi i} \int_\Gamma \mu R_\mu(M)\, d\mu, \tag{2.6}$$

$$I = \frac{1}{2\pi i} \int_\Gamma R_\mu(M) \, d\mu. \qquad (2.7)$$

If, moreover, f and g are elements of $\mathcal{F}(M)$ with domains U_f and U_g resp. and if, in addition to the above requirements, $\Gamma \in U_f \cap U_g$, then

$$f(M) \cdot g(M) = \frac{1}{2\pi i} \int_\Gamma f(\mu) g(\mu) R_\mu(M) \, d\mu. \qquad (2.8)$$

For the proof we refer the reader to the proof of Theorem 10, p. 468 in [18], which contains the assertions of the present lemma.

2.10. DEFINITION. A bounded linear map $P: Z \to Z$ is called a projection if

$$P^2 = P. \qquad (2.9)$$

2.11. LEMMA. Let σ_0 be a spectral set of M. Let Γ_0 be as in Lemma 2.7. Then the operator P_{σ_0} defined by

$$P_{\sigma_0} = \frac{1}{2\pi i} \int_{\Gamma_0} R_\mu(M) \, d\mu \qquad (2.10)$$

is a projection.

PROOF. Let $\sigma_0^* = \sigma - \sigma_0$. By the definition of spectral sets, there exist bounded open sets V_0, V_0^*, U_0, U_0^* such that $\sigma_0 \subset V_0 \subset \overline{V}_0 \subset U_0$, $\sigma_0^* \subset V_0^* \subset \overline{V}_0^* \subset U_0^*$ with $\overline{U}_0$ and $\overline{U}_0^*$ disjoint and such that $\Gamma_0 = \partial V_0$ and $\Gamma^* = \partial V^*$ are rectifiable Jordan curves. Let $\Gamma = \Gamma_0 \cup \Gamma^*$.

Then (2.8) holds if we set

$$f(\mu) = g(\mu) = \begin{cases} 1, & \mu \in U_0, \\ 0, & \mu \in U_0^*. \end{cases} \qquad (2.11)$$

But with this choice of f and g we see from (2.10) that the right member of (2.8) equals P_{σ_0} since $f(\mu) = g(\mu) = 0$ for $\mu \in \Gamma_0^*$, and taking account also of (2.5), we see that $f(M) = g(M) = P_{\sigma_0}$. Thus the left member of (2.8) equals $P_{\sigma_0}^2$.

2.12. LEMMA. Let σ_0 and σ_1 be two disjoint spectral sets for M. Let Γ_0 and P_{σ_0} be defined as in subsection 2.11, and let Γ_1 and P_{σ_1} be defined correspondingly with respect to σ_1. Then

$$P_{\sigma_0} \cdot P_{\sigma_1} = P_{\sigma_1} \cdot P_{\sigma_0} = \theta_1 \qquad (2.12)$$

where $\theta_1 x = \theta$ for every $x \in E$.

PROOF. Since σ_0, σ_1, and $\sigma_2 = \sigma - (\sigma_1 \cup \sigma_2)$ are closed bounded sets, there exist for $i = 0, 1, 2$ bounded open sets V_i, U_i such that $\sigma_i \subset V_i \subset \overline{V}_i \subset U_i$ with the sets $\overline{U}_1, \overline{U}_2, \overline{U}_3$ being pairwise disjoint and such that $\Gamma_i = \partial V_i$ is a rectifiable Jordan curve which we assume to be oriented in the counterclockwise sense. If we set

$$f(\mu) = \begin{cases} 1 & \text{for } \mu \in U_0, \\ 0 & \text{for } \mu \in U_1 \cup U_2, \end{cases} \qquad g(\mu) = \begin{cases} 1 & \text{for } \mu \in U_1, \\ 0 & \text{for } \mu \in U_0 \cup U_2, \end{cases}$$

then with $\Gamma = \Gamma_1 \cup \Gamma_2 \cup \Gamma_3$

$$P_{\sigma_0} = \frac{1}{2\pi i} \int_\Gamma f(\mu) R_\mu(M)\, d\mu = f(M),$$

$$P_{\sigma_1} = \frac{1}{2\pi i} \int_\Gamma g(\mu) R_\mu(M)\, d\mu = g(M),$$

and application of (2.8) yields the assertion (2.12) since $f(\mu) \cdot g(\mu) = 0$ for all $\mu \in \Gamma$.

2.13. LEMMA. *Let σ_0 be a spectral set of M, let P_{σ_0} be the projection (2.10), and let Z_0 be the range of P_{σ_0}, i.e., $Z_0 = P_{\sigma_0} Z$. Then $M Z_0 \subset Z_0$ and σ_0 is the spectrum of the restriction of M to Z_0.*

For a proof see [**18**, pp. 574, 575].

2.14. LEMMA. *The restriction M_{σ_0} of M to $Z_0 = P_{\sigma_0} Z$ is given by*

$$M_{\sigma_0} = \int_{\Gamma_0} \mu R_\mu(M)\, d\mu \tag{2.13}$$

where Γ_0 is as in subsection 2.7.

PROOF. Let Γ be as in Lemma 2.9, and let V_0 and V be the bounded open sets whose boundaries are Γ_0 and Γ resp. Then M is given by (2.6), i.e., by (2.5) with $f(\mu) = \mu$. Now $z = P_{\sigma_0}(z)$ if and only if $z \in Z_0 = P_{\sigma_0} Z$. Therefore $M(z) = M P_{\sigma_0}(z)$ for $z \in Z_0$. But P_{σ_0} is given by (2.10) or what is the same

$$P_{\sigma_0} = \frac{1}{2\pi i} \int_\Gamma g(\mu) R_\mu(M)\, d\mu, \tag{2.14}$$

with $g(\mu)$ as in (2.11). The assertion (2.13) follows now from (2.5) and (2.14) on application of (2.8).

2.15. So far we assumed of the linear operator M only boundedness or what is the same continuity. In the remainder of this section we require complete continuity, i.e., we make the additional assumption that for any bounded set $\beta \subset E$ the closure of $M(\beta)$ is compact. The definitions and assertions stated below are classical.

Proofs may be found, e.g., in [**18**, p. 577ff.].

2.16. The spectrum $\sigma(M)$ of a completely continuous linear operator M on E is at most countable. If $\sigma(M)$ contains infinitely many points, then 0 is the only accumulation point of $\sigma(M)$. It follows that each $\lambda_0 \in \sigma(M)$ which is different from 0 is a spectral set σ_0 of M. For the projection P_{σ_0} defined by (2.10) we write:

$$P_{\lambda_0} = \frac{1}{2\pi i} \int_{\Gamma_0} R_\mu(M)\, d\mu. \tag{2.15}$$

For Γ_0 we may and will take a circle with center λ_0 and radius smaller than the distance of λ_0 from the rest of $\sigma(M)$. Γ_0 is supposed to be oriented in the counterclockwise sense. We recall that the points of $\sigma(M)$ are also referred to as eigenvalues of M.

2.17. Let $\lambda_0 \in \sigma(M)$ and $\lambda_0 \neq 0$. Then

$$P_{\lambda_0} Z = \{z \in Z \mid (\lambda_0 I - M)^n = 0\} \tag{2.16}$$

for some positive integer n. The dimension $\nu(\lambda_0)$ of the linear subspace $P_{\lambda_0} Z$ of Z is finite and is called the generalized multiplicity of λ_0 as an eigenvalue of M.

2.18. Let $\lambda_0 \in \sigma(M)$ and $\lambda_0 \neq 0$. Then the equation

$$\lambda_0 z - M(z) = \theta \tag{2.17}$$

has nontrivial solutions (i.e., solutions $z \neq 0$) called eigenelements of M to the eigenvalue λ_0. These eigenelements obviously form a linear subspace of E. The dimension $r(\lambda_0)$ of this subspace satisfies the inequality $1 \leq r(\lambda_0) \leq \nu(\lambda_0)$ and is called the multiplicity of λ_0 as an eigenvalue of M.

2.19. LEMMA. *Let M be a completely continuous linear map $Z \to Z$ and suppose that the linear map $m = I - M$ mapping Z onto Z is one-to-one. Then the inverse $n = m^{-1}$ exists and is of the form $m(\xi) = \xi - N(\xi)$, where N is completely continuous on Z.*

PROOF. By assumption m is a linear, continuous, and therefore bounded, one-to-one map of Z onto Z. By a well-known theorem (see, e.g., [**18**, p. 57]) these properties imply that the linear map $n = m^{-1}$ is bounded. Now if $\xi = z - M(z)$, then $n(\xi) = z = \xi + M(z) = \xi + M(n(\xi))$. Since n is bounded, i.e., continuous, and M is completely continuous, the map $N(\xi) = -M(n(\xi))$ is completely continuous.

§3. The complexification Z of a real Banach space E

3.1. The construction of Z from E is analogous to the construction of complex numbers as ordered pairs of real numbers: the elements z of Z are ordered pairs (x, y) of points of the real Banach space E. If $z_i = (x_i, y_i)$, $i = 1, 2$, are two elements of Z, then their sum is defined by $z_1 + z_2 = (x_1 + x_2, y_1 + y_2)$, and if $c = \alpha + i\beta$, α, β real, is a complex number, then cz_1 is defined by

$$cz_1 = (\alpha x_1 - \beta y_1, \alpha y_1 + \beta x_1). \tag{3.1}$$

Finally if as usual $\|\cdot\|$ denotes the norm in E, we set for the point $z = (x, y) \in Z$

$$\|z\|_1 = \sup_{\phi} \|x \cos \phi + y \sin \phi\|. \tag{3.2}$$

3.2. In the case that E is the real line, then Z is the complex plane. The reader may verify that then $\|z\|_1$ is the absolute value of the complex number z.

3.3. The number $\|z\|_1$ defined by (3.2) is a norm, i.e., for points z, z_1, z_2 in Z and complex number c

$$\|z\|_1 \geq 0 \tag{3.3}$$

with the equality holding if and only if $z = \theta$, the zero element of Z, and

$$\|z_1 + z_2\| \leq \|z_1\| + \|z_2\|, \tag{3.4}$$

$$\|cz\|_1 = |c| \cdot \|z\|. \tag{3.5}$$

3.4. With the norm $\|z\|_1$ the linear space Z is complete and thus a Banach space.

The verification of the assertions contained in subsections 3.3 and 3.4 is of course based on the corresponding properties of the Banach space E of which Z is the complexification. We leave this verification to the reader.

3.5. It is easily verified from our definitions that $\|(x, \theta_1)\|_1 = \|x\|$. Thus the linear one-to-one map E into Z given by

$$x \to (x, \theta) \tag{3.6}$$

is norm preserving. Identifying $x \in E$ with $(x, \theta) \in Z$, we consider E as a subset of Z. This identification also allows us to write $\|z\|$ instead of $\|z\|_1$ without ambiguity.

3.6. From the multiplication rule (2.1) we see that $(\theta, y) = i(y, \theta)$ for any $y \in E$ where as usual $i = (0, 1)$, the imaginary unit of the complex plane. Therefore $z = (x, y) = (x, \theta) + (\theta, y) = (x, \theta) + i(y, \theta)$, and by our above identification we see that every $z \in Z$ may be written uniquely in the form

$$z = x + iy, \qquad x \in E, \ y \in E. \tag{3.7}$$

3.7. For later use we note the inequality

$$\tfrac{1}{2}(\|x\| + \|y\|) \le \|z\| \le \|x\| + \|y\| \quad \text{for } z = x + iy. \tag{3.8}$$

The right part of this inequality is a restatement of the triangle inequality (3.4). To prove the left part we note that $\|x\| = \|x \cos 0 + y \sin 0\| \le \sup_\phi \|x \cos \phi + y \sin \phi\| = \|z\|$ by (3.2). In the corresponding way we see that $\|y\| \le \|z\|$. Thus $\|x\| + \|y\| \le 2\|z\|$.

3.8. DEFINITION. Let L be a linear bounded operator $E \to E$. Then the map $M: Z \to Z$ defined for $z = x + iy$ by

$$M(z) = L(x) + iL(y) \tag{3.9}$$

is called the complex extension of L.

3.9. LEMMA. (i) M is linear; (ii) M is bounded; (iii) M is completely continuous if and only if L is completely continuous; (iv) if L is continuous, then $m = I - M$ is nonsingular if and only if $l = I - L$ is not singular (I denotes the identity map in the respective space); (v) $m(x) = \theta$ has a solution $z \ne \theta$ if and only if $l(x) = \theta$ has a solution $x \ne \theta$; (vi) m is not singular (i.e., $m(Z) = Z$) if and only if m is one-to-one, i.e., if $z = \theta$ is the only root of $m(z) = \theta$.

PROOF. The proof of (i) consists of an obvious verification.

Proof of (ii). Let μ be a bound for L. Then by (3.9), (3.4), and (3.8)

$$\|M(z)\| \le \|L(x)\| + \|L(y)\| \le \mu(\|x\| + \|y\|) \le 2\mu\|z\|.$$

Proof of (iii). Suppose L is completely continuous. Then, by (ii), M is continuous and it remains to show that if $z_n = x_n + iy_n$ where x_n and y_n are bounded sequences in E, then the $M(z_{n_i})$ converge for some subsequence z_{n_i} of the z_n. Now by (3.8) the boundedness of the z_n implies the boundedness of the x_n and y_n. It therefore follows from the complete continuity of L that for some subsequence n_i of the integers the $L(x_{n_i})$ and $L(y_{n_i})$ converge. But then for $z_{n_i} = x_{n_i} + iy_{n_i}$ the sequence $M(z_{n_i}) = L(x_{n_i}) + iL(y_{n_i})$ converges. The converse is obvious since L is a restriction of M to E.

Proof of (iv): If $z = x + iy$, $\varsigma = \xi + i\eta$, with x, y, ξ, η elements of E, then the equation $m(z) = \varsigma$ is equivalent to the couple of equations $l(x) = \xi$, $l(y) = \eta$. This obviously implies that l maps E onto E if and only if m maps Z onto Z. We obtain a proof of assertion (v) if in the proof of (iv) we set $\varsigma = \xi = \eta = 0$. Finally assertion (vi) follows from assertions (iv) and (v) in conjunction with part (iv) of Lemma 12 in Chapter 1, the latter being valid also for complex Banach spaces.

3.10. DEFINITION. With E considered as subset of Z (cf. subsection 3.5), the elements of E are called the real elements of Z. If $z = x + iy$ with x, y real, the element $\bar{z} = x - iy$ is called conjugate to z and the map $z \to \bar{z}$ is called a conjugation. A subspace Z_1 of Z is called invariant under conjugation if $z \in Z_1$ implies that $\bar{z} \in Z_1$.

3.11. LEMMA. *Let M be a bounded linear map $Z \to Z$. Then M is the complex extension of a linear bounded map $L: E \to E$ if and only if for all $z \in Z$*

$$M(\bar{z}) = \overline{M(z)}. \tag{3.10}$$

PROOF. The necessity is obvious from Definition 3.8. Then let M be a linear bounded map $Z \to Z$ satisfying (3.10). Now for each $x \in E$, $M(x)$ is an element of Z. Therefore

$$M(x) = L_1(x) + iL_2(x), \tag{3.11}$$

where $L_1(x)$ and $L_2(x)$ are elements of E.

It will be sufficient to show that $L_2(x) = 0$, for then M will be the complex extension of L_1. Now by (3.11) for $z = x + iy$

$$M(z) = M(x) + iM(y) = L_1(x) - L_2(y) + i(L_2(x) + L_1(y)). \tag{3.12}$$

This holds for all $z \in Z$. Therefore

$$M(\bar{z}) = L_1(x) + L_2(y) + i(L_2(x) - L_1(y)). \tag{3.13}$$

On the other hand, we see from (3.12) that

$$\overline{M(z)} = L_1(x) - L_2(y) - i(L_2(x) + L_1(y)). \tag{3.14}$$

By assumption (3.10) the left members of (3.13) and (3.14) are equal. Comparing the imaginary parts of the latter equalities, we see that $L_2(x) = -L_2(x)$, i.e., $L_2(x) = 0$ as we wanted to prove.

3.12. LEMMA. *Let Z_1 be a linear subspace of Z and let E_1 be the set of real elements in Z_1. It is asserted:*

(i) *if Z_1 is invariant under conjugation, then for each $z = x + iy \in Z_1$ the elements x and y belong to E_1;*

(ii) *if, besides satisfying the assumption of (i), Z_1 is finite dimensional, then there exists a base for Z_1 which consists of elements of E_1;*

(iii) *a finite-dimensional subspace Z_1 of Z is invariant under conjugation if and only if there exists a projection $P: Z$ onto Z_1, such that*

$$P(\bar{z}) = \overline{P(z)} \quad \text{for all } z \in Z. \tag{3.15}$$

PROOF. (i) If $z = x + iy \in Z_1(x, y \text{ real})$, then by assumption $\bar{z} = x - iy \in Z_1$. Consequently $x = (z + \bar{z})/2$ and $y = (z - \bar{z})/2i$ lie in $Z_1 \cap E = E_1$.

Proof of (ii). E_1 is obviously a linear subspace of the real space E.

Moreover, if $b_1, b_2, \ldots, b_r$ are linearly independent elements of this subspace, they are also linearly independent as elements of Z_1. For if

$$\sum_{j=1}^{r} c_j b_j = \theta,$$

with $c_j = \alpha_j + i\beta_j$, a_j, β_j real, then

$$\sum_{j=1}^{r} \alpha_j b_j + i \sum_{j=1}^{r} \beta_j b_j = \theta,$$

and thus

$$\sum_{j=1}^{r} \alpha_j b_j = \sum_{j=1}^{r} \beta_j b_j = \theta.$$

Therefore $\alpha_j = \beta_j = 0$, $j = 1, 2, \ldots, r$, on account of the independence of the b_j as elements of E_1. Thus $c_j = \alpha_j + i\beta_j = 0$. This proves the independence of the b_j as elements of Z_1. It follows that the subspace E_1 of E is finite dimensional since Z_1 is finite dimensional. Then let $b_1, \ldots, b_n$ be a base for E_1. Since the b_j are independent also as elements of Z_1, it remains to show that they span Z_1. Now let $z = x + iy \in Z_1$. Then there exist real numbers α_j, β_j such that

$$x = \sum_{j=1}^{n} \alpha_j b_j, \qquad y = \sum_{j=1}^{n} \beta_j b_j, \tag{3.16}$$

and thus

$$z = x + iy = \sum_{j=1}^{n} (\alpha_j + i\beta_j) b_j.$$

Proof of (iii). The coefficients α_j in the first sum in (3.16) are obviously linear in x for x in the finite-dimensional subspace E_1 of E. They are therefore also continuous (see, e.g., [**18**, p. 245]). Consequently, by the Hahn-Banach theorem, these bounded linear functions $\alpha_j = \alpha_j(x)$ can be extended to bounded linear functions $A_j(x)$ defined for all $x \in E$. We now define for $j = 1, 2, \ldots, n$ continuous linear functionals C_j on Z by setting for $z = x + iy$

$$C_j(z) = A_j(x) + iA_j(y). \tag{3.17}$$

Then

$$C_j(\bar{z}) = \overline{C_j(z)} \tag{3.18}$$

since the $A_j(x)$ are real-valued. We now set

$$P(z) = \sum_{j=1}^{n} C_j(z)b_j, \tag{3.19}$$

where the b_j form a real base for Z_1. We then see from (3.18) that (3.15) is satisfied. We now show that P is a projection, i.e., that (2.9) is satisfied. Since the b_j are real, we see from (3.19) and (3.17) that

$$P^2(z) = P(P(z)) = \sum_{j=1}^{n} b_j C_j \left(\sum_{l=1}^{n} C_l(z)b_l \right) = \sum_{j=1}^{n} b_j \sum_{l=1}^{n} C_l(z)C_j(b_l). \tag{3.20}$$

Since $b_l \in E_1$, $C_j(b_l) = A_j(b_l) = \alpha_j(b_l)$. But $\alpha_j(b_l) = 0$ if $j \neq l$, and 1 for $j = l$ as is seen from (3.16) with $x = b_l$. Therefore

$$C_j(b_l) = \begin{cases} 1 & \text{if } j = l, \\ 0 & \text{if } j \neq l, \end{cases}$$

and we see from (3.20) and (3.19) that

$$P^2(z) = \sum_{j=1}^{n} b_j C_j(z) = P(z).$$

This finishes the proof that the invariance of Z_1 under conjugation implies the existence of a projection P satisfying (3.15).

Conversely if such projection P exists, then for $z \in Z_1$ we see that $z = P(z)$, and therefore by (3.15) that $\bar{z} = \overline{P(z)} = P(\bar{z})$ which shows that $\bar{z} \in Z_1$.

§4. On the index j of linear nonsingular L.-S. maps
on complex and real Banach spaces

4.1. DEFINITION. Let Z be a complex Banach space. Let

$$m(z) = z - M(z) \tag{4.1}$$

be a nonsingular L.-S. map $Z \to Z$. Then with $\mathcal{R}(\lambda)$ denoting the real part of the complex number λ, the index $j(m)$ of m is defined as follows: if M has no eigenvalues λ with $\mathcal{R}(\lambda) > 1$ we set $j(m) = 1$. If M does have eigenalues λ with $\mathcal{R}(\lambda) \geq 1$, we denote them by $\lambda_1, \lambda_2, \ldots, \lambda_s$ and define

$$j(m) = (-1)^{\sum_{i=1}^{s} \nu_i}, \tag{4.2}$$

where $\nu_i = \nu(\lambda_i)$ is the generalized multiplicity of λ_i as an eigenvalue of M (see subsection 2.17 for the definition of "generalized multiplicity," and see §1 for a motivation for the definition of $j(m)$). Note that by subsection 2.16 the number of eigenvalues λ of M with $\mathcal{R}(\lambda) \geq 1$ is finite and by subsection 2.17 the generalized multiplicity by $\nu(\lambda)$ of an eigenvalue λ of M is finite.

4.2. DEFINITION. Let E be a real Banach space, and let

$$l(x) = x - L(x) \tag{4.3}$$

be a nonsingular L.-S. map $E \to E$. An eigenvalue λ_0 of L is then a real number such that for some point $x \in E$ different from θ

$$L(x) = \lambda_0 x. \tag{4.4}$$

If Z is the complexification of E and M the complex extension of L, it is clear from §3 that λ_0 is an eigenvalue of L if and only if λ_0 is a real eigenvalue of M. We define the generalized multiplicity of an eigenvalue $\lambda_0 \neq 0$ of L as the dimension $\mu(\lambda_0)$ of the space

$$E_{\lambda_0} = \{x \in E \mid (\lambda_0 I - L)^n x = \theta\} \tag{4.5}$$

for some integer $n \geq 1$. $\mu(\lambda_0)$ is finite, for it is clear that $\mu(\lambda_0) \leq \nu(\lambda_0)$, the generalized multiplicity of λ_0 as an eigenvalue of M (see subsection 2.17; actually $\mu(\lambda_0) = \nu(\lambda_0)$ as we will see in subsection 4.11).

4.3. DEFINITION. Let l and L be as in subsection 4.2. We define the index $j(l)$ to be 1 if L has no eigenvalues > 1. If L has eigenvalues > 1, we denote them by $\lambda_1 > \lambda_2 > \cdots > \lambda_s$ and define the index $j(l)$ by

$$j(l) = (-1)^{\sum_{\rho=1}^{s} \mu_\rho}, \tag{4.6}$$

where $\mu_\rho = \mu(\lambda_\rho)$, $\rho = 1, 2, \ldots, s$. Note that this definition agrees with (1.5), which was shown to agree in the finite-dimensional case with (1.2). Note also that $\lambda = 1$ is not an eigenvalue of L since l is not singular.

4.4. DEFINITION. A linear map l of the real Banach space E into itself is said to be type I^- if it can be constructed as follows: let E^1 be a one-dimensional subspace of E and let E^2 be a complementary subspace such that every $x \in E$ can be written in the form (cf. Lemma 4 of Chapter 1)

$$x = x_1 + x_2, \qquad x_1 \in E^1, \; x_2 \in E^2. \tag{4.7}$$

Then l maps $x = x_1 + x_2$ into $-x_1 + x_2$ (cf. §8 of the introduction). Maps of type I^- in a complex Banach space Z are defined correspondingly.

Note that $l(x) = x - 2x_1$, i.e., $l(x)$ is of the form (4.3) with $L(x) = 2x_1$. Since x_1 lies in the one-dimensional space E^1, $L(x)$ is completely continuous. Thus l is an L.-S. map. Since obviously l maps E onto E, the map l is not singular. Thus $j(l)$ is defined for a map l of type I^-.

4.5. LEMMA.

$$j(l) = \begin{cases} +1 & \text{if } l = I, \text{ the identity map,} \\ -1 & \text{if } l \text{ is of type } I^-. \end{cases} \tag{4.8}$$

PROOF. If $l = I$ then the $L(x)$ in (4.3) equals θ for all x. Therefore (4.4) for $x \neq \theta$ is satisfied only with $\lambda_0 = 0$. Thus L has no eigenvalues ≥ 1 which by Definition 4.3 proves the first part of assertion (4.8). Suppose now l is of the I^- type. Then with the notation used in (4.7), $l(x) = -x_1 + x_2 = x - 2x_1$, i.e., $L(x) = 2x_1$ by (4.3) and the eigenvalue equation (4.4) reads $2x_1 = \lambda_0 x = \lambda_0(x_1 + x_2)$. This implies $x_2 = \theta$ and $\lambda_0 = 2$. Thus 2 is the only eigenvalue of L, and

$$j(l) = (-1)^{\mu(2)}. \tag{4.9}$$

Here $\mu(2)$ is the dimension of the space given by (4.5) with $\lambda_0 = 2$ and $L(x) = 2x_1$. Now for $i = 1, 2$ let P_i be the projection $x \to x_i$ (cf. (4.7)). Then $\lambda_0 I - L = 2(P_1 + P_2) - 2P_1 = 2P_2$. Thus, by (4.5), E_2 consists of those x for which $P_2^n x = 0$ for some integer $n \geq 1$. But for these integers $P_2^n = P_2$ since P_2 is a projection, and we see that $E_2 = \{x \in E \mid P_2(x) = \theta\} = E^1$. Thus $\mu(2) = \dim E^1 = 1$. This together with (4.9) proves the second part of the assertion (4.8).

Since L.-S. homotopy (see subsection 1.2 of Chapter 2) is transitive, it follows from Lemma 4.5 that the linear homotopy theorem (stated in subsection 1.3 of Chapter 2) is equivalent to the following one.

4.6. THEOREM. *Let l_0 and l_1 be two linear nonsingular L.-S. maps $E \to E$. Then l_0 and l_1 are linearly L.-S. homotopic (see Definition 1.2 in Chapter 2) if and only if*

$$j(l_0) = j(l_1). \tag{4.10}$$

Now for questions of continuity it is preferable to deal with the complexification Z of E and the complex extension M of the operator L given by (4.3). For if M changes continuously, a real eigenvalue of M may become complex. But in this case the eigenvalue of L as an operator on E "disappears" since the concept of a complex eigenvalue makes no sense in the real space E. We therefore state the following theorem whose proof will be given in §5.

4.7. THEOREM. *Let l_0 and l_1 be as in Theorem 4.6. Let Z be the complexification of E, and for $i = 0, 1$ let M_i be the complex extension of $L_i = I - l_i$ on E, and let $m_i(z) = z - M_i(z)$. It is asserted that the equality*

$$j(m_0) = j(m_1) \tag{4.11}$$

(see Definition 4.1) holds if and only if there exists an L.-S. linear homotopy $m(x, t) = z - M(z, t)$, $0 \leq t \leq 1$, between $m_0(z) = m(z, 0)$ and $m_1(z) = m(z, 1)$ which satisfies for $z \in Z$ and $t \in [0, 1]$

$$M(\bar{z}, t) = \overline{M(z, t)}. \tag{4.12}$$

The remainder of §4 is devoted to showing

4.8. THEOREM. *Theorems 4.6 and 4.7 are equivalent.*

The proof of this theorem requires some lemmas.

4.9. LEMMA. *Let E and Z be as in Theorems 4.7 and 4.8. Let L be a continuous linear map $E \to E$, and let M be its complex extension. Then* (i) *the spectrum $\sigma(M)$ of M is symmetric with respect to the real axis of the complex λ-plane, and* (ii) *for $\lambda \in \rho(M)$*

$$R_{\overline{\lambda}}(M)(z) = \overline{R_\lambda(M)(\overline{z})} \tag{4.13}$$

(see Definition 2.1).

PROOF. (i) $\sigma(M)$ and the resolvent set $\rho(M)$ are disjoint subsets of the λ-plane whose union is the whole λ-plane. Therefore assertion (i) is equivalent to the assertion that $\rho(M)$ is symmetric to the real λ-axis. Let $\lambda_0 \in \rho(M)$.

We have to prove

$$\overline{\lambda}_0 \in \rho(M). \tag{4.14}$$

Now by the definition of $R_\lambda(M)$

$$(\lambda_0 I - M)R_{\lambda_0}(M)(z) = z \tag{4.15}$$

for all $z \in Z$. But by (3.10) for $u \in Z$

$$\overline{(\lambda_0 I - M)(u)} = (\overline{\lambda}_0 I - M)(\overline{u}).$$

Applying this relation with $u = R_{\lambda_0}(M)(z)$ we see from (4.15) that

$$(\overline{\lambda}_0 I - M)\overline{R_{\lambda_0}(M)(z)} = \overline{z}.$$

Replacing z here by $\overline{z}$ we see that

$$(\overline{\lambda}_0 I - M)\overline{R_{\lambda_0}(M)(\overline{z})} = z$$

for all $z \in Z$. This shows that

$$S(z) = \overline{R_{\lambda_0}(M)(\overline{z})} \tag{4.16}$$

is a right inverse of $(\overline{\lambda}_0 I - M)$. But interchanging the factors in (4.15) we conclude that $S(z)$ is also a left inverse of $\overline{\lambda}_0 I - M$. Thus S is an inverse of $\overline{\lambda}_0 I - M$. Since S is bounded, the assertion (4.14) follows by definition of $\rho(M)$.

Proof of (ii). By definition, $R_{\overline{\lambda}_0}(M)$ is the inverse of $\overline{\lambda}_0 I - M$. Therefore by (4.16)

$$R_{\overline{\lambda}_0}(M)(z) = S(z) = \overline{R_{\lambda_0}(M)(\overline{z})}.$$

This proves assertion (ii).

4.10. LEMMA. *Let E, Z, L, M be as in Lemma 4.9. Let λ_0 be an isolated point of the spectrum $\sigma(M)$ of M such that by Lemma 4.9 $\overline{\lambda}_0$ is also an isolated point of $\sigma(M)$. Let c_0 be a counterclockwise-oriented circle with center λ_0 and radius r_0 where r_0 is such that the closure of the circular disk $B(\lambda_0, r_0)$ contains no point of $\sigma(M)$ other than λ_0. Let c_0' be the counterclockwise-oriented circle with center $\overline{\lambda}_0$ and radius r_0. Let P_{λ_0} and $P_{\overline{\lambda}_0}$ be the linear operators defined by*

$$P_{\lambda_0} = \frac{1}{2\pi i} \int_{c_0} R_\mu(M)\, d\mu, \qquad P_{\overline{\lambda}_0} = \frac{1}{2\pi i} \int_{c_0'} R_\mu(M)\, d\mu. \tag{4.17}$$

Then the ranges $Z_{\lambda_0} = P_{\lambda_0} Z$ and $Z_{\overline{\lambda}_0} = P_{\overline{\lambda}_0} Z$ are conjugate to each other, i.e., if z varies over Z_{λ_0}, then $\overline{z}$ varies over $Z_{\overline{\lambda}_0}$.

PROOF. λ_0 and $\overline{\lambda}_0$ are spectral sets (see subsections 2.1 and 2.3). Therefore, by subsection 2.11, P_{λ_0} and $P_{\overline{\lambda}_0}$ are projections. We will show first our lemma follows from the relation

$$P_{\overline{\lambda}_0}(\overline{z}) = \overline{P_{\lambda_0}(z)} \quad \text{for all } z \in Z. \tag{4.18}$$

Indeed, since P_{λ_0} is a projection, $z \in Z_{\lambda_0} = P_{\lambda_0} Z$ if and only if $z = P_{\lambda_0}(z)$ or $\overline{z} = \overline{P_{\lambda_0}(z)}$, or, by (4.18), $\overline{z} = P_{\overline{\lambda}_0}(\overline{z})$. But since $P_{\overline{\lambda}_0}$ is a projection, the last equality is a necesary and sufficient condition for $\overline{z}$ to be an element of $P_{\overline{\lambda}_0} Z = Z_{\overline{\lambda}_0}$.

Thus for the proof of our lemma it remains to verify (4.18). For this purpose we set $\mu = \overline{\lambda}_0 + r_0 e^{i\phi}$ and $\lambda = \overline{\mu} = \lambda_0 + r_0 e^{-i\phi}$ such that $d\mu = i r_0 e^{i\phi} \, d\phi$ and $d\lambda = -i r_0 e^{-i\phi} \, d\phi$. We then see from (4.17) and (4.13) that

$$P_{\overline{\lambda}_0}(\overline{z}) = \frac{1}{2\pi} \int_0^{2\pi} R_\mu(M)(\overline{z}) r_0 e^{i\phi} \, d\phi = \frac{1}{2\pi} \int_0^{2\pi} \overline{R_\lambda(M)(z)} r_0 e^{i\phi} \, d\phi$$

$$= \overline{\frac{1}{2\pi} \int_0^{2\pi} R_\lambda(M)(z) r_0 e^{-i\phi} \, d\phi} = \overline{-\frac{1}{2\pi i} \int_0^{2\pi} R_\lambda(M)(z)(-i r_0 e^{-i\phi}) \, d\phi}$$

$$= \overline{-\frac{1}{2\pi i} \int_{-c_0} R_\lambda(M)(z) \, d\lambda} = \overline{\frac{1}{2\pi i} \int_{c_0} R_\lambda(M)(z) \, d\lambda} = \overline{P_{\lambda_0}(z)}.$$

This proves (4.18), and thus our lemma.

4.11. COROLLARY TO LEMMA 4.10. (i) *The spaces Z_{λ_0} and $Z_{\overline{\lambda}_0}$ have the same dimension; thus $\nu(\lambda_0) = \nu(\overline{\lambda}_0)$ (cf. subsection 2.17).*

(ii) *If λ_0 is real, then Z_{λ_0} is invariant under conjugation.*

(iii) *If λ_0 is real, then the generalized multiplicity $\nu(\lambda_0)$ of λ_0 as an eigenvalue of M equals the generalized multiplicity $\mu(\lambda_0)$ of λ_0 as an eigenvalue of L (cf. subsections 2.17 and 4.2).*

PROOF. Assertions (i) and (ii) are obvious consequences of Lemma 4.10. As to assertion (iii) we noted already in subsection 4.2 that $\mu(\lambda_0) \leq \nu(\lambda_0)$. It remains to prove that

$$\nu(\lambda_0) \leq \mu(\lambda_0). \tag{4.19}$$

Now it follows from assertion (ii) of our corollary and Lemma 3.12 that $Z_{\lambda_0} = P_{\lambda_0} Z$ has a base $b_1, b_2, \ldots, b_r$ consisting of elements of E. Then

$$r = \nu(\lambda_0),$$

since r and $\nu(\lambda_0)$ both equal the dimension of Z_{λ_0} (cf. subsection 2.17). Since each $b_j \in Z_{\lambda_0}$, we have (also by subsection 2.17)

$$(\lambda_0 I - M)^n b_j = \theta, \qquad j = 1, 2, \ldots, \nu(\lambda_0) \tag{4.20}$$

for some n. But by definition (3.9) of the extension M of L, we see that $M(b_j) = L(b_j)$ since $b_j \in E$. Thus by (4.20) $(\lambda_0 I - L)^n b_j = 0$, $j = 1, 2, \ldots, \nu(\lambda_0)$. The asserted inequality (4.19) now follows from the definition of $\mu(\lambda_0)$ given in 4.2 since λ_0 is real.

4.12. LEMMA. *We use the notation of Lemma 4.10. We assume now that the linear map $l(x) = x - L(x)$ is a nonsingular L.-S. map $E \to E$. Let $m(z)$ be its complex extension. Then*

$$j(m) = j(l). \tag{4.21}$$

PROOF. By Lemma 3.9 the nonsingularity of l implies that $m(z) = z - M(z)$ is a nonsingular L.-S. map $Z \to Z$. Thus both members of (4.21) are defined. To prove their equality we note first: if λ_0 is a complex eigenvalue of M, then so is $\bar{\lambda}_0$ (see Lemma 4.9), and $\nu(\lambda_0) = \nu(\bar{\lambda}_0)$ by part (i) of Corollary 4.11. Thus $\nu(\lambda_0) + \nu(\bar{\lambda}_0)$ is an even number. Therefore in Definition 4.1 (see in particular (4.2)) we may omit those $\nu_i = \nu(\lambda_i)$ for which λ_i is complex. But for a real eigenvalue λ_i of M we know from part (i) of the corollary 4.11 that $\nu(\lambda_i) = \mu(\lambda_i)$, and the definition of $j(l)$ in subsection 4.3 (see in particular (4.6)) shows that $j(m) = j(l)$.

4.13. We are now ready to prove Theorem 4.8. In the present section we show that Theorem 4.7 implies Theorem 4.6.

(a) Suppose (4.10) is satisfied. We have to show that l_0 and l_1 are linearly L.-S. homotopic. Let m_0 and m_1 be the complex extensions of l_0 and l_1 resp. Then, by Lemma 4.12, the equality (4.10) implies (4.11). Therefore, by Theorem 4.7 there exists a homotopy $m(z, t) = z - M(z, t)$ of the properties asserted in that theorem. But by Lemma 3.11 the relation (4.12) implies that for each $t \in [0, 1]$ the map $M(z, t)$ is the complex extension of a bounded linear map $L(x, t) \colon E \to E$ which by Lemma 3.9 is completely continuous and such that $l(x, t) = x - L(x, t)$ is nonsingular. This $l(x, t)$ gives the desired homotopy between l_0 and l_1.

(b) Assume now that l_0 and l_1 are linearly L.-S. homotopic. We have to prove (4.10). By assumption there exists a linear L.-S. homotopy $l(x, t) = x - L(x, t)$ connecting l_0 with l_1. For each $t \in [0, 1]$ let $m(x, t) = z - M(z, t)$ be the complex extension of $l(x, t)$. Then $m(z, t)$ is an L.-S. homotopy connecting $m_0(z) = m(z, 0)$ with $m_1 = m(z, 1)$ which by Lemma 3.11 satisfies (4.12). The assertion (4.10) follows now from Theorem 4.7 in conjunction with Lemma 4.12.

4.14. In this section we show that Theorem 4.6 implies Theorem 4.7.

(a) For $i = 0, 1$ let $m_i(z) = z - M_i(z)$ be a linear nonsingular L.-S. map $Z \to Z$ which is the complex extension of the linear nonsingular L.-S. map $l_i(x) = x - L_i(x) \colon E \to E$, and suppose that (4.11) holds. Now by Lemma 4.12 the relation (4.11) implies (4.10). Consequently by Theorem 4.6 there exists a linear L.-S. homotopy $l(x, t) = x - L(x, t)$ connecting $l_0 = l(x, 0)$ with $l_1 = l(x, 1)$. Then the complex extension $m(z, t) = z - M(z, t)$ of $l(x, t)$ is a linear L.-S. homotopy which connects m_0 with m_1 and which, by Lemma 3.11, satisfies (4.12).

(b) Again, for $i = 0, 1$, let $m_i(z) = z - M_i(z)$ be a nonsingular linear L.-S. map $Z \to Z$ which is the extension of a linear nonsingular L.-S. map $l_i(x) = x - L_i(x) \colon E \to E$. Suppose now there exists a linear L.-S. homotopy which connects m_0 with m_1 and satisfies (4.12). From the latter relation and from Lemma 3.11 we conclude that there exists a linear L.-S. homotopy which connects l_0 and l_1 and that therefore by Theorem 4.6, $j(l_0) = j(l_1)$. The asserted relation (4.11) follows now from Lemma 4.12.

§5. Proof of the linear homotopy theorem

5.1. The linear homotopy theorem (see subsection 1.3 of Chapter 2) is, as already stated in the last paragraph of subsection 4.5, equivalent to Theorem 4.6, which in turn, by Theorem 4.8, is equivalent to Theorem 4.7. It is thus sufficient to prove Theorem 4.7. We start by proving that if a homotopy $m_t(z) = m(x,t)$ of the properties stated in that theorem exists, then (4.11) is true. To do this we will show that $j(m_t)$ is constant for $t \in [0,1]$. Since $j(m_t)$ is integer-valued it will be sufficient to prove the "continuity" of that number or, more precisely, to prove the following theorem.

5.2. THEOREM. *Let Z be the complexification of the real Banach space E, and let $m_0(z) = z - M_0(z)$ be the complex extension of the nonsingular linear L.-S. map $l_0(x) = x - L_0(x)$ mapping E into E. Then there exists a positive number ε_0 of the following property: if $m(z) = z - M(z)$ is the complex extension of a linear nonsingular L.-S. map $l: E \to E$ and if*

$$\|m - m_0\| = \|M - M_0\| < \varepsilon < \varepsilon_0, \tag{5.1}$$

then

$$j(m) = j(m_0). \tag{5.2}$$

For the proof we need the following well-known elementary lemma.

5.3. LEMMA. *Let Z be an arbitrary complex Banach space, and let n_0 be a bounded linear map $Z \to Z$ which has a bounded everywhere-defined inverse n_0^{-1}. Then every linear bounded map $n: Z \to Z$ satisfying*

$$\|n - n_0\| < \|n_0^{-1}\|^{-1} \tag{5.3}$$

has a bounded everywhere-defined inverse. (As the following proof shows, the lemma is also valid in a real Banach space E.)

PROOF. Suppose first that $n_0 = I$. Then the assumption (5.3) reads

$$\|n - I\| < 1, \tag{5.4}$$

from which it easily follows that

$$n_1 = \sum_{i=0}^{\infty} (I - n)^i \tag{5.5}$$

converges (with $(I - n)^0 = I$). Moreover, writing $n = I - (I - n)$, one sees that $nn_1 = n_1 n = I$, and thus $n_1 = n^{-1}$.

In the general case we see from (5.3) that

$$\|I - n_0^{-1} n\| = \|n_0^{-1}(n_0 - n)\| \le \|n_0^{-1}\| \|n_0 - n\| < 1. \tag{5.6}$$

Thus (5.4) is satisfied if n is replaced by $n_0^{-1}n$. Thus $n_2 = (n_0^{-1}n)^{-1}$ exists and $(n_2 n_0^{-1})n = n_2(n_0^{-1}n) = I$. This shows that $n_2 n_0^{-1}$ is a left inverse of n. Similarly, we see from the inequality $\|I - nn_0^{-1}\| < 1$, which is proved the same way as (5.6), that n has a left inverse. But the existence of a left and a right inverse proves the existence of a unique inverse.

5.4. PROOF OF THEOREM 5.2. We distinguish two cases: (A) M_0 has no eigenvalues λ with real part $\mathcal{R}(\lambda) \geq 1$; (B) M_0 does have such eigenvalues.

Proof in case (A). We define the subset $\Delta = \Delta(M_0)$ of the λ-plane by $\Delta = \{\lambda \mid \mathcal{R}(\lambda) \geq 1\}$. Then Δ is a closed subset of $\rho(M_0)$. Therefore by Lemma 2.5 there exists a positive number μ_0 such that

$$\|R_\lambda(M_0)\| < \mu_0/2 \quad \text{for } \lambda \in \Delta. \tag{5.7}$$

We now claim that if M is a linear completely continuous map $Z \to Z$ satisfying

$$\|M - M_0\| < \tfrac{1}{2}(\mu_0/2)^{-1}, \tag{5.8}$$

then

$$\Delta \subset \rho(M). \tag{5.9}$$

Indeed if $n_0(z) = \lambda z - M_0(z)$ and $n(z) = \lambda z - M(z)$ with $\lambda \in \Delta$, then by (5.8), by (5.7), and by the definition of R_λ

$$\|n - n_0\| = \|M - M_0\| < (\mu_0/2)^{-1} < \|R_\lambda(M_0)\|^{-1} = \|n_0^{-1}\|^{-1}. \tag{5.10}$$

But by Lemma 5.3 this inequality implies that $n^{-1} = R_\lambda(M)$ exists, as a bounded operator, i.e., that $\lambda \in \rho(M)$ as asserted.

It is now easy to see that (5.8) implies the asserted equality (5.2). Indeed $j(m_0) = 1$ by Definition 4.1. But by definition of Δ we see from (5.9) that no eigenvalue λ of M has a real part ≥ 1. Thus, again by Definition 4.1, $j(m) = 1$.

Proof in case (B). $\sigma(M_0)$ is closed and bounded (see subsection 2.1). Moreover by Lemma 3.9 M_0 is completely continuous and therefore (see subsection 2.16) zero is the only accumulation point of $\sigma(M)$. It follows that $\sigma(M_0)$ contains only a finite number of points λ with $R(\lambda) \geq 1$. Moreover, if we denote them by $\lambda_1^0, \lambda_2^0, \ldots, \lambda_s^0$, then there exists a positive number r of the following properties: if for $i = 1, 2, \ldots, s$, d_i is the open circular disk with center λ_i^0 and radius r, then (a) λ_i^0 is the only eigenvalue of M_0 in the closed disk $\bar{d}_i$; (b) the $\bar{d}_i$ are disjoint; (c) for those i for which $\mathcal{R}(\lambda_i^0) > 1$ the disk $\bar{d}_i$ is contained in the open half plane $\mathcal{R}(\lambda) > 1$; (d) for those i for which λ_i^0 is not real the disk $\bar{d}_i$ does not intersect the real axis; (e) for those $\lambda \in \sigma(M_0)$ for which $\mathcal{R}(\lambda) < 1$ we have $\mathcal{R}(\lambda) < 1 - r$; (f) $0 < r < 1$. In our present case (B) we define

$$\Delta = \Delta(M_0) = \{\lambda \mid \mathcal{R}(\lambda) \geq 1 - r\} - \bigcup_{j=1}^{s} d_j. \tag{5.11}$$

Then Δ is a closed subset of $\rho(M_0)$. As in case (A) we conclude from this the existence of a positive μ_0 such that (5.7) holds and for completely continuous linear M satisfying (5.8) the inclusion (5.9) holds.

To prove that (5.1) with small enough ε_0 implies the assertion (5.2), we choose our notation in such a way that $\lambda_1^0, \lambda_2^0, \ldots, \lambda_r^0$ are real and $\lambda_{r+1}^0, \lambda_{r+2}^0, \ldots, \lambda_s^0$ are complex, $0 \leq r \leq s$ ($r = 0$ means that there are no real numbers among the λ_i^0, and $r = s$ that all of them are real; in any case $s \geq 1$ by assumption). Since M_0 is the complex extension of a completely continuous operator $L_0 \colon E \to E$, we see from Lemma 4.9 and the corollary to Lemma 4.11 that if λ is one of the complex eigenvalues $\lambda_{r+1}^0, \ldots, \lambda_s^0$ of M_0, then so is $\overline{\lambda}$ and $\nu^0(\lambda) = \nu^0(\overline{\lambda})$, where $\nu^0(\lambda)$ denotes the generalized multiplicity of λ as an eigenvalue of M_0. Consequently by definition (4.2) of $j(m)$

$$j(m_0) = (-1)^{\sum_{\rho=1}^{r} \nu(\lambda_\rho^0)}, \tag{5.12}$$

where it is understood that the sum stands for zero if $r = 0$. Suppose now that M is the complex extension of a linear completely continuous operator $L \colon E \to E$ and that M satisfies (5.8). Then (5.9) is true and therefore by definition (5.11) of Δ

$$\sigma(M) \cap \{\lambda \mid \mathcal{R}(\lambda) \geq 1 - r\} \subset \bigcup_{j=1}^{s} d_j = \bigcup_{j=1}^{r} d_j + \bigcup_{j=r+1}^{s} d_j.$$

Now by property (d) of the radius r of the disks d_j, these disks do not intersect the real axis for $j = r + 1, \ldots, s$. Thus the eigenvalues of M contained in these disks are complex, and the argument used to establish (5.12) shows that

$$j(m) = (-1)^{\sum \nu(\lambda)},$$

where the sum is extended over those eigenvalues λ of M which lie in one of the d_j for $j = 1, 2, \ldots, r$ and where $\nu(\lambda)$ denotes the generalized multiplicity of λ as an eigenvalue of M. Thus if $\lambda_1^j, \lambda_2^j, \ldots, \lambda_{r_j}^j$ are those eigenvalues of M which lie in d_j for $j = 1, 2, \ldots, r$, then

$$j(m) = (-1)^{\sum_{j=1}^{r} \sum_{i=1}^{r_j} \nu(\lambda_i^j)}. \tag{5.13}$$

Comparison of (5.12) with (5.13) shows that our assertion (5.2) will be proved once it is shown that

$$\nu(\lambda_j^0) = \sum_{i=1}^{r_j} \nu(\lambda_i^j), \qquad j = 1, 2, \ldots, r. \tag{5.14}$$

We recall that here the multiplicity ν at the left refers to M_0 while the ν's at the right refer to M.

Now by subsection 2.17 and (2.15)

$$\nu(\lambda_j^0) = \dim P_j^0 Z, \qquad j = 1, 2, \ldots, r, \tag{5.15}$$

where

$$P_j^0 = \frac{1}{2\pi i} \int_{\partial d_j} R_\mu(M_0) \, d\mu \tag{5.16}$$

with the circle ∂d_j oriented in the counterclockwise sense. In the same way we see that

$$\nu(\lambda_i^j) = \dim P_i^j, \qquad i = 1, 2, \ldots, r_j, \tag{5.17}$$

where

$$P_i^j = \frac{1}{2\pi i} \int_{\partial d_i^j} R_\mu(M)\, d\mu \tag{5.18}$$

and where d_i^j is a circular disk with center λ_i^j and a radius so small that the closures $\bar{d}_i^j$ are disjoint and lie in d_j.

The assertion (5.14) now reads

$$\dim P_j^0 Z = \sum_{i=1}^{r_j} \dim P_i^j Z, \qquad j = 1, 2, \ldots, r. \tag{5.19}$$

If we set

$$P^j = \frac{1}{2\pi i} \int_{\partial d_j} R_\mu(M)\, d\mu, \qquad j = 1, 2, \ldots, r, \tag{5.20}$$

we see from the "Cauchy theorem" (2.4) that

$$P^j(Z) = \sum_{i=1}^{r_j} P_i^j Z \tag{5.21}$$

and from this equality together with Lemma 2.12 that $P^j Z$ is the direct sum of the spaces $P_1^j Z, P_2^j Z, \ldots, P_{r_j}^j Z$, and that therefore

$$\dim P^j Z = \sum_{i=1}^{r_j} \dim P_i^j Z. \tag{5.22}$$

Thus the assertion (5.19) may be written as

$$\dim P_j^0 Z = \dim P^j Z, \qquad j = 1, 2, \ldots, r. \tag{5.23}$$

We will prove that if μ_0 is as in (5.7), if $\varepsilon_1 = (\mu_0/2)^{-1}$ (cf. (5.8)), if

$$\varepsilon_2 = \mu_0^{-2} \min_{j=1,2,\ldots,r} \|P_j^0\|^{-1}, \tag{5.24}$$

and if

$$\varepsilon_0 = \min(\varepsilon_1/2, \varepsilon_2), \tag{5.25}$$

then (5.1) implies (5.23) (and thus (5.2)). For the proof we need two lemmas; the first is related to Lemma 5.3 while the second is due to J. Schwartz [50, p. 424].

5.5. LEMMA. *Let $n_0(z)$ be as in Lemma 5.3, and let n be a linear bounded map $Z \to Z$ satisfying*

$$\|n - n_0\| < (2\|n_0^{-1}\|)^{-1}. \tag{5.26}$$

Then n^{-1} exists and satisfies

$$\|n^{-1} - n_0^{-1}\| < 2\|n_0^{-1}\|^2 \|n - n_0\|. \tag{5.27}$$

5.6. LEMMA. *Let Q_0 and Q_1 be projections $Z \to Z$. Suppose that*

$$\|Q_1 - Q_0\| \le \tfrac{1}{2}\|Q_0\|^{-1}. \tag{5.28}$$

Then the dimension δ_0 of $Q_0 Z$ equals the dimension δ_1 of $Q_1 Z$.

5.7. We postpone the proof of the two preceding lemmas and prove that (5.1) implies (5.23). Let $n_0(z) = \mu z - M_0(z)$ and $n(z) = \mu z - M(z)$. Then by (5.8) and (5.7)

$$\|n(z) - n_0(z)\| = \|M(z) - M_0(z)\| < \tfrac{1}{2}(\mu_0/2)^{-1}$$
$$< \tfrac{1}{2}\|R_\mu(M_0)\|^{-1} = \tfrac{1}{2}\|n_0^{-1}\|^{-1},$$

i.e., (5.26) is satisfied. Therefore by Lemma 5.5 the inequality (5.27) holds:

$$\|R_\mu(M) - R_\mu(M_0)\| < 2\|R_\mu(M_0)\|^2\|M - M_0\|.$$

Thus by (5.1), (5.25) and (5.24)

$$\|R_\mu(M) - R_\mu(M_0)\| < 2\|R_\mu(M_0)\|^2\mu_0^{-2}\min_{j=1,2,\dots,r}\|P_j^0\|^{-1},$$

and by (5.7)

$$\|R_\mu(M) - R_\mu(M_0)\| < \tfrac{1}{2}\min_{j=1,2,\dots,r}\|P_j^0\|^{-1}. \qquad (5.29)$$

But by (5.20) and (5.16)

$$\|P^j - P_j^0\| = \frac{1}{2\pi}\left\|\int_{\partial d_j} R_\mu(M) - R_\mu(M_0)\,d\mu\right\|$$

$$\leq \frac{1}{2\pi}\int_0^{2\pi}\|R_\mu(M) - R_\mu(M_0)\|\,|r e^{i\phi}|\,d\phi.$$

Therefore by (5.29) (recalling that $0 < r < 1$) we see that $\|P^j - P_j^0\| < \tfrac{1}{2}\|P_j^0\|^{-1}$. But by Lemma 5.6 this inequality implies the assertion (5.23).

It remains to prove Lemmas 5.5 and 5.6.

5.8. PROOF OF LEMMA 5.5. That the inequality (5.26) implies the existence of n^{-1} follows from Lemma 5.3. Moreover it follows easily from the proof for that lemma (see in particular (5.5) and (5.6)) that

$$n^{-1}n_0 = (n_0^{-1}n)^{-1} = \sum_{j=0}^{\infty}(I - n_0^{-1}n)^j$$

or

$$n^{-1} = \sum_{j=0}^{\infty}(I - n_0^{-1}n)^j n_0^{-1}.$$

Thus

$$n^{-1} - n_0^{-1} = \sum_{j=1}^{\infty}(I - n_0^{-1}n)^j n_0^{-1} = (I - n_0^{-1}n)\sum_{k=0}^{\infty}(I - n_0^{-1}n)^k n_0^{-1}.$$

Therefore

$$\|n^{-1} - n_0^{-1}\| \leq \|n_0^{-1}\|\cdot\|I - n_0^{-1}n\|\cdot\sum_{k=0}^{\infty}\|I - n_0^{-1}n\|^k$$

$$= \frac{\|n_0^{-1}\|\cdot\|I - n_0^{-1}n\|}{1 - \|I - n_0^{-1}n\|} \leq \frac{\|n_0^{-1}\|^2\|n_0 - n\|}{1 - \|n_0^{-1}\|\cdot\|n_0 - n\|}.$$

This inequality together with the assumption (5.26) proves the assertion (5.27).

5.9. PROOF OF LEMMA 5.6. We will establish the inequalities

$$\left.\begin{array}{r}\|Q_0 - Q_0Q_1\| \\ \|Q_1 - Q_1Q_0\|\end{array}\right\} < 1, \tag{5.30}$$

and then show that they imply our lemma. Since $Q_0^2 = Q_0$ we see that

$$\|Q_0 - Q_0Q_1\| = \|Q_0(Q_0 - Q_1)\| \le \|Q_0\| \cdot \|Q_0 - Q_1\|.$$

By assumption (5.28) this inequality implies the first of the two inequalities (5.30). To prove the second one, we note that $\|Q_0\| \ge 1$ since Q_0 is the identity on its range Q_0Z. Using this and again the assumption (5.28), we see that

$$\|Q_1\| = \|Q_0 + Q_1 - Q_0\| \le \|Q_0\| + \|Q_1 - Q_0\|$$
$$\le \|Q_0\| + 1/(2\|Q_0\|) \le \|Q_0\| + \|Q_0\|/2.$$

Thus

$$\|Q_1\| < 2\|Q_0\|. \tag{5.31}$$

Since $Q_1^2 = Q_1$ we see that

$$\|Q_1 - Q_1Q_0\| = \|Q_1(Q_1 - Q_0)\| \le \|Q_1\| \cdot \|Q_1 - Q_0\|.$$

But if follows from (5.31) and from (5.28) that here the right member is less than 1. This proves the second of the equalities (5.30).

To derive the lemma from (5.30), we set $Z_0 = Q_0Z$, $Z_1 = Q_1Z$, and denote by $(Q_0Q_1)_0$ the restriction of Q_0Q_1 to Z_0. Then $(Q_0Q_1)_0$ maps $Z_0 \to Z_0$, while Q_0 restricted to Z_0 is the identity on Z_0. Therefore the first of the inequalities (5.30) implies by Lemma 5.3 that $(Q_0Q_1)_0$ has an inverse on Z_0. Thus $Z_0 \subset$ range of$(Q_0Q_1)_0 \subset$ range of $Q_0Q_1 \subset$ range of Q_1, and $Z_0 \subset Q_1Z = Z_1$. Therefore $\delta_0 = \dim Z_0 \le \delta_1 = \dim Z_1$. In a similar way one derives from the second of the inequalities (5.30) that $\delta_1 \le \delta_0$. Thus $\delta_1 = \delta_0$ as asserted.

5.10. We finished the proof that the equality (4.11) is necessary for the existence of the homotopy described in Theorem 4.7. We now turn to the sufficiency proof. We recall that $\lambda = 1$ is not an eigenvalue of the complex extension M_0 of L_0 since $m_0 = I - M_0$ is not singular. We consider first the special case that M_0 has no real eigenvalues > 1. In this case it follows from Lemma 4.12 and Definition 4.3 that $j(m_0) = j(l_0) = 1 = j(I)$. Therefore in the special case considered we have to prove

5.11. LEMMA. *Let M_0 satisfy the assumption of Theorem 4.7. In addition it is assumed that M_0 has no real eigenvalues $\lambda > 1$. Then there exists a linear L.-S. homotopy which connects m_0 with I and which satisfies (4.12).*

PROOF. For $t \in [0,1]$ we set $m(z,t) = z - (1 - t)M_0(z)$. That $M(z,t) = (1 - t)M_0(z)$ satisfies (4.12) is clear from Lemma 3.11. It remains to show that $m_t(z) = m(z,t)$ is nonsingular for all $t \in [0,1]$. For $t = 0$ and $t = 1$

the nonsingularity is obvious since $m(z,0) = m_0(z)$ and $m(z,1) = I$. Suppose then $m_{t_0}(z)$ to be singular for some t_0 in the open interval $(0,1)$. Then $z_0 - (1-t_0)M_0(z_0) = z_0$ for some $z_0 \neq \theta$, or $M(z_0) = \lambda_0 z_0$ with $\lambda_0 = (1-t_0)^{-1} > 1$. This contradicts our assumption that M_0 has no real eigenvalues > 1.

5.12. We suppose now that M_0 has real eigenvalues greater than 1. We denote them by $\lambda_1 > \lambda_2 > \cdots > \lambda_s$. Let $\sigma_0 = \sigma_0(M_0)$ be their union, and let $\sigma_1 = \sigma_1(M_0)$ be the complement of σ_0 in $\sigma(M_0)$:

$$\sigma(M_0) = \sigma_0 \cup \sigma_1. \tag{5.32}$$

Let

$$P_0 = P_{\sigma_0}, \qquad P_1 = P_{\sigma_1} \tag{5.33}$$

be the projections defined in Lemma 2.11 (cf. also subsection 2.12), and let

$$Z_0 = P_0 Z, \qquad Z_1 = P_1 Z. \tag{5.34}$$

Then as will be shown in subsection 5.13

$$Z = Z_0 \dotplus Z_1 \quad \text{(direct sum)} \tag{5.35}$$

with Z_0 being finite-dimensional. Corresponding to the decomposition (5.34) we set

$$M_0^0 = P_0 M_0, \qquad M_0^1 = P_1 M_0. \tag{5.36}$$

We will show that

$$M_0 = \begin{pmatrix} M_0^0 & 0 \\ 0 & M_0^1 \end{pmatrix} \tag{5.37}$$

(see subsection 5.15) and that σ_0 is the spectrum of M_0^0 and σ_1 the spectrum of M_0^1 (see subsection 5.16). Thus M_0^1 has no real eigenvalues > 1, and Lemma 5.11 may be applied to $m_0' = I_1 - M_0^1$, where I_1 denotes the identity map on Z_1. Discussion of the "finite-dimensional" part M_0^0 of M_0 will begin in subsection 5.20.

5.13. PROOF OF (5.35). Let V_0 and V_1 be open sets in the λ-plane containing σ_0 and σ_1 resp. with $\overline{V}_0$ and $\overline{V}_1$ disjoint. Let $\Gamma_0 = \partial V_0$ and $\Gamma_1 = \partial V_1$, where Γ_0 and Γ_1 are supposed to be rectifiable Jordan curves oriented in the counterclockwise sense. Then by subsections 2.11 and 2.12, by (5.33) and by Lemma 2.9

$$I = P_0 + P_1 = \frac{1}{2\pi i} \int_\Gamma R_\mu(M_0)\,d\mu, \qquad \Gamma_0 + \Gamma_1. \tag{5.38}$$

This shows that every $z \in Z$ can be represented by

$$z = z_0 + z_1, \qquad z_0 \in Z_0,\ z_1 \in Z_1. \tag{5.39}$$

The uniqueness of this representation follows from (2.12).

To establish (5.37) we prove the following lemma.

5.14. LEMMA.

$$M_0^0 = \frac{1}{2\pi i} \int_{\Gamma_0} \mu R_\mu(M_0)\, d\mu = M_0 P_0, \tag{5.40}$$

$$M_0^1 = \frac{1}{2\pi i} \int_{\Gamma_1} \mu R_\mu(M_0)\, d\mu = M_0 P_1, \tag{5.41}$$

where M_0^0 and M_0^1 are the operators defined in (5.36). Moreover, for $z \in Z$,

$$M_0(z) = M_0^0(z_0) + M_0^1(z), \tag{5.42}$$

$$M_0^0(z_1) = \theta, \qquad M_0^1(z_0) = \theta, \tag{5.43}$$

with z_0, z_1 as in (5.39).

PROOF. Let

$$f(\mu) = \begin{cases} 1 & \text{for } \mu \in \overline{V}_0, \\ 0 & \text{for } \mu \in \overline{V}_1, \end{cases}$$

where V_0 and V_1 are as in subsection 5.13. Then by (2.10) and (5.33)

$$P_0 = \frac{1}{2\pi i} \int_{\Gamma_0} R_\mu(M_0)\, d\mu = \frac{1}{2\pi i} \int_{\Gamma_0 + \Gamma_1} f(\mu) R_\mu(M_0)\, d\mu,$$

and from (5.32) and Lemma 2.9 we see that

$$M_0 = \frac{1}{2\pi i} \int_{\Gamma_0 + \Gamma_1} \mu R_\mu(M_0)\, d\mu.$$

Therefore from (2.8) with $g(\mu) = \mu$ on $\overline{V}_0 \cup \overline{V}_1$

$$P_0 M_0 = M_0 P_0 = \frac{1}{2\pi i} \int_{\Gamma_0 \cup \Gamma_1} f(\mu)\mu R_\mu(M_0)\, d\mu = \frac{1}{2\pi i} \int_{\Gamma_0} \mu R_\mu(M_0)\, d\mu. \tag{5.44}$$

By definition (5.36) this equality proves (5.40). Assertion (5.41) is proved correspondingly.

To prove (5.42) we note that by (5.39) and (5.34)

$$M_0(z) = M_0(z_0 + z_1) = M_0(z_0) + M_0(z_1)$$
$$= M_0(P_0 z) + M_0(P_1 z) = M_0 P_0(z_0) + M_0 P_1(z_1).$$

By (5.40) and (5.41) this proves (5.42). Finally by (5.36), (5.44), and (2.12)

$$M_0^0(z_1) = P_0 M_0(z_1) = M_0 P_0(z_1) = M_0 P_0 P_1(z) = 0.$$

This proves the first of the relations (5.43). The second one follows similarly.

5.15. PROOF OF (5.37). Since $P_0 M_0 = M_0 P_0$ by (5.44) and since $P_1 M_0 = M_0 P_1$ is established the same way,

$$M_0(z) = \begin{pmatrix} P_0 M_0(z) \\ P_1 M_0(z) \end{pmatrix} = \begin{pmatrix} M_0 P_0(z) \\ M_0 P_1(z) \end{pmatrix}.$$

Therefore by the direct decomposition (5.39)

$$M_0(z) = \begin{pmatrix} M_0 P_0(z_0) & M_0 P_0(z_1) \\ M_0 P_1(z_0) & M_0 P_1(z_1) \end{pmatrix}.$$

This relation proves (5.37) as we see from (5.40), (5.41), and (5.43).

5.16. Proof that σ_0 is the spectrum of M_0^0 and σ_1 is the spectrum of M_0^1. It follows from (5.42) and (5.43) that, restricted to Z_0, $M_0 = M_0^0$, and, restricted to Z_1, $M_0 = M_0^1$. Our assertion now follows from subsection 2.13.

This finishes the proof of the assertions contained in subsection 5.12. We next prove

5.17. LEMMA. (i) Z_0 and Z_1 are invariant under conjugation (see Definition 3.10); (ii) Z_0 is finite dimensional.

PROOF. (i) Let λ_j, $j = 1, 2, \ldots, s$, be as in subsection 5.12. Let d_j be a circular disk with center λ_j of radius so small that the closures $\bar{d}_j$ are disjoint and lie in V_0 (cf. subsection 5.13). Let the boundary ∂d_j of d_j be oriented in the counterclockwise sense, and let

$$Q_j = \frac{1}{2\pi i} \int_{\partial d_j} R_\mu(M_0) \, d\mu, \qquad j = 1, 2, \ldots, s. \tag{5.45}$$

Since the λ_j are the only eigenvalues of M_0 in V_0, it follows from the Cauchy theorem 2.7 and from 5.12 that

$$P_0 = \sum_{j=1}^{s} Q_j \tag{5.46}$$

and

$$Z_0 = P_0 Z = \sum_{j=1}^{s} Q_j Z. \tag{5.47}$$

Since each λ_j is a spectral set for M_0, each Q_j is a projection by Lemma 2.11, and by the Corollary 4.11 to Lemma 4.10 each of the spaces $Q_j Z$ is invariant under conjugation since λ_j is real. Thus we see from (5.47) that Z_0 is invariant under conjugation. Since Z also is invariant under conjugation, being the complexification of a real Banach space E, it now follows that Z_1 also is invariant under conjugation.

(ii) By Lemma 2.12 $Q_i Q_j = 0$ for $i \neq j$. It follows that the decomposition (5.47) is direct. Therefore

$$\dim Z_0 = \sum_{j=1}^{s} \dim Q_j Z_0.$$

But by 2.17 the dimension of $Q_j Z_0$ is the finite number $\nu(\lambda_j)$. Thus

$$\dim Z_0 = \sum_{j=1}^{s} \nu(\lambda_j). \tag{5.48}$$

5.18. Let I_0 and I_1 be the identity maps on Z_0 and Z_1 resp., and let, as always, I denote the identity map on Z. Then

$$m_0 = I - M_0 \begin{pmatrix} m_0^0 & 0 \\ 0 & m_0^1 \end{pmatrix}, \tag{5.49}$$

where

$$m_0^0 = I_0 - M_0^0, \qquad m_0^1 = I_1 - M_0^1. \tag{5.50}$$

This is an immediate consequence of (5.37).

5.19. LEMMA. *There exists an L.-S. homotopy $m_0^1(z_1, t) = z_1 - M_0^1(z_1, t)$, $z_1 \in Z_1$, $t \in [0, 1]$, connecting m_0^1 with I_1 and satisfying*

$$M_0^1(\bar{z}, t) = \overline{M_0^1(z, t)}.$$

PROOF. The lemma will be proved once it is shown that M_0^1 satisfies the assumption made in Theorem 5.11 for M_0. That is, we have to verify: (i) Z_1 is the complexification of a real Banach space E_1; (ii)$m_0^1 = I_1 - M_0^1$ and I_1 are the complex extensions of linear nonsingular L.-S. maps on E_1; (iii) M_1^0 has no real eigenvalues > 1. Now (i) follows from the fact that Z_1 is invariant under conjugation (Lemma 5.17). The assertion of (ii), that I_1 is the complex extension of the identity on E_1, is obvious. To prove that M_0^1 is the complex extension of a linear continuous map L_0^1 on E_1, it is by Lemma 3.11 sufficient to prove the relation

$$M_0^1(\bar{z}) = \overline{M_0^1(z)}, \qquad z \in Z_1. \tag{5.51}$$

But this relation follows from the fact that $M_0^1(z) = M_0(z)$ for $z \in Z_1$, by (5.41) and that $M_0(\bar{z}) = \overline{M_0(z)}$ since M_0 is by assumption the extension of a linear continuous map on E. That L_0^1 is completely continuous follows from Lemma 3.9 since $M_0^1(z) = M_0(z)$ is completely continuous by assumption. Finally, that $l_0^1 = I_1 - L_0^1$ is nonsingular follows also from Lemma 3.9 since otherwise, by that lemma, m_0^1 and therefore m_0 would be singular. Finally that assertion (iii) is true was already noticed in the lines directly following (5.37).

5.20. Our next goal is to prove that $m_0 = I - M_0$ is L.-S. homotopic (with a homotopy satisfying (4.12)) to either I or to a map of type I^- (see Definition 4.4). It follows from subsection 5.19 and (5.49) that m_0 is L.-S. homotopic (with the condition (4.12)) to

$$\tilde{m}_0 = \begin{pmatrix} m_0^0 & 0 \\ 0 & I_1 \end{pmatrix}, \qquad m_0^0 = I_0 - M_0^0. \tag{5.52}$$

It will therefore be sufficient to prove the corresponding assertion for the map m_0^0 which maps the finite-dimensional space Z_0 onto Z_0 (cf. Lemma 5.17). Our first step in this direction will be to prove

5.21. LEMMA. *There exists a real base $b_1, b_2, \ldots, b_N$ of Z_0 such that*

$$m_0^0(b_j) = -b_j, \qquad j = 1, 2, \ldots, N, \tag{5.53}$$

where N is the dimension of Z_0 (cf. (5.48)).

PROOF. As shown in the proof of part (ii) of Lemma 5.17 the decomposition of Z_0 given by (5.47) is direct. Therefore a basis for Z_0 is obtained by putting together the bases for each of the spaces $Q_j Z_0$, $j = 1, 2, \ldots, s$. But by subsection 2.17 and by assertion (iii) of Corollary 4.11

$$\dim Q_j Z_0 = \nu(\lambda_j) = \mu(\lambda_j). \tag{5.54}$$

Now let λ_0 be one of the λ_j, and let Q_0 be the corresponding Q_j. $Q_0 Z_0$ is invariant under conjugation (see Corollary 4.11). It follows that $Q_0 Z_0$ is the complexification of a real Banach space E_0 of the same dimension, and every base of E_0 is a real base of Z_0 (cf. Lemma 3.12). Therefore if L_0 is the linear map $E_0 \to E_0$ of which the map M_0^0 (restricted to $Q_0 Z_0$) is the complex extension, it will for the proof of our lemma be sufficient to show that there exists a base $b_1, \ldots, b_{N_0}$ of E_0 such that

$$l_0(b_j) = -b_j, \qquad j = 1, 2, \ldots, N_0, \tag{5.55}$$

where

$$l_0(x) = x - L_0(x), \qquad x \in E_0. \tag{5.56}$$

Let n_1 be the "index of nilpotency" of the map $Q_0 Z_0 \to Q_0 Z_0$ given by

$$l_{\lambda_0} = \lambda_0 I_0 - L_0, \tag{5.57}$$

i.e., the smallest positive integer n for which (4.5) holds for all $x \in Q_0 Z_0$. Then by a well-known theorem of linear algebra (see, e.g., [**26**, p. 111]) there exist integers $r, n_2, n_3, \ldots, n_r$ and elements $\beta_1, \beta_2, \ldots, \beta_r$ such that $n_1 \geq n_2 \geq \cdots \geq n_r$ and such that the elements

$$\beta_1, l_{\lambda_0} \beta_1, \ldots, l_{\lambda_0}^{n_1 - 1} \beta_1, \ldots, \beta_\rho, l_{\lambda_0} \beta_\rho, \ldots, l_{\lambda_0}^{n_\rho - 1} \beta_\rho, \ldots, \beta_r, l_{\lambda_0} \beta_r, \ldots, l_{\lambda_0}^{n_r - 1} \beta_r \tag{5.58}$$

form a base for E_0, while

$$l_{\lambda_0}^{n_1} \beta_1 = l_{\lambda_0}^{n_2} \beta_2 = \cdots = l_{\lambda_0}^{n_r} \beta_r = \theta. \tag{5.59}$$

Thus E_0 is the direct sum of spaces $E_{0\rho}$ with bases

$$\beta_\rho^1 = \beta_\rho, \quad \beta_\rho^2 = l_{\lambda_0} \beta^\rho, \ldots, \beta^{n_\rho} = l_{\lambda_0}^{n_\rho - 1} \beta_\rho. \tag{5.60}$$

We then see from (5.58) that

$$l_{\lambda_0} \beta_\rho^1 = \beta_\rho^2, \quad l_{\lambda_0} \beta_\rho^2 = \beta_\rho^3, \ldots, l_{\lambda_0} \beta_\rho^{n_\rho - 1} = \beta_\rho^{n_\rho}, \tag{5.61}$$

while by (5.59)

$$l_{\lambda_0}^{n_\rho} \beta_\rho^1 = \theta. \tag{5.62}$$

But from (5.56) and (5.57)

$$l_0 = (1 - \lambda_0) I_0 + l_{\lambda_0},$$

and therefore from (5.61), (5.62),

$$l_0 \beta_\rho^1 = (1 - \lambda_0) \beta_\rho^1 + \beta_\rho^2, \ldots, l_0 \beta_\rho^{n_\rho - 1} = (1 - \lambda_0) \beta_\rho^{n_\rho - 1} + \beta_\rho^{n_\rho},$$

$$l_0 \beta_\rho^{n_\rho} = (1 - \lambda_0) \beta_\rho^{n_\rho}. \tag{5.63}$$

We now define for $x \in E_{0\rho}$ and $t \in [0,1]$ a homotopy $l_t = l(x,t)$ by setting

$$l_t \beta_\rho^1 = [(1 - \lambda_0)(1 - t) + (-1)t]\beta_\rho^1 + (1 - t)\beta_\rho^2,$$

$$\vdots \tag{5.64}$$

$$l_t \beta_\rho^{n_\rho - 1} = [(1 - \lambda_0)(1 - t) + (-1)t]\beta_\rho^{n_\rho - 1} + (1 - t)\beta_\rho^{n_\rho},$$
$$l_t \beta_\rho^{n_\rho} = [(1 - \lambda_0)(1 - t) + (-1)t]\beta_\rho^{n_\rho}.$$

Obviously (5.63) and (5.64) agree for $t = 0$, while for $t = 1$ by (5.64)

$$l_1 \beta_\rho^j = -\beta_\rho^j, \qquad j = 1, 2, \ldots, n_\rho. \tag{5.65}$$

Recalling that $\lambda_0 > 1$ we see that $[(1 - \lambda_0)(1 - t) + (-1)t]$ is the linear convex combination of the points -1 and $1 - \lambda_0 < 0$, and therefore does not contain the point 0. This shows that l_t is not singular for all $t \in [0,1]$. Thus l_0 and l_1 are homotopic and (5.65) proves our lemma

5.22. We will now prove the assertion stated in subsection 5.20 as our "next goal." By subsection 5.20 and Lemma 5.21 this assertion obviously follows from the following lemma.

5.23. LEMMA. *Let Z_0 be a complex Banach space of finite dimension N which is the complexification of a real Banach space E_0. Let m_0^0 be a nonsingular linear map $Z_0 \to Z_0$. Suppose there exists a real base $b_1, b_2, \ldots, b_N$ in Z_0 for which (5.54) holds. Then m_0^0 is L.-S. homotopic if (4.12) holds to the identity I if N is even and to a map of type I^- if N is odd.*

PROOF. If N is even, we set $N = 2p$ and define $m_0^0(x, t)$ for $t \in [0, 1]$ and for $i = 1, 2, \ldots, p$ by

$$m_0^0(b_{2i-1}, t) = b_{2i-1} \cos(1 - t)\pi - b_{2i} \sin(1 - t)\pi,$$
$$m_0^0(b_{2i}, t) = b_{2i-1} \sin(1 - t)\pi + b_{2i} \cos(1 - t)\pi. \tag{5.66}$$

Then $m_0^0(z, 0) = m_0^0(z)$ by (5.54), while $m_0^0(z, 1)$ sends b_j into b_j for $j = 1, 2, \ldots, N$, i.e., $m_0^0(z, 1)$ is the identity map. Moreover $m_0^0(z, t)$ is not singular, the determinant of (5.66) being 1.

If N is odd, we change our notation by denoting the given base of Z_0 by $b_0, b_1, \ldots, b_N$. We then set $m_0^0(b_0, t) = -b_0$, while $m_0^0(b_j, t)$ for $j = 1, 2, \ldots, N = 2p$ is given by (5.66). We thus obtain a nonsingular homotopy $m_0^0(z, t)$ connecting the map $m_0^0(z)$ given by $m_0^0(b_j) = -b_j$ for $j = 0, 1, \ldots, N$ with the map given by $m_0^0(b_0, 1) = -b_0$, $m_0^0(b_j, 1) = b_j$, $j = 1, 2, \ldots, N$. The latter map is by definition of the I^- type.

5.23. We thus proved the assertion of subsection 5.20. It is clear that the identity map I on Z is not homotopic to a map m of type I^- since for such m the equality $j(m) = j(I)$ would hold (see 5.10). But by (4.21) and Lemma 4.5 this leads to the contradiction $+1 = -1$. Thus to finish the proof of Theorem 4.7 it remains to prove that two maps of the I^- type are L.-S. homotopic.

5.24. LEMMA. *Let $m_0(z) = z - M_0(z)$ and $M_1(z) = z - M_1(z)$ be linear L.-S. maps $Z \to Z$ of type I^- which are complex extensions of the real maps l_0 and l_1 resp. Then there exists an L.-S. homotopy $m(z, t) = z - M(z, t)$ connecting m_0 with m_1 and satisfying the condition* (4.12).

PROOF. Again let E denote the real Banach space of which Z is the complexification. l_0 and l_1 are then L.-S. linear maps $E \to E$ of type I^-. It is easily seen that it will be sufficient to show that l_0 and l_1 are L.-S. homotopic.

By Definition 4.4 there exists for $i = 0, 1$ a direct decomposition $E = E_1^i \dotplus E_2^i$, where E_1^i is a one-dimensional subspace, and there exists a corresponding unique representation for $x \in E$

$$x = x_1^i + x_2^i, \qquad x_1^i \in E_1^i, \ x_2^i \in E_2^i,$$

such that

$$l_0(x) = -x_1^0 + x_2^0 = x - 2x_1^0, \qquad l_1(x) = x - 2x_1^1. \tag{5.67}$$

Now let e^0 and e^1 be unit elements in E_1^0 and E_1^1 resp.:

$$\|e^0\| = \|e^1\| = 1. \tag{5.68}$$

Then by (5.67)

$$\begin{aligned}
l_0(e^0) &= -e^0, & l_1(e^1) &= -e^1, \\
l_0(x) &= x - 2\alpha(x)e^0, & l_1(x) &= x - 2\beta(x)e^1,
\end{aligned} \tag{5.69}$$

where α and β are real-valued continuous functionals on E satisfying

$$\alpha(e^0) = \beta(e^1) = 1. \tag{5.70}$$

For the proof that the maps (5.69) are L.-S. homotopic, we distinguish three cases:

(A) $e^0 = e^1$;
(B) $e^0 = -e^1$;
(C) e^0 and e^1 are linearly independent.

Case (A). We set

$$\gamma_t(x) = (1 - t)\alpha(x) + t\beta(x), \qquad t \in [0, 1], \tag{5.71}$$

and

$$l_t(x) = x - 2\gamma_t(x)e^0. \tag{5.72}$$

It is clear that for $t = 0$ and $t = 1$ this definition agrees with the one given by (5.69). It remains to verify that l_t is nonsingular for each $t \in [0, 1]$. If this were not true, then $l_{\bar{t}}(\bar{x}) = \theta$ for some $\bar{t} \in [0, 1]$ and $\bar{x} \in E$ different from θ, i.e., by (5.72), $\bar{x} = ae^0$ for some real $a \neq 0$. Since $l_{\bar{t}}$ is linear, we see that $l_{\bar{t}}(e^0) = \theta$, and therefore from (5.72) that $e^0(1 - 2\gamma_{\bar{t}}(e^0)) = 0$. Thus

$$2\gamma_{\bar{t}}(e^0) = 1. \tag{5.72a}$$

But since $e^0 = e^1$ by assumption, we see from (5.70) and (5.71) that $\gamma_{\bar{t}}(e^0) = (1 - \bar{t})\alpha(e^0) + \bar{t}\beta(e^1) = 1$ in contradiction to (5.72a).

Case (B). This case reduces to case (A) since, by (5.69), $l_1(x) = x - 2\tilde{\beta}(x)e^0$ with $\tilde{\beta}(x) = \beta(-x)$, and since $\tilde{\beta}(e^0) = \beta(-e^0) = \beta(e^1) = 1$.

Case (C). Let E_2 be the two-dimensional subspace of E spanned by e^0 and e^1, and let E_3 be a complementary subspace to E_2 such that every $x \in E$ has the unique representation

$$x = x_2 + x_3, \qquad x_2 \in E_2, \ x_3 \in E_3. \tag{5.73}$$

Now let $\tilde{r}_t$ be the rotation in E_2 determined by

$$\begin{aligned} \tilde{r}_t(e^0) &= e^0 \cos(t\pi/2) + e^1 \sin(t\pi/2), \\ \tilde{r}_t(e^1) &= -e^0 \sin(t\pi/2) + e^1 \cos(t\pi/2). \end{aligned} \qquad 0 \le t \le 1. \tag{5.74}$$

We note that

$$\tilde{r}_1(e^0) = e^1. \tag{5.75}$$

We extend $\tilde{r}_t$ to a map r_t on E by setting, for $x \in E$ and $t \in [0,1]$,

$$r_t(x) = x - P_2(x) + \tilde{r}_t(P_2(x)), \tag{5.76}$$

where P_2 is the projection $x = x_2 + x_3 \rightarrow x_2$ (cf. (5.73)). This is indeed an extension of $\tilde{r}_t$ since $x = P_2(x)$ for $x \in E_2$. (5.76) is obviously an L.-S. map. To show that it is nonsingular we will prove that r_t^{-1} exists: let $y = r_t(x)$. Since $P_2^2 = P_2$ and since $r_t(P_2(x)) \in E_2$, we see from (5.76) that $P_2(y) = \tilde{r}_t(P_2(x))$, and since $\tilde{r}_t$ obviously has an inverse, it follows that

$$P_2(x) = \tilde{r}_t^{-1}(P_2(y)). \tag{5.77}$$

On the other hand, if P_3 is the projection $x = x_2 + x_3 \rightarrow x_3$, we see from (5.76) that $P_3(x) = P_3(y)$. Thus by (5.77) $x = P_2(x) + P_3(x) = \tilde{r}_t^{-1}(P_2(y)) + P_3(y)$ which shows that the inverse r_t^{-1} exists. We therefore may define for $x \in E$ and $t \in [0,1]$

$$l(x,t) = x - 2\alpha(r_t^{-1}(x))r_t(e^0) \tag{5.78}$$

with α as in (5.69). Then

$$l(x,0) = l_0(x), \tag{5.79}$$

since r_0 is the identity map. Moreover by (5.75)

$$l(x,1) = x - 2\alpha(r_1^{-1}(x))e^1 = x - 2\alpha_1(x)e^1, \tag{5.80}$$

where

$$\alpha_1(x) = \alpha(r_1^{-1}(x)). \tag{5.81}$$

To show that $l_0(x)$ and $l(x,1)$ are L.-S. homotopic we have, because of (5.80), only to show that $l(x,t)$ is not singular. If this were not true, then by (5.78)

$$x = 2\alpha(r_t^{-1}(x))r_t(e^0) \tag{5.82}$$

for some $x \in E$ with $x \ne \theta$ and some $t \in [0,1]$. Thus $x = ar_t(e^0)$ with $a \ne 0$. Substitution in (5.82) and cancelling a shows that $1 = 2\alpha(r_t^{-1}r_t(e_0)) = 2\alpha(e^0)$ which contradicts (5.70). Now we want to prove that l_0 and l_1 (see (5.69) and (5.70)) are L.-S. homotopic. Since we just proved that l_0 and $l(\cdot,1)$ are

L.-S. homotopic, it remains to show that l_1 and $l(\cdot, 1)$ are L.-S. homotopic. But comparing (5.80) with the definition (5.69) of l_1 we see that we are in case (A) provided that

$$\alpha_1(e^1) = 1. \tag{5.83}$$

Now by (5.81), (5.75) and (5.70), $\alpha_1(e^1) = \alpha(r_1^{-1}(e^1)) = \alpha(\tilde{r}_1^{-1}(e^1)) = \alpha(e^0) = 1$.

This finishes the proof of the linear homotopy theorem.

§6. Two multiplication theorems for the indices

6.1. THEOREM. *Let E be a real Banach space, and l_0 and l_1 be two nonsingular L.-S. maps $E \to E$. Then*

$$j(l_0 l_1) = j(l_0) \cdot j(l_1). \tag{6.1}$$

PROOF. Let I be the identity map on E and let l^- be a fixed L.-S. map of type I^- on E. Then by Lemma 4.5

$$j(I) = +1, \qquad j(l^-) = -1. \tag{6.2}$$

By the linear homotopy theorem each of the maps l_0 and l_1 is L.-S. homotopic to exactly one of the maps I and l^-. Denoting L.-S. homotopy by the symbol "$\sim$" there are four cases:

$$
\begin{array}{lll}
\text{(i)} & l_0 \sim I, & l_1 \sim I; \\
\text{(ii)} & l_0 \sim I, & l_1 \sim l^-; \\
\text{(iii)} & l_0 \sim l^-, & l_1 \sim I; \\
\text{(iv)} & l_0 \sim l^-, & l_1 \sim l^-.
\end{array} \tag{6.3}
$$

Then

$$l_0 l_1 \sim I \cdot I = I \quad \text{in case (i)},$$
$$l_0 l_1 \sim I \cdot l^- = l^- \quad \text{in case (ii)},$$
$$l_0 l_1 \sim l^- I = l^- \quad \text{in case (iii)},$$
$$l_0 l_1 \sim l^- l^- = I \quad \text{in case (iv)}.$$

By Theorem 4.6 and by (6.2) this implies

$$
\begin{aligned}
j(l_0 l_1) &= +1 \quad \text{in case (i)}, \\
j(l_0 l_1) &= -1 \quad \text{in case (ii)}, \\
j(l_0 l_1) &= -1 \quad \text{in case (iii)}, \\
j(l_0 l_1) &= +1 \quad \text{in case (iv)}.
\end{aligned} \tag{6.4}
$$

On the other hand we see from (6.3), from Theorem 4.6 and from (6.2) that

$$
\begin{array}{lll}
j(l_0) = 1, & j(l_1) = 1 & \text{in case (i)}, \\
j(l_0) = 1, & j(l_1) = -1 & \text{in case (ii)}, \\
j(l_0) = -1, & j(l_1) = +1 & \text{in case (iii)}, \\
j(l_0) = -1, & j(l_1) = -1 & \text{in case (iv)}.
\end{array} \tag{6.5}
$$

Comparison of (6.4) with (6.5) makes the assertion (6.1) evident.

6.2. THEOREM. *Let l be a nonsingular L.-S. map $E \to E$. Let $E = E_1 \dotplus E_2$ such that every $h \in E$ has the unique representation*

$$h = h_1 + h_2, \qquad h_1 \in E_1, \ h_2 \in E_2. \tag{6.6}$$

Let l_1 and l_2 be the restrictions of l to E_1 and E_2 resp., and suppose that

$$l(h) = \begin{pmatrix} l_1(h_1) & 0 \\ 0 & l_2(h_2) \end{pmatrix}. \tag{6.7}$$

Then

$$j(l) = j(l_1) \cdot j(l_2). \tag{6.8}$$

PROOF. It is clear that l_1 and l_2 are linear nonsingular L.-S. maps $E_1 \to E$ and $E \to E_2$ resp. Thus the right member of (6.8) is defined. Now let

$$m_1 = \begin{pmatrix} l_1 & 0 \\ 0 & I_2 \end{pmatrix}, \qquad m_2 = \begin{pmatrix} I_1 & 0 \\ 0 & l_2 \end{pmatrix}, \tag{6.9}$$

where I_1 and I_2 are the identity maps on E_1 and E_2 resp. Then $l = m_1 m_2$. Therefore, by Theorem 6.1, $j(l) = j(m_1) \cdot j(m_2)$. It remains to verify that

$$j(l_1) = j(m_1), \qquad j(l_2) = j(m_2). \tag{6.10}$$

By Definition 4.3 of the index j, it will for the proof of the first of these equalities be sufficient to show:

(i) a real number $\lambda \neq 0$ is an eigenvalue of $M_1 = I - m_1$ if and only if λ is an eigenvalue of $L_1 = I_1 - l_1$;

(ii) for such an eigenvalue $\lambda \neq 0$

$$\mu_{l_1}(\lambda) = \mu_{m_1}(\lambda), \tag{6.11}$$

where $\mu_{l_1}(\lambda)$ and $\mu_{m_1}(\lambda)$ denote the generalized multiplicity of λ as an eigenvalue of L_1 and M_1 resp. (see Definition 4.2).

Proof of (i). Let $\lambda \neq 0$ be an eigenvalue of $M_1 = I - m_1$. Then there exists an $h \neq 0$ such that $(I - m_1)h = \lambda h$ or by (6.9) and (6.6) such that

$$L_1 h_1 = \lambda h_1 \tag{6.12}$$

and $0 h_2 = \lambda h_2$. This implies $h_2 = \theta$ since $\lambda \neq 0$. Thus $h = h_1$, and (6.12) shows that λ is an eigenvalue of L_1. The converse follows even more easily.

Proof of (ii). Let $\lambda \neq 0$ be an eigenvalue of L_1 and therefore, by (i), an eigenvalue of $M_1 = I - m_1$. By (6.9)

$$\lambda I - M_1 = \begin{pmatrix} \lambda I_1 - L_1 & 0 \\ 0 & \lambda I_2 \end{pmatrix},$$

and for any positive integer n

$$(\lambda I - M_1)^n = \begin{pmatrix} (\lambda I_1 - L_1)^n & 0 \\ 0 & \lambda^n I_2 \end{pmatrix}.$$

From this, from (6.6) and from $\lambda \neq 0$ it follows that for $h \in E$, $(\lambda I - M_1)^n h = \theta$ if and only if $(\lambda I_1 - L_1)^n h_1 = \theta$ and $h_2 = \theta$. This implies (6.11) by Definition 4.2, and therefore, by Definition 4.3, proves the first of our assertions (6.10). The second one is proved correspondingly.

6.3. COROLLARY TO THEOREM 6.2. *Let E, E_1, E_2, h, h_1, h_2 be as in Theorem 6.2, and for $i = 1, 2$ let π_i be the projection $h = h_1 + h_2 \to h_i$. Let m be a nonsingular L.-S. map $E \to E$. Let m_1 denote the restriction of $\pi_1 m$ to E_1, and let m_2 denote the restriction of $\pi_2 m$ to E_2. We assume*

$$\pi_2 m(E_1) = 0. \tag{6.13}$$

Then

$$j(m) = j(m_1) \cdot j(m_2). \tag{6.14}$$

PROOF. Let $m_{12}(h_2) = \pi_1 m(h_2)$. Then

$$m(h) = \begin{pmatrix} m_1(h_1) & m_{12}(h_2) \\ \theta & m_2(h_2) \end{pmatrix}. \tag{6.15}$$

We note first that m_2 is nonsingular, for otherwise $m_2(h_2) = 0$ for some $h_2 \neq \theta$. Then $m(h) = 0$ for $h = h_2$ since then $h_1 = \theta$ and $m_{12}(h_2) = \pi_1 m(h_2) = \theta$. This contradicts the assumed nonsingularity of m. Similarly one sees that m_1 is not singular.

Now let for $t \in [0, 1]$

$$m^t(h) = \begin{pmatrix} m_1(h_1) & t m_{12}(h_2) \\ \theta & m_2(h_2) \end{pmatrix}. \tag{6.16}$$

We prove that for each t this map is not singular by showing that for arbitrary $k \in E$ the equation

$$m^t(h) = k \tag{6.17}$$

has a solution. Now if $k = k_1 + k_2$ with $k_1 \in E_i$, then by (6.16) the equation (6.17) is equivalent to

$$m_1(h_1) + t m_{12}(h_2) = k_1, \qquad m(h_2) = k_2.$$

Since m_1 and m_2 are nonsingular, this system obviously has a solution. We thus see from (6.16) and (6.15) that $m = m^1$ is L.-S. homotopic to the map

$$m^0 = \begin{pmatrix} m_1(h_1) & \theta \\ \theta & m_2(h_2) \end{pmatrix}.$$

Consequently by Theorem 4.6, $j(m) = j(m^0)$. But $j(m^0) = j(m_1) \cdot j(m_2)$ by Theorem 6.2. This proves the assertion (6.14).

APPENDIX B

Proof of the Sard-Smale Theorem 4.4 of Chapter 2

1. THE THEOREM OF SARD. *Let R^n and R^m be real Euclidean spaces of finite dimension n and m resp. Let U be an open subset of R^n, and let $f \in C^r(U) \colon U \to R^m$, where r is a positive integer satisfying*

$$r > \max(0, n - m). \tag{1}$$

Then the set of critical values of f (see Definition 17 in Chapter 1) is of Lebesgue measure 0.

For a proof of and further literature on Sard's theorem we refer the reader to [1, §15].

2. COROLLARY TO SARD'S THEOREM. *Let E^n and E^m be (real) Banach spaces of finite dimension n and m resp. Let Ω be a bounded open set in E^n and $f \in C^r(\overline{\Omega})$ (cf. §1.16) where r is a positive integer satisfying (1). Then the set of points in E^m which are regular values for f (see §1.17) is dense in E^m.*

This is an obvious consequence of Theorem 1 since on the one hand a subset of R^m of measure 0 contains no open set and therefore its complement contains a point in every open subset of R^m and is thus dense in R^m, while on the other hand every finite-dimensional Banach space is linearly isomorphic to a Euclidean space of the same dimension (see, e.g., [18, p. 245]).

3. The proof given below for the Sard-Smale theorem is an adaptation to the simpler Banach *space* case of the proof given in [1, §16] for Smale's generalization of Sard's theorem to certain Banach *manifolds*. Using the preceding corollary we will prove a "local version" of the Sard-Smale theorem in Lemma 4 for maps of a very special form. The lemma in §5 will allow us to show that the local theorem holds also for the more general maps treated in §6. The latter result implies the Sard-Smale theorem as shown in §7.

4. LEMMA. *Let Π_1, K, and K^* be Banach spaces, where K is finite dimensional and where*

$$\dim K^* = \dim K - p, \tag{2}$$

231

with p being a positive integer. Let Π and Σ be the Banach spaces defined as the direct products

$$\Pi = \Pi_1 \dotplus K, \qquad \Sigma = \Pi_1 \dotplus K^* \tag{3}$$

(cf. §1.3). Let V be a bounded open subset of Π. Let ψ be a C^s-map $V \to \Sigma$ where s is an integer satisfying

$$s \geq p + 1. \tag{4}$$

We suppose moreover that the map $\psi \colon V \to W = \psi(V)$ is of the special form

$$\psi(\varsigma) = \varsigma_1 + \chi(\varsigma), \tag{5}$$

where

$$\varsigma = \varsigma_1 + \varsigma_2, \qquad \varsigma_1 \in \Pi_1, \quad \varsigma_2 \in K, \tag{6}$$

and where the map $\chi \colon V \to K^*$ is completely continuous.

Now if ς^0 is a point of V, then there exists an open neighborhood V_0 of ς^0 such that

$$V_0 \subset \overline{V}_0 \subset V \tag{7}$$

and such that the set $R(\overline{\psi}_0)$ of those points of Σ which are regular values for the restriction $\overline{\psi}_0$ of ψ to V_0 is dense and open.

PROOF. Let $\varsigma^0 = \varsigma_1^0 + k^0$, $\varsigma_1^0 \in \Pi_1$, $k^0 \in K$. Since ς^0 is an element of the open set V, there obviously exist positive numbers ε_1 and ε_2 such that the "rectangular" neighborhood $V_0 = V_1 \times V_2$ satisfies (7) if

$$V_1 = \{\varsigma_1 \in \Pi_1 \mid \|\varsigma_1 - \varsigma_1^0\| < \varepsilon_1\}, \qquad V_2 = \{k \in K \mid \|k - k^0\| < \varepsilon_2\}. \tag{8}$$

We will show first that the set $R(\psi_0)$ of regular values for the restriction ψ_0 of ψ to V_0 is dense in Σ. For this purpose we consider an arbitrary point $\overline{\sigma} = \overline{\varsigma}_1 + \overline{k}^*$ in Σ where $\overline{\varsigma}_1 \in \Pi_1$ and $\overline{k}^* \in K^*$. We have to show that every neighborhood N of $\overline{\sigma}$ contains a point which is a regular value for ψ_0. Now this is obvious if the equation

$$\psi_0(\varsigma) = \overline{\sigma} = \overline{\varsigma}_1 + \overline{k}^* \tag{9}$$

has no solution in V_0 since then, by definition, $\overline{\sigma}$ is a regular value of ψ_0. Suppose now (9) has a solution $\tilde{\varsigma}_1 + \tilde{k}$, $\tilde{\varsigma}_1 \in V_1$, $\tilde{k} \in V_2$. Then by (5) and (9), $\tilde{\varsigma}_1 = \overline{\varsigma}_1$ and

$$\chi(\overline{\varsigma}_1 + \tilde{k}) = \overline{k}^*. \tag{10}$$

Now for $\overline{\varsigma}_1 \in V_1$ we define a map $\chi_{\overline{\varsigma}_1} \colon V_2 \subset K$ into K^* by setting

$$\chi_{\overline{\varsigma}_1}(k) = \chi(\overline{\varsigma}_1 + k). \tag{11}$$

From (2) and (4) we see that this map satisfies the assumption of the Corollary 2 (with $E^m = K^*$, $E^n = K$). Therefore by that corollary every neighborhood $N_2 \subset K^*$ of $\overline{k}^*$ contains a regular value k^* for $\chi_{\overline{\varsigma}_1}$. We will show that for such k^* the point $\sigma = \overline{\varsigma}_1 + k^* \in \Sigma$ is a regular value for ψ_0. This will obviously prove our assertion that every neighborhood N of $\overline{\sigma} = \overline{\varsigma}_1 + \overline{k}^*$ contains a regular value for ψ_0. Now if the equation

$$\chi_{\overline{\varsigma}_1}(k) = k^* \tag{12}$$

has no solution $k \subset V_2$, then the equation

$$\psi_0(\varsigma) = \bar{\varsigma}_1 + k^* \tag{13}$$

has no solution $\varsigma \in V_0$ as is seen from (5) and (11). Thus in this case $\bar{\varsigma}_1 + k^*$ is a regular value for ψ_0.

Suppose now (12) has a solution $k \in V_2$. Then since k^* is a regular value for $\chi_{\bar{\varsigma}_1}$, the differential $D\chi_{\bar{\varsigma}_1}(k; \kappa)$, $\kappa \in K$, is nonsingular. On the other hand, $\varsigma = \bar{\varsigma}_1 + k$ is the corresponding solution of (13), and with $h = \lambda + \kappa$, $\lambda \in \Pi_1$, $\kappa \in K$, we see from (5) that

$$D\psi_0(\varsigma; h) = \begin{pmatrix} \lambda & \theta \\ D\chi(\varsigma; \lambda) & D\chi(\varsigma; \kappa) \end{pmatrix}. \tag{14}$$

But $D\chi(\varsigma; \kappa) = D\chi_{\bar{\varsigma}_1}(k; \kappa)$ by (11). This shows that the differential (14) is not singular and therefore that $\bar{\varsigma}_1 + k^*$ is a regular value for ψ_0 for every solution $\varsigma \in V_1 \times V_2$ of (13).

This finishes the proof that $R(\psi_0)$ is dense in Σ. Now let U_0 be a rectangular neighborhood of ς^0 whose closure $\bar{U}_0$ is a subset of V_0. As follows directly from the definition of "regular value," the relation $\bar{U}_0 \subset V_0$ implies that the set of regular values for the restriction $\bar{\psi}_0$ of ψ to $\bar{U}_0$ contains the set $R(\psi_0)$ and is therefore dense in Σ. In other words, changing our notation, we can choose the rectangular neighborhood V_0 of ς^0 satisfying (7) in such a way that $R(\bar{\psi}_0)$ is dense in Σ.

To prove our lemma it remains to show that $R(\bar{\psi}_0)$ is open or, what is the same, that its complement, the set $S(\bar{\psi}_0)$ of singular values for $\bar{\psi}_0$, is closed. Now it follows directly from the definition of "singular value" that $S(\bar{\psi}_0) = \bar{\psi}_0(S_1)$ if S_1 denotes the set of singular *points* of $\bar{\psi}_0$ in $\bar{V}_0$. But to prove that S_1 is closed, it will be sufficient to show that S_1 is closed as follows from the special form (5) of ψ (cf. the proof of part (ii) of §1.10). The proof that S_1 is closed is quite similar to the proof of part (ii) of Lemma 19 in Chapter 1: let $\varsigma^1, \varsigma^2, \ldots$ be a convergent sequence of points in S_1. We have to prove that

$$\bar{\varsigma} = \lim_{i \to \infty} \varsigma^i \in S_1.$$

Since $\varsigma^i \in S_1 \subset \bar{V}_0$ and since $\bar{V}_0$ is a closed set, we see that

$$\bar{\varsigma} \in \bar{V}_0 \subset V.$$

Now if $\bar{\varsigma} \notin S_1$ the differential $D\psi(\bar{\varsigma}; h)$ would not be singular, i.e., $\bar{\varsigma} \subset V$ would be a regular point for ψ. But then, by the continuity of the differential and by Lemma 5.3 of Appendix A, all points of some neighborhood of $\bar{\varsigma}$ would be regular points for ψ. This contradicts the fact that every neighborhood of $\bar{\varsigma}$ contains infinitely many of the singular points ς^i.

5. LEMMA. *Let* Π *be a Banach space, let* E *be a subspace of* Π *of finite codimension* p, *let* Z *be an open bounded subset of* Π, *and let* $\phi = \phi(z)$ *be an*

L.-S. map $\overline{Z} \to E$. *We suppose that* $\phi \in C^s(\overline{z})$ *for some integer* $s \geq 1$. *Let* z_0 *be a point of* Z *and*

$$l_0(\eta) = D\phi(z_0; \eta). \tag{15}$$

Moreover, let K *denote the kernel of* l_0, *i.e.*, $K = \{\eta \in \Pi \mid l_0(\eta) = \theta\}$. *Then the following assertions* (i) *and* (ii) *are made:*

(i) K *is finite dimensional, and there exists a subspace* Π_1 *of* Π *such that*

$$\Pi = \Pi_1 \dotplus K. \tag{16}$$

Moreover if $R \subset E$ *is the range of* l_0, *then there exists a finite-dimensional subspace* K^* *of* E *with the properties:*

(a) $E = R \dotplus K^*$;   (b) $\dim K^* = \dim K - p.$ (17)

Thus every $y \in E$ *has the unique representation*

$$y = r + k^* = \pi_1(y) + \pi_2(y), \qquad r \in R, \ k^* \in K^*, \tag{18}$$

where π_1 *and* π_2 *are the projections* $y \to r$ *and* $y \to k^*$ *resp.*

(ii) *Let* Σ *be the direct sum of* Π_1 *and* K^* (*see part* (ii) *of* §1.3):

$$\Sigma = \Pi_1 \dotplus K^* \tag{19}$$

such that each point $\sigma \in \Sigma$ *has the unique representation*

$$\sigma = \sigma_1 + \sigma_2 = \pi_1^\Sigma(\sigma) + \pi_2^\Sigma(\sigma), \qquad \sigma_1 \in \Pi_1, \ \sigma_2 \in K^*, \tag{20}$$

where π_1^Σ *and* π_2^Σ *are the projections* $\sigma \to \sigma_1$ *and* $\sigma \to \sigma_2$ *resp. It is asserted: there exists a linear L.-S. isomorphism* h *of* E *onto* Σ *and an L.-S.* C^s *isomorphism* h_1 *of some neighborhood* $U \subset Z \subset \Pi$ *of* z_0 *onto some neighborhood* $V \subset \Pi$ *of* $h_1(z_0)$ *such that the map*

$$\psi(\varsigma) = h\phi(h_1^{-1}(\varsigma)), \qquad \varsigma \in V, \tag{21}$$

is of the form (5) *where* $\varsigma, \varsigma_1, \varsigma_2$ *and* χ *are as described in the lines following* (5).

PROOF. (i) That K is finite dimensional follows from Lemmas 12 and 18 in Chapter 1, and the existence of a Π_1 satisfying (16) follows from Lemma 4 in Chapter 1. Since ϕ is an L.-S. map whose domain is the subset $\overline{Z}$ of Π and whose range lies in the subspace E of Π, it may also be considered as an L.-S. map $\overline{Z} \to \Pi$. Therefore, by Lemma 12 of Chapter 1 there exists a finite-dimensional subspace K_1^* of Π such that

$$\Pi = R \dotplus K_1^*. \tag{22}$$

Now $E \cap R = R$ since $R \subset E$. Therefore intersecting both members of (22) with E we see that the finite-dimensional space $K^* = E \cap K_1^*$ satisfies (17a). Since p is the codimension of E in Π, comparison of (22) with (17a) shows that $\dim K^* = \dim K_1^* - p$. But by §1.12, $\dim K_1^* = \dim K$. This proves (17b).

Proof of (ii). We note first that the restriction of l_0 to Π_1 maps Π_1 onto R. Indeed, if r is a given element of R, there exists by definition of R a $z \in \Pi$ such that $l_0(z) = r$. But by (16)

$$z = z_1 + k, \qquad z_1 \in \Pi_1, \ k \in K. \tag{23}$$

Since $l_0(k) = \theta$ by definition of K, we see that $l_0(z_1) = r$. Thus l_0 as a map $\Pi_1 \to R$ is a nonsingular L.-S. map Π_1 onto R. It follows from the argument given for the proof of part (v) of §1.12 that this map has an inverse $m \colon R \to \Pi_1$ which is a nonsingular L.-S. map. Now every $y \in E$ has the decomposition (18). If for $r = \pi_1(y)$, $k^* = \pi_2(y)$ we define

$$h(y) = m(r) + k^*, \tag{24}$$

we obtain a map of E onto the space Σ defined by (19). Since $m(r)$ is linear and L.-S. and since K^* is finite dimensional, it follows that h is a linear nonsingular L.-S. map.

Using the decomposition (16) we set for $z = z_1 + k \in Z$, $z_1 \in \Pi_1$, $k \in K$,

$$h_1(z) = m\pi_1\phi(z) + k. \tag{25}$$

Since $\phi(z) \in E$, it follows from (18) that $\pi_1\phi(z) \in R$. Therefore the first term of the right member of (25) is an element of Π_1, and $h_1(z) \in \Pi$ by (16). Moreover since ϕ and m are L.-S. mappings and since K is finite dimensional, we see from (25) that h_1 is an L.-S. map.

We prove next that h_1 maps some open neighborhood U of z_0 onto some open neighborhood V of $h_1(z_0)$ and that $h_1^{-1} \in C^s(V)$. For this purpose it will by the inversion theorem in §1.21 be sufficient to show that the differential of h_1 at z_0 is the identity map on Π, i.e., that

$$Dh_1(z_0; \eta) = \eta \tag{26}$$

for all

$$\eta = \eta_1 + \kappa \in \Pi, \qquad \eta_1 \in \Pi_1, \ \kappa \in K \tag{27}$$

(cf. (16)). Now since m and π_1 are linear, we see from (25) that

$$Dh_1(z; \eta) = m\pi_1 D\phi(z; \eta) + \kappa \tag{28}$$

for all $z \in Z$. But for $z = z_0$ we see from (15) and from (27) that $D\phi(z_0; \eta) = l_0(\eta) = l_0(\eta_1)$ since $\kappa \in K$ and, therefore, $l_0(\kappa) = \theta$. Since R is the range of l_0, we see from (18) that $\pi_1 D\phi(z_0, \eta) = \pi_1 l_0(\eta_1) = l_0(\eta_1)$. But $l_0(\eta_1)$ is the restriction of l_0 to Π_1 and, by definition, m is its inverse. Therefore $m\pi_1 D\phi(z_0; \eta) = ml_0(\eta_1) = \eta_1$. Thus by (28) $Dh_1(z_0; \eta) = \eta_1 + \kappa$, which by (27) proves (26).

Then let U and V be sets of the properties stated above. We set

$$\varsigma = h_1(z). \tag{29}$$

If z varies over U, then ς varies over V. Since $\varsigma \in \Pi$ we may write $\varsigma = \varsigma_1 + \varsigma_2$, $\varsigma_1 \in \Pi_1$, $\varsigma_2 \in K$ (cf. (16)). Then by (25) and (20)

$$\varsigma_1 = m\pi_1\phi(z) \in \Pi_1, \qquad \varsigma_2 = k \in K. \tag{30}$$

Now if

$$\psi(\varsigma) = h\phi h_1^{-1}(\varsigma), \qquad \varsigma \in V, \tag{31}$$

then with $z = h_1^{-1}(\varsigma)$ by (24), (18) and (30)

$$\psi(\varsigma) = h\phi(z) = m\pi_1\phi(z) + \pi_2\phi(z) = \varsigma_1 + \pi_2\phi h_1^{-1}(\varsigma).$$

With $\chi(\varsigma) = \pi_2 \phi h_1^{-1}(\varsigma)$ this equality shows that ψ is of the asserted form (5). Obviously $\chi(\varsigma)$ is bounded and lies in K^*. Since K^* is finite dimensional, we see that χ is completely continuous. Thus assertion (ii) of Lemma 4 is proved.

6. LEMMA. *With the notation used in Lemma 5 we suppose that the assumptions of that lemma are satisfied. In addition we strengthen the assumption that $\phi \in C^s(\overline{Z})$ with $s \geq 1$ by requiring that*

$$s \geq p + 1. \tag{32}$$

It is asserted that corresponding to each point $z_0 \in Z$ there exists an open set U_0 which in addition to satisfying

$$z_0 \in U_0 \subset \overline{U}_0 \subset Z \tag{33}$$

has the following property: if $\overline{\phi}_0$ is the restriction of ϕ to $\overline{U}_0$, then the set $R(\overline{\phi}_0)$ of points in E which are regular values for $\overline{\phi}_0$ is dense and open.

PROOF. It follows from Lemma 5 together with the assumption (32) that all assumptions of Lemma 4 are satisfied. Consequently, the point

$$\varsigma^0 = h_1(z_0) \in \Pi \tag{34}$$

has an open neighborhood V_0 with the properties asserted in that lemma. We claim that then the set

$$U_0 = h_1^{-1}(V_0) \tag{35}$$

has the properties asserted in the present lemma. In the first place U_0 is open and $\overline{U}_0 = h_1^{-1}(\overline{V}_0)$ as follows from Theorem 2.3 in Chapter 5. Next the following two assertions hold:

(a) for $y_0 \in E$ the equation $\phi(z) = y_0$ has a solution $z \in \overline{U}_0$ if and only if for $\sigma_0 = h(y_0) \in \Sigma$ the equation $\psi(\varsigma) = \sigma_0$ has a solution $\varsigma \in \overline{V}_0$;

(b) for $z_0 \in \overline{U}_0$ the differential $D\phi(z_0; \cdot)$ is not singular if and only if the differential $D\psi(h_1(z_0); \cdot)$ is not singular.

(a) follows directly from the relation (21) between ψ and ϕ and the properties of h and h_1.

(b) holds for the same reasons in conjunction with the chain rule. But (a) and (b) together imply that $R(\overline{\phi}_0) = h^{-1}R(\overline{\psi}_0)$.

Now $R(\overline{\psi}_0)$ is dense and open by Lemma 4. Therefore $h^{-1}R(\overline{\psi}_0)$ is dense and open since h^{-1} is a linear one-to-one L.-S. map Σ onto E.

7. *Proof of the Sard-Smale theorem 4.4 of Chapter 2.* Let y_0 be an arbitrary point of E. We have to prove that every neighborhood (in E) of y_0 contains a regular value for ϕ. Now if the equation

$$\phi(z) = y_0, \qquad y_0 \in E - \phi(\partial Z), \tag{36}$$

has no solution, then, by definition, y_0 is a regular value for ϕ and every neighborhood of y_0 contains a regular value, e.g., y_0.

Suppose now that the set C_0 of roots of (36) is not empty. C_0 is compact by Lemma 10 in Chapter 1, and every z_0 has a neighborhood U_0 of the property asserted in Lemma 6. As z_0 varies over C_0, these neighborhoods form an open covering of this compact set. Consequently there exists a finite subcovering, say, $U_0^1, U_0^2, \ldots, U_0^q$ such that (cf. (33)) for $i = 1, 2, \ldots, q$

$$U_0^i \subset \overline{U}_0^i \subset Z; \qquad C_0 \subset U = \sum_{i=1}^{q} U_0^i. \tag{37}$$

If $\overline{\phi}_i$ is the restriction of ϕ to $\overline{U}_0^i$, then by Lemma 6 the set $R(\overline{\phi}_i)$ of regular values for $\overline{\phi}_i$ is dense and open in E. Consequently by Baire's density theorem (see e.g. [**2**; p. 108, Theorem V′]) the set

$$\Delta = \bigcap_{i=1}^{q} R(\overline{\phi}_i) \tag{38}$$

is dense in E, and therefore in every neighborhood of y_0. We show next the existence of an open neighborhood N_0 of y_0 such that

$$\phi^{-1}(N_0) \subset U. \tag{39}$$

Indeed, otherwise, there would exist a sequence of positive numbers δ_ν converging to 0, of points $y_\nu \in B(y_0, \delta_\nu)$, and of points x_ν such that

$$\phi(x_\nu) = y_\nu, \qquad x_\nu \in \overline{Z}, \tag{40}$$

but

$$x_\nu \notin U. \tag{41}$$

Now since the y_ν converge to y_0 and since ϕ is an L.-S. map, it is easy to see that a subsequence of the x_ν converges to a point x_0 which by (40) satisfies $\phi(x_0) = y_0$, i.e.,

$$x_0 \in \phi^{-1}(y_0) = C_0. \tag{42}$$

But by (41)

$$x_0 \notin U \tag{43}$$

since U is open. But by (42) and (37), $x_0 \in C_0 \subset U$ which contradicts (43).

Then let N_0 be a neighborhood of y_0 satisfying (39), and let $\Delta_1 = \Delta \cap N_0$. Δ_1 is not empty since Δ is dense and N_0 open in E. We now show that every point y_1 of Δ_1 is a regular value for ϕ. If the equation

$$\phi(z) = y_1 \tag{44}$$

has no solutions, there is nothing to prove. But if it has a solution, then every solution lies in U by (39), and therefore, by (37), in some of the sets U_0^i, say for $i = 1, 2, \ldots, q_1$ with $i \leq q_1 \leq q$. Consider now for such i a solution $z \in U_0^i$. Then $\phi(z) = \phi_i(z) = y_1 \in \Delta = \bigcap_{j=1}^{q} R(\phi_j) \subset R(\phi_i)$, and $D\phi(z; h)$ is not singular.

This finishes the proof that every point of $N_0 \cap \Delta$ is a regular value for ϕ, and therefore it finishes the proof of the Sard-Smale theorem since an N_0 satisfying (39) can be chosen as a subset of a given neighborhood of y_0.

References

1. R. Abraham and J. Robbins, *Transversal mappings and flows*, Benjamin, New York-Amsterdam, 1967.

2. P. Alexandroff and H. Hopf, *Topologie*, Springer-Verlag, 1935.

3. H. Amann and S. A. Weiss, *On the uniqueness of the topological degree*, Math. Z. **130** (1973), 39–54.

4. S. Banach, *Théorie des opérations linéaires*, Warsaw, 1932.

5. R. Bonic and J. Frampton, *Smooth functions on Banach manifolds*, J. Math. Mech. **15** (1966), 877–898.

6. Yu. G. Borisovich, V. G. Zvyagin, and Y. I. Sapronov, *Nonlinear Fredholm maps and the Leray-Schauder theory*, Uspeki Mat. Nauk. **32** (1977), 3–54; English transl. in Russian Math. Surveys **32** (1977).

7. K. Borsuk, *Drei Sätze über die n-dimensionale Sphäre*, Fund. Math. **20** (1933), 177–190.

8. L. E. J. Brouwer, *Beweis der Invarianz der Dimensionzahl*, Math. Ann. **70** (1911), 161–165.

9. ____, *Über Abbildung von Mannigfaltigkeiten*, Math. Ann. **71** (1912), 97–115.

10. F. Browder, *Nonlinear operators and nonlinear equations of evolution in Banach spaces*, Proc. Sympos. Pure Math., vol. 18, pt. 2, Amer. Math. Soc., Providence, R. I., 1976.

11. F. Browder and R. Nussbaum, *The topological degree for noncompact nonlinear mappings in Banach spaces*, Bull. Amer. Math. Soc. **74** (1968), 671–676.

12. F. Browder and W. Petryshyn, *Approximation methods and the generalized topological degree for nonlinear mappings in Banach spaces*, J. Funct. Anal. **3** (1969), 217–245.

13. R. Courant and D. Hilbert, *Methods of mathematical physics*. Vol. 1, Interscience, 1953.

14. J. Cronin, *Fixed points and topological degree in nonlinear analysis*, Math. Surveys, no. 11, Amer. Math. Soc., Providence, R. I., 1964.

15. K. Deimling, *Nichtlineare Gleichungen und Abbildungsgrade*, Springer-Verlag, 1974.

16. J. Dieudonné, *Foundations of modern analysis*, Academic Press, 1960.

17. J. Dugundji, *An extension of Tietze's theorem*, Pacific J. Math. **1** (1951), 353–367.

18. N. Dunford and J. T. Schwartz, *Linear operators*. Part I: *General theory*, Interscience, 1958.

19. G. Eisenack and C. Fenske, *Fixpunkttheorie*, Bibliographisches Institut, Mannheim, 1978.

20. K. Elworthy and A. Tromba, *Differential structures and Fredholm maps on Banach manifolds*, Proc. Sympos. Pure Math, vol. 15, Amer. Math. Soc., Providence, R. I., 1970.

21. C. Fenske, *Analytische Theorie des Abbildungsgrades in Banach Rämmen*, Math. Nachr. **48** (1971), 279–290.

22. K. Geba and A. Granas, *Infinite dimensional cohomology theories*, J. Math. Pures Appl. **52** (1973), 145–270.

23. A. Granas, *Homotopy extension theorem in Banach spaces and some of its applications to the theory of nonlinear equations*, Bull. Acad. Polon. Sci. Math. Astr. Phys. **7** (1959), 387–394. (Russian; English summary).

24. ____, *The theory of compact vector fields and some of its applications to topology of functional spaces*. (I). Rozprawy Mat., Warsaw, 1962.

25. ____, *Points fixes pour les applications compactes: espaces de Lefschetz et la théorie de l'indice*, Presses de l'Université de Montréal, Montréal, 1980.

26. P. Halmos, *Finite dimensional vector spaces*, Van Nostrand, Princeton, N. J., 1958.

27. H. Hopf, *Topologie*, lectures given at the University of Berlin, 1926/27, (unpublished). Notes by E. Pannwitz.

28. S. T. Hu, *Theory of retracts*, Wayne State Univ. Press, Detroit, Mich., 1965.

29. W. Hurewicz and H. Wallman, *Dimension theory*, Princeton Univ. Press, Princeton, N. J., 1941.

30. N. Jacobson, *Lectures in abstract algebra*. Vol. II: *Linear algebra*, Van Nostrand, Princeton, N. J., 1953.

31. M. Krasnoselskiĭ, *Topological methods in the theory of nonlinear integral equations*, Gosudorstv. Izdat. Tehn.-Teor. Lit., Moscow, 1956; English transl., Macmillan, 1964.

32. L. Kronecker, *Uber die Charakteristik von Funktionen Systemen*, Monatsberichte der Kgl. Preussischen Akademie der Wissenschaften, 1878, 145–152.

33. J. Kurzweil, *On approximation in real Banach spaces*, Studia Math. **14** (1954), 213–231.

34. J. Leray, *Topologie des espaces de Banach*, C. R. Acad. Sci. Paris, **200** (1935), 1082–1084.

35. ——, *La théorie des points fixes et ses applications en analyse*, (Proc. Internat. Congr. of Math., Vol. 2, Cambridge, Mass., 1950) Amer. Math. Soc., Providence, R. I., 1952, pp. 202–208.

36. J. Leray and J. Schauder, *Topologie et équations fonctionnelles*, Ann. École Norm. **51** (1934), 45–78.

37. N. G. Lloyd, *Degree theory*, Cambridge Univ. Press, New York, 1978.

38. J. Mawhin, *Equivalence theorems for nonlinear operator equations and coincidence theory for some mappings in locally convex topological vector spaces*, J. Differential Equations **12** (1972), 610–636.

39. J. W. Milnor, *Topology from the differentiable viewpoint*, The University Press of Virginia, Charlottesville, Va., 1965.

40. H. Minkowski, *Geometrie der Zahlen*, Teubner, Leipzig und Berlin, 1910.

41. N. Moulis, *Approximations de fonctions différentiable sur certain espaces de Banach*, Thesis, Université de Paris, 1970.

42. M. Nagumo, *A theory of degree of mapping based on infinitesimal analysis*, Amer. J. Math **73** (1951), 485–496.

43. ——, *Degree of mapping in convex linear topological spaces*, Amer. J. Math. **73** (1951), 497–511.

44. L. Nirenberg, *Topics in nonlinear functional analysis*, Lecture Notes, 1973–74, Courant Inst. of Math. Sciences, New York University, New York, 1974. Notes by R. A. Artin.

45. E. H. Rothe, *Über Abbildungsklassen von Kugeln des Hilbertschen Raumes*, Compositio Math. **4** (1937), 294–307.

46. ——, *Zur Theorie der topologischen Ordnung und der Vektorfelder in Banachschen Räumen*, Compositio Math. **5** (1937), 177–197.

47. ——, *The theory of topological order in some linear topological spaces*, Iowa State College J. Sci. **13** (1939), 373–390.

48. ——, *Mapping degree in Banach spaces and spectral theory*, Math. Z. **63** (1955), 195–218.

49. A. Sard, *The measure of critical values of differentiable maps*, Bull. Amer. Math. Soc. **48** (1942), 883–890.

50. J. T. Schwartz, *Perturbations of spectral operators and applications. I: Bounded perturbations*, Pacific J. Math. **4** (1954), 415–458.

51. ——, *Nonlinear functional analysis*, Lecture Notes, Courant Inst. of Math. Sciences, New York University, New York, 1963–1964.

52. H. W. Siegberg, *Brouwer degree; history and numerical computation* (Sympos. Fixed point algorithms and complementarity problems, Univ. of Southampton, Southampton, 1979), North-Holland, 1980, pp. 389–411.

53. S. Smale, *An infinite dimensional version of Sard's theorem*, Amer. J. Math. **87** (1965), 861–866.

54. E. Zeidler, *Vorlesungen uber nichtlineare Funktional Analysis. I: Fixpunktsätze*, Teubner, Leipzig, 1976.

Index

ABCDEFGHIJ — 89876